U.S.NRC
United States Nuclear Regulatory Commission

Protecting People and the Environment

NUREG-0498
Supplement 2, Vol. 2

Final Environmental Statement

Related to the Operation of Watts Bar Nuclear Plant, Unit 2

Supplement 2

Final Report

Office of Nuclear Reactor Regulation

AVAILABILITY OF REFERENCE MATERIALS
IN NRC PUBLICATIONS

U.S.NRC

United States Nuclear Regulatory Commission

Protecting People and the Environment

NUREG-0498
Supplement 2, Vol. 2

Final Environmental Statement

Related to the Operation of Watts Bar Nuclear Plant, Unit 2

Supplement 2

Final Report

Manuscript Completed: May 2013
Date Published: May 2013

Office of Nuclear Reactor Regulation

Abstract

The U.S. Nuclear Regulatory Commission (NRC) prepared this supplemental final environmental statement in response to the Tennessee Valley Authority (TVA) application for a facility operating license. The proposed action requested is for the NRC to issue an operating license for a second light-water nuclear reactor at the Watts Bar Nuclear (WBN) Plant in Rhea County, Tennessee.

TVA received construction permits (CPs) for two units at the WBN site and began construction in 1973. In 1978, the NRC issued a final environmental statement related to the operating license for WBN Units 1 and 2. On March 4, 2009, the NRC received an update to the application from TVA for a facility operating license to possess, use, and operate WBN Unit 2. The NRC published the notice of the receipt of application and the opportunity for hearing in the *Federal Register* on May 1, 2009. NRC regulations in Title 10 of the *Code of Federal Regulations* (10 CFR) 51.92, "Supplement to the Final Environmental Impact Statement," require the NRC staff to prepare a supplement to the final environmental statement if there are substantial changes in the proposed action relevant to environmental concerns or if there are significant new circumstances or information relevant to environmental concerns and bearing on the proposed action or its impacts. The same regulation permits the NRC staff to prepare a supplement when, in its opinion, preparation of a supplement will further the interests of the National Environmental Policy Act of 1969.

This supplement documents the NRC staff's environmental review related to the operating license for WBN Unit 2. The NRC staff evaluated a full scope of environmental topics, including land and water use, air quality and meteorology, terrestrial and aquatic ecology, radiological and nonradiological impacts on humans and the environment, historic and cultural resources, socioeconomics, and environmental justice. The NRC staff's evaluations are based on (1) the application submitted by TVA, including the environmental report and previous environmental impact statements and historical documents, (2) consultation with other Federal, State, Tribal, and local agencies, (3) the NRC staff's independent review, and (4) the NRC staff's consideration of comments related to the environmental review received during the public scoping process.

Table of Contents

Table of Contents

Table of Contents

Table of Contents

Table of Contents

Figures

Tables

Table of Contents

Executive Summary

On March 4, 2009, the Tennessee Valley Authority (TVA) submitted to the U.S. Nuclear Regulatory Commission (NRC) a request to reactivate its application for a license to operate a second light-water nuclear reactor at the Watts Bar Nuclear (WBN) Plant in Rhea County, Tennessee. The NRC published a notice of receipt of the application and the opportunity for hearing in the *Federal Register* on May 1, 2009 (74 FR 20350). The proposed action is NRC issuance of a 40-year facility operating license for WBN Unit 2. WBN Unit 2, a pressurized-water reactor, could produce up to 3,425 megawatts thermal. The reactor-generated heat would be used to produce steam to drive steam turbines, providing 1,160 megawatts electric of net electrical power capacity to the region.

Section 102 of the National Environmental Policy Act of 1969, as amended (NEPA) (42 U.S.C. 4321), directs that an environmental impact statement (EIS) be prepared for major Federal actions that significantly affect the quality of the human environment. In 1978, the NRC issued a final environmental statement related to the operating license for WBN Units 1 and 2 (NUREG-0498, "Final Environmental Statement Related to Operation of Watts Bar Nuclear Plant Units Nos. 1 and 2," December 1978, 1978 FES-OL) for operating Units 1 and 2 at the WBN site (the final environmental statement [FES] is an EIS equivalent). Because TVA did not operate WBN Unit 2 as scheduled, the NRC's regulations in Title 10 of the *Code of Federal Regulations* (10 CFR) 51.92, "Supplement to the Final Environmental Impact Statement," require the NRC staff to prepare a supplement to the 1978 FES-OL. The purpose of this supplement is to determine if there are substantial changes in the proposed action relevant to environmental concerns or if significant new circumstances or information exist related to environmental concerns that bear on the proposed action or its impacts.

Upon acceptance of the TVA application, the NRC began the environmental review process described in 10 CFR Part 51, "Environmental Protection Regulations for Domestic Licensing and Related Regulatory Functions," by publishing a notice of intent in the *Federal Register* to prepare a supplemental final environmental statement (SFES) and conduct scoping. On October 6, 2009, the NRC held two scoping meetings in Sweetwater, Tennessee, to obtain public input on the scope of the environmental review. To gather information and become familiar with the WBN site and its environs, the NRC and its contractor, Pacific Northwest National Laboratory, visited the WBN site and environs in Rhea County, Tennessee, October 6–8, 2009.

During the site visit, the NRC team met with TVA staff, public officials, and the public. The NRC reviewed the comments received during the scoping process and contacted Federal, State, Tribal, regional, and local agencies to solicit comments. This SFES includes (1) the results of the NRC staff's analyses, which consider and weigh the environmental effects of the NRC's proposed action, issuance of a facility operating license for WBN Unit 2, (2) mitigation measures for reducing or avoiding adverse effects, and (3) the NRC staff's recommendation on the proposed action.

The NRC's standard of significance for impacts was established using the Council on Environmental Quality terminology for "significant" as defined in 40 CFR 1508.27. In addition, NRC guidance states that "Information in the GEIS [Generic Environmental Impact Statement] for license renewal, for example, the impact categorization approach (i.e., SMALL, MODERATE, and LARGE), may also be used in the preparation of NEPA documents prepared in conjunction with other types of applications such as ESPs [early site permits] and COLs [combined licenses] when it is appropriate to do so." The NRC staff used the impact categorization approach in this SFES. Impact categories include:

- SMALL—Environmental effects are not detectable or are so minor that they will neither destabilize nor noticeably alter any important attribute of the resource.

- MODERATE—Environmental effects are sufficient to alter noticeably, but not to destabilize, important attributes of the resource.

- LARGE—Environmental effects are clearly noticeable and sufficient to destabilize important attributes of the resource.

The NRC staff considered potential mitigation measures for each resource category only if adverse impacts were identified.

In preparing this SFES for WBN Unit 2, the NRC staff reviewed the TVA "Final Supplemental Environmental Impact Statement for Completion and Operation of Watts Bar Nuclear Plant Unit 2," dated February 15, 2008, which TVA submitted to the NRC as the environmental report portion of its application. The NRC staff also consulted with other Federal, State, Tribal, regional, and local agencies and followed the guidance set forth in NUREG-1555, "Standard Review Plans for Environmental Reviews of Nuclear Power Plants," dated October 1999. In addition, the NRC staff considered public comments related to the environmental review received during the scoping process. Appendix D to this SFES includes these scoping comments and the NRC staff's responses to them.

The draft SFES was published in October 2011. The U.S. Environmental Protection Agency Notice of Filing in the *Federal Register* (76 FR 70130) indicated a 75-day comment period, commencing on November 10, 2011, to allow members of the public to comment on the results of the NRC staff's review. This was amended in the *Federal Register* on November 18, 2011 to a 45-day comment period (76 FR 71560). The NRC issued a Notice of Availability (76 FR 70169) of the draft SFES in the *Federal Register* that provided a 45-day comment period and announced the date and location of the public meetings. On December 8, 2011, two public meetings were held in Sweetwater, Tennessee. At the meetings, the NRC staff described the results of the NRC environmental review, answered questions related to the review, and provided members of the public with information to assist them in formulating their comments. Based on comments received at the public meetings, the comment period was extended by the NRC to January 24, 2012 (76 FR 80409). When the comment period ended on January 24, 2012, the NRC staff considered and addressed all the comments received. All comments received on the draft SFES are included in Appendix E.

In this SFES, the NRC staff concludes that impacts from the operation of WBN Unit 2 associated with water use, terrestrial resources, aquatic ecology, design-basis accidents, socioeconomics, the radiological and nonradiological environments, decommissioning, air quality, and land use are generally consistent with those reached in the 1978 FES-OL and Supplement No. 1 to the "Final Environmental Statement Related to the Operation of Watts Bar Nuclear Plant, Units 1 and 2," dated April 1995 (1995 SFES-OL-1). In some cases, the impacts were less than those identified in the 1978 FES-OL.

Groundwater quality, public services, noise, socioeconomic transportation, cultural and historical resources, environmental justice, greenhouse gas emissions, severe accidents, severe accident mitigation alternatives, and cumulative impacts were not addressed in the 1978 FES-OL but are addressed in this SFES. The NRC staff concludes that impacts associated with the operation of WBN Unit 2 on groundwater quality, public services, noise, socioeconomic transportation, cultural and historical resources, greenhouse gas emissions, and severe accidents would be SMALL. In addition, the NRC staff concludes that the operation of WBN Unit 2 would not result in a disproportionately high and adverse human health or environmental effect on any of the low-income communities near the WBN site.

The NRC staff also considered cumulative impacts from past, present, and reasonably foreseeable future actions. The NRC staff concludes that, although some of the cumulative impacts are LARGE as the result of other activities that affected the environment, the incremental impact from operation of WBN Unit 2 would in all cases be minor.

The NRC staff's recommendation to the Commission related to the environmental aspects of the proposed action is that the operating license for WBN Unit 2 be issued as proposed. This recommendation is based on (1) the application, including the February 15, 2008 final EIS submitted by TVA as the ER, and responses to staff requests for additional information submitted by TVA; (2) the NRC staff's review conducted for the 1978 FES-OL; (3) consultation with Federal, State, Tribal, and local agencies; (4) the NRC staff's own independent review of information available since the preparation and publication of the 1978 FES-OL; and (5) the assessments summarized in this SFES, including consideration of public comments received during scoping and on the draft SFES.

The NRC's final safety evaluation report, anticipated to be published in 2014, will address the NRC staff's evaluation of the site safety and emergency preparedness aspects of the proposed action.

Abbreviations/Acronyms

χ/Q	atmospheric dispersion value
°C	degree(s) Celsius
°F	degree(s) Fahrenheit
ac	acre(s)
ACRS	Advisory Committee on Reactor Safeguards
A.D.	Anno Domini
ADAMS	Agencywide Documents Access and Management System (NRC)
ADEM	Alabama Department of Environmental Management
ADTV	average daily traffic volume
AEC	U.S. Atomic Energy Commission
ALARA	as low as is reasonably achievable
AOC	averted offsite costs
AOE	averted occupational exposure
AOSC	averted onsite costs
APE	area of potential effect or averted public exposure
AQCR	Air Quality Control Region
ATWS	anticipated transient without scram
B.C.	Before Christ
BLS	Bureau of Labor Statistics
BMP	best management practice
Bq	becquerel
Btu	British thermal unit(s)
Btu/hr	British thermal unit(s) per hour
CAFTA	Computer Aided Fault Tree Analysis
CCP	centrifugal charging pump
CCS	component cooling water system
CCW	condenser circulating water or condenser cooling water
CDC	Centers for Disease Control
CDF	core damage frequency

Abbreviations/Acronyms

CDWE	condensate demineralizer waste evaporator
CEQ	Council on Environmental Quality
CET	containment event trees
CFR	*Code of Federal Regulations*
cfs	cubic (foot) feet per second
Ci	curies
cm	centimeter(s)
CO_2	carbon dioxide
CORMIX	Cornell Mixing Zone Expert System
CPI	Consumer Price Index
CPPR	construction permit power reactor
CTBD	cooling-tower blowdown
CWA	Clean Water Act
CWS	Circulating Water System
dBA	decibels on the A-weighted scale
DBA	design basis accident
DC	design certification
D.C.	District of Columbia
DOE	U.S. Department of Energy
DSM	demand-side management
EAB	exclusion area boundary
ECCS	emergency core cooling system
EDG	emergency diesel generator
EIS	environmental impact statement
ELF	extremely low frequency
EMF	electromagnetic field
EO	Executive Order
EPA	U.S. Environmental Protection Agency
EPACT	1992 National Energy Policy Act
EPRI	Electric Power Research Institute
ER	environmental report
ERCW	essential raw cooling water
ESFAS	emergency safety features actuation system

ESRP	Environmental Standard Review Plan
FCC	Federal Communications Commission
FERC	Federal Energy Regulatory Commission
FES	final environmental statement
FES-CP	final environmental statement related to the construction permit for WBN Units 1 and 2
FES-OL	final environmental statement related to the operating license for WBN Units 1 and 2
FHA	fuel handling accident
FIVE	fire-induced vulnerability evaluation
FONSI	finding of no significant environmental impact
FR	*Federal Register*
FSAR	Final Safety Analysis Report
ft	foot (feet)
ft^3	cubic foot (feet)
FWS	U.S. Fish and Wildlife Service
gal	gallon(s)
GAO	U.S. General Accounting Office
GC	gaseous centrifuge
GCRP	U.S. Global Change Research Program
GD	gaseous diffusion
GEIS	Generic Environmental Impact Statement
GEIS-DECOM	Generic Environmental Impact Statement on Decommissioning of Nuclear Facilities Regarding the Decommissioning of Nuclear Power Reactors
GHG	greenhouse gas
gpd	gallon(s) per day
gpm	gallon(s) per minute
GWPP	Ground Water Protection Program
Gy	gray(s)
ha	hectare(s)
HCLPF	high confidence of low probability of failure
HFO	high winds, floods, and other
HLW	high-level waste

Abbreviations/Acronyms

HPFP	high pressure fire protection
HPI	high pressure injection
hr	hour(s)
HRA	human reliability analysis
HVAC	heating, ventilation, and air conditioning
Hz	hertz
I	Interstate
IAEA	International Atomic Energy Agency
ICRP	International Commission on Radiological Protection
IMP	internal monitoring point
in.	inch(es)
in.^{2}	square inch(es)
IPE	Individual Plant Examination
IPEEE	Individual Plant Examination of External Events
IPS	intake pumping station
IRP	Integrated Resource Plan
ISFSI	independent spent fuel storage installation
ISLOCA	interfacing system loss-of-coolant accidents
kg	kilogram(s)
km	kilometer(s)
km^2	square kilometer(s)
kV	kilovolt(s)
L/d	liter(s) per day
L/s	liter(s) per second
L/yr	liter(s) per year
lb	pound(s)
LCV	level control valves
LERF	large early release frequency
LLW	low-level waste
LM	log mile(s)
LOCA	loss-of-cooling accident
LOOP	loss of offsite power

LPZ	Low Population Zone
LVWTP	Low Volume Waste Treatment Pond
LWR	light water reactor
m	meter(s)
m^3/s	cubic meter(s) per second
MACR	maximum averted cost risk
MACCS2	MELCOR Accident Consequence Code System
MCR	main control room
MEI	maximally exposed individual
MGD	million gallons per day
mg/L	milligram(s) per liter
mGy	milligray(s)
mGy/yr	milligray(s) per year
MHz	megahertz
mi	mile(s)
mi^2	square mile(s)
MIT	Massachusetts Institute of Technology
MMACR	modified maximum averted cost risk
mo	month(s)
mrad	millirad(s)
mrad/d	millirad(s) per day
mrem	millirem(s)
mrem/yr	millirem(s) per year
msl	mean sea level
mSv	millisievert(s)
mSv/yr	millisievert(s) per year
MT	metric ton(s)
MW	megawatt(s)
MW(e)	megawatt(s) electric
MW(t)	megawatt(s) thermal
NCRP	National Council on Radiation Protection and Measurements
NEI	Nuclear Energy Institute
NEPA	National Environmental Policy Act of 1969, as amended

Abbreviations/Acronyms

NERC	North American Electric Reliability Corporation
NESC	National Electrical Safety Code
NHPA	National Historic Preservation Act of 1966, as amended
NPDES	National Pollutant Discharge Elimination System
NPF	nuclear power facility
NRC	U.S. Nuclear Regulatory Commission
NSSS	Nuclear Steam Supply System
O&M	operation and maintenance
ODCM	Offsite Dose Calculation Manual
OL	Operating License
OSHA	Occupational Safety and Health Administration
PCB	polychlorinated biphenyl
pCi/L	picocurie(s) per liter
PDS	plant damage state
PNNL	Pacific Northwest National Laboratory
PORV	power-operated relief valves
PPA	purchased power arrangement
ppm	parts per million
PRA	probabilistic risk assessment
PWR	pressurized-water reactor
RAI	Request for Additional Information
RCP	reactor coolant pump
RCS	reactor coolant system
RCRA	Resource Conversation and Recovery Act
RCW	raw cooling water
rem	roentgen equivalent man
REMP	radiological environmental monitoring program
RLE	review level earthquake
ROI	region of influence
ROS	Reservoir Operations Study
RPS	reactor protection system
RRS	Reliability Review Subcommittee

RRW	risk-reduction worth
RWST	refueling water storage tank
Ryr	reactor-year
s/m^2	second(s) per square meter
SACE	Southern Alliance for Clean Energy
SAMA	severe accident mitigation alternative
SAMDA	severe accident mitigation design alternative
SBO	station blackout
SCCW	Supplemental Condenser Cooling Water
SEIS	supplemental environmental impact statement
SERC	Southeastern Electric Reliability Corporation
SFES	supplemental final environmental statement
SFES-OL-1	NRC 1995 Supplement No. 1 to the Final Environmental Statement related to the operating license
SFES-OL-2	NRC 2011 Supplement No. 2 to the Final Environmental Statement related to the operating license
SGTR	steam generator tube rupture
SHPO	State Historic Preservation Officer
SPCC plan	Spill, Prevention, Control, and Countermeasure Plan
SRP	Standard Review Plan
Sv	sievert(s)
TACIR	Tennessee Advisory Committee on Intergovernmental Relations
TDEC	Tennessee Department of Environment and Conservation
TDOH	Tennessee Department of Health
TDOT	Tennessee Department of Transportation
TDS	total dissolved solids
TN	Tennessee State Route
TOSHA	Tennessee Occupational Safety and Health Administration
tpy CO_2e	tons per year of carbon dioxide equivalent
TRM	Tennessee River Mile
TRO	Total Residual Oxident
TVA	Tennessee Valley Authority
TWRA	Tennessee Wildlife Resource Agency

Abbreviations/Acronyms

USGS	U.S. Geological Survey
V	volt(s)
WCD	Waste Confidence Decision
WBN	Watts Bar Nuclear
WNA	World Nuclear Association
WOG	Westinghouse Owners Group
yd^3	cubic yard(s)
YHP	Yard Holding Pond
yr	year(s)

Appendix A

Contributors to the Supplement

Appendix A

Contributors to the Supplement

The overall responsibility for the preparation of this supplemental final environmental statement (SFES) was assigned to the Office of Nuclear Reactor Regulation, U.S. Nuclear Regulatory Commission (NRC). Members of the Office of Nuclear Reactor Regulation prepared the SFES with assistance from other NRC organizations and the Pacific Northwest National Laboratory.

Name	Affiliation	Function or Qualifications
NUCLEAR REGULATORY COMMISSION		
Joel Wiebe	Nuclear Reactor Regulation	Project Manager
Patrick Milano	Nuclear Reactor Regulation	Project Manager
Justin Poole	Nuclear Reactor Regulation	Project Manager
Laurel Bauer	Nuclear Reactor Regulation	Branch Chief
Andrew Imboden	Nuclear Reactor Regulation	Branch Chief, Air Quality
Dennis Beissel[a]	Nuclear Reactor Regulation	Environmental Team Lead, Hydrology, Nonradiological Human Health
Dennis Logan	Nuclear Reactor Regulation	Environmental Team Lead, Ecology
Joseph Giacinto	Nuclear Reactor Regulation	Environmental Team Lead
Elaine Keegan	Nuclear Reactor Regulation	Environmental Team Lead
Jeffrey Rikhoff	Nuclear Reactor Regulation	Socioeconomics, Benefit-cost, Environmental Justice
Kevin Folk	Nuclear Reactor Regulation	Hydrology
Alice Erickson	Nuclear Reactor Regulation	Hydrology
Ray Galluci	Nuclear Reactor Regulation	Severe Accident Mitigation Alternatives
Andrew Stuyvenberg	Nuclear Reactor Regulation	Need for Power, Alternatives, Benefit-Cost
Jennifer Davis	Federal and State Materials and Environmental Management	Cultural Resources
Allison Travers	Nuclear Reactor Regulation	Cultural Resources
Stephen Klementowicz	Nuclear Reactor Regulation	Health Physics, Nonradiological Human Health
PACIFIC NORTHWEST NATIONAL LABORATORY[b]		
Rebekah Krieg		Project Manager, Aquatic Ecology
Amanda Stegen		Task Leader
Tonya Keller		Deputy Task Leader
Beverly Miller		Deputy Task Leader
Joanne Duncan		Deputy Task Leader, References
Robert Bryce		Hydrology, Groundwater, Surface Water

Appendix A

Name	Affiliation	Function or Qualifications
Corey Duberstein		Terrestrial Ecology
Katherine Cort		Land Use, Socioeconomics, Environmental Justice, Benefit-Cost, Need for Power, Alternatives
Tara O'Neil		Cultural Resources
Eva Eckert Hickey		Health Physics
Richard Traub		Health Physics
James V. Ramsdell, Jr		Air Quality, Meteorology, Design Basis Accidents, Severe Accidents
Steve Short		Severe Accidents
Amoret Bunn		Nonradiological Human Health
Lara Aston		Nonradiological Human Health
Tom Anderson		Cumulative Impacts
Terri Miley		Cumulative Impacts
Georganne O'Connor		Program Specialist
Nancy Kohn		Comment Binning
Lubov Lavrentiev		Assistant
Dave Payson		Technical Editing
Susan Ennor		Technical Editing
Michael Parker		Technical Editing, Document Design
Tomiann Parker		References
Sharon Johnson		References
Kathy Neiderhiser		Document Design
Rose Zanders		Graphics
Donna Austin-Workman		Graphics
Susan Loper		Graphics

(a) Staff member is no longer with the NRC Office of Nuclear Reactor Regulation.
(b) Pacific Northwest National Laboratory is operated by Battelle for the U.S. Department of Energy.

Appendix B

Organizations Contacted

Appendix B

Organizations Contacted

The following Federal, State, Tribal, regional, and local organizations were contacted during the course of the U.S. Nuclear Regulatory Commission staff's independent review of potential environmental impacts from operation of one new nuclear unit at the Watts Bar Nuclear site in Rhea County, Tennessee:

Absentee Shawnee Tribe of Oklahoma, Shawnee, Oklahoma

Advisory Council on Historic Preservation, Washington, D.C.

Alabama-Coushatta Tribe of Texas, Wetumka, Oklahoma

Alabama-Quassarte Tribal Town, Wetumka, Oklahoma

Cherokee Nation, Tahlequah, Oklahoma

Choctaw Nation of Oklahoma, Durant, Oklahoma

Dayton City School System, Dayton, Tennessee

Department of Interior, Office of Environmental Policy and Compliance, Atlanta, Georgia

Eastern Band of Cherokee Indians, Cherokee, North Carolina

Eastern Band of the Cherokee Indians, Bryson City, North Carolina

Eastern Shawnee Tribe of Oklahoma, Seneca, Missouri

Harmon, Curran, Spielberg & Eisenberg, L.L.P., Washington, D.C.

Jena Band of Choctaw Indians, Jena, Louisiana

Kialegee Tribal Town, Wetumka, Oklahoma

Meigs County School System, Decatur, Tennessee

Muscogee (Creek) Nation of Oklahoma, Okmulgee, Oklahoma

Appendix B

Shawnee Tribe, Miami, Oklahoma

Southeast Tennessee Development District, Chattanooga, Tennessee

Southern Alliance for Clean Energy, Washington, D.C.

Tennessee Department of Agriculture, Nashville, Tennessee

Tennessee Department of Economic and Community Development, Nashville, Tennessee

Tennessee Department of Environment and Conservation, Chattanooga Field Office, Chattanooga, Tennessee

Tennessee Department of Environment and Conservation, Nashville, Tennessee

Tennessee Department of Transportation, Nashville, Tennessee

Tennessee Historical Commission, Nashville, Tennessee

Tennessee Wildlife Resource Agency, Nashville, Tennessee

The Chickasaw Nation, Ada, Oklahoma

Thlopthlocco Tribal Town, Okemah, Oklahoma

United Keetoowah Band Headquarters, Tahlequah, Oklahoma

U.S. Army Corps of Engineers, Nashville, Tennessee

U.S. Environmental Protection Agency, Atlanta, Georgia

U.S. Fish and Wildlife Service, Cookeville, Tennessee

Watts Bar Utility District, Kingston, Tennessee

Appendix C

Chronology of NRC Staff Environmental Review Correspondence Related to Tennessee Valley Authority Application for an Operating License for Watts Bar Nuclear Plant Unit 2

Appendix C

Chronology of NRC Staff Environmental Review Correspondence Related to Tennessee Valley Authority Application for an Operating License for Watts Bar Nuclear Plant Unit 2

This appendix contains a chronological list of correspondence between the U.S. Nuclear Regulatory Commission (NRC) and the Tennessee Valley Authority (TVA) and other correspondence related to the NRC staff's environmental review, under Title 10 of the Code of Federal Regulations (CFR) Part 51, for the TVA application for an operating license (OL) at the Watts Bar Nuclear (WBN) Unit 2 site in Rhea County, Tennessee.

All documents, with the exception of those containing proprietary information or otherwise exempt from disclosure, have been placed in the NRC's Public Document Room, at One White Flint North, 11555 Rockville Pike (first floor), Rockville, Maryland, and are available electronically from the Public Electronic Reading Room found on the Internet at the following web address: http://www.nrc.gov/reading-rm.html. The public can use this site to gain access to the NRC's Agencywide Document Access and Management System (ADAMS), which provides text and image files of NRC's publicly available documents. The ADAMS accession numbers for each document are included below.

Author	Recipient	Date of Letter/Email
Tennessee Valley Authority (M. Bajestani)	U.S. Nuclear Regulatory Commission	February 15, 2008 (ML080510469)
U.S. Nuclear Regulatory Commission (J.F. Williams)	Tennessee Valley Authority (A. Bhatnagar)	June 3, 2008 (ML081210270)
U.S. Nuclear Regulatory Commission (J.F. Williams)	Tennessee Valley Authority (A. Bhatnagar)	June 20, 2008 (ML081500030)
Tennessee Valley Authority (M. Bajestani)	U.S. Nuclear Regulatory Commission	July 2, 2008 (ML081850460)
Tennessee Valley Authority (M. Bajestani)	U.S. Nuclear Regulatory Commission	January 27, 2009 (ML090360588)

Author	Recipient	Date of Letter/Email
U.S. Nuclear Regulatory Commission (J. Wiebe)	U.S. Fish and Wildlife Service (M. Jennings)	September 2, 2009 (ML092100088)
U.S. Nuclear Regulatory Commission (J. Wiebe)	Southern Alliance for Clean Energy (D. Curran and M. Fraser)	September 4, 2009 (ML092440217)
U.S. Nuclear Regulatory Commission (J. Wiebe)	Advisory Council on Historic Preservation (D. Klima)	September 10, 2009 (ML092120105)
U.S. Nuclear Regulatory Commission (J. Wiebe)	Cherokee Nation (R. Allen)	September 10, 2009 (ML092110475)
U.S. Nuclear Regulatory Commission (J. Wiebe)	Eastern Band of the Cherokee Indians (T. Howe)	September 10, 2009 (ML092110475)
U.S. Nuclear Regulatory Commission (J. Wiebe)	Eastern Band of the Cherokee Indians (R. Townsend)	September 10, 2009 (ML092110475)
U.S. Nuclear Regulatory Commission (J. Wiebe)	United Keetoowah Band Headquarters (L. Larue-Stopp)	September 10, 2009 (ML092110475)
U.S. Nuclear Regulatory Commission (J. Wiebe)	The Chickasaw Nation (V. (Gingy) Nail)	September 10, 2009 (ML092110475)
U.S. Nuclear Regulatory Commission (J. Wiebe)	Choctaw Nation of Oklahoma (T. Cole)	September 10, 2009 (ML092110475)
U.S. Nuclear Regulatory Commission (J. Wiebe)	Choctaw Nation of Oklahoma (G. Pyle)	September 10, 2009 (ML092110475)
U.S. Nuclear Regulatory Commission (J. Wiebe)	Jena Band of Choctaw Indians (L. Strange)	September 10, 2009 (ML092110475)
U.S. Nuclear Regulatory Commission (J. Wiebe)	Muscogee (Creek) Nation of Oklahoma (J. Bear)	September 10, 2009 (ML092110475)
U.S. Nuclear Regulatory Commission (J. Wiebe)	Alabama-Coushatta Tribe of Texas (B. Battise)	September 10, 2009 (ML092110475)

Author	Recipient	Date of Letter/Email
U.S. Nuclear Regulatory Commission (J. Wiebe)	Alabama-Quassarte Tribal Town (A. Asbury)	September 10, 2009 (ML092110475)
U.S. Nuclear Regulatory Commission (J. Wiebe)	Kialegee Tribal Town (E. Bucktrot and G. Bucktrot)	September 10, 2009 (ML092110475)
U.S. Nuclear Regulatory Commission (J. Wiebe)	Thlopthlocco Tribal Town (C. Coleman)	September 10, 2009 (ML092110475)
U.S. Nuclear Regulatory Commission (J. Wiebe)	Absentee Shawnee Tribe of Oklahoma (K. Kaniatobe)	September 10, 2009 (ML092110475)
U.S. Nuclear Regulatory Commission (J. Wiebe)	Eastern Shawnee Tribe of Oklahoma (R. DuShane)	September 10, 2009 (ML092110475)
U.S. Nuclear Regulatory Commission (J. Wiebe)	Eastern Shawnee Tribe of Oklahoma (G.J. Wallace)	September 10, 2009 (ML092110475)
U.S. Nuclear Regulatory Commission (J. Wiebe)	Shawnee Tribe (R. Sparkman)	September 10, 2009 (ML092110475)
U.S. Nuclear Regulatory Commission (J. Wiebe)	Shawnee Tribe (B. Pryor)	September 10, 2009 (ML092110475)
U.S. Nuclear Regulatory Commission (J. Wiebe)	Tennessee Historical Commission (J.Y. Garrison)	September 10, 2009 (ML092120097)
U.S. Nuclear Regulatory Commission (J. Wiebe)	U.S. Army Corps of Engineers (R. Gatlin)	September 10, 2009 (ML092110147)
U.S. Nuclear Regulatory Commission (J. Wiebe)	Office of Environment Policy and Compliance, Department of Interior (G.L. Hogue)	September 10, 2009 (ML092110147)
U.S. Nuclear Regulatory Commission (J. Wiebe)	Sam Nunn Atlanta Federal Center (A.S. Meiburg and S. Gordon)	September 10, 2009 (ML092110147)
U.S. Nuclear Regulatory Commission (J. Wiebe)	Tennessee Department of Environment and Conservation (M. Apple)	September 10, 2009 (ML092110147)

Author	Recipient	Date of Letter/Email
U.S. Nuclear Regulatory Commission (J. Wiebe)	Tennessee Department of Environment and Conservation (S. Baxter)	September 10, 2009 (ML092110147)
U.S. Nuclear Regulatory Commission (J. Wiebe)	Tennessee Department of Environment and Conservation (B. Bowen)	September 10, 2009 (ML092110147)
U.S. Nuclear Regulatory Commission (J. Wiebe)	Tennessee Department of Economic and Community Development (M. Atchinson)	September 10, 2009 (ML092110147)
U.S. Nuclear Regulatory Commission (J. Wiebe)	Environment and Planning Environmental Division (E. Cole)	September 10, 2009 (ML092110147)
U.S. Nuclear Regulatory Commission (J. Wiebe)	Tennessee Department of Agriculture (K. Givens)	September 10, 2009 (ML092110147)
U.S. Nuclear Regulatory Commission (J. Wiebe)	Tennessee Department of Environment and Conservation (P. Davis)	September 10, 2009 (ML092110147)
U.S. Nuclear Regulatory Commission (J. Wiebe)	Water Supply (R. Foster)	September 10, 2009 (ML092110147)
U.S. Nuclear Regulatory Commission (J. Wiebe)	Tennessee Department of Environment and Conservation (J. Fyke)	September 10, 2009 (ML092110147)
U.S. Nuclear Regulatory Commission (J. Wiebe)	Division of Radiological Health (L.E. Nanney)	September 10, 2009 (ML092110147)
U.S. Nuclear Regulatory Commission (J. Wiebe)	Tennessee Department of Environment and Conservation (B. Stephens)	September 10, 2009 (ML092110147)
U.S. Nuclear Regulatory Commission (J. Wiebe)	Tennessee Department of Environment and Conservation (M. Tummons)	September 10, 2009 (ML092110147)
U.S. Nuclear Regulatory Commission (J. Wiebe)	Groundwater (A. Schwendimann)	September 10, 2009 (ML092110147)
U.S. Nuclear Regulatory Commission (J. Wiebe)	Tennessee Wildlife Resource Agency (E. Carter)	September 10, 2009 (ML092110147)

Author	Recipient	Date of Letter/Email
U.S. Nuclear Regulatory Commission (J. Wiebe)	Resource Management Division (A. Marshall)	September 10, 2009 (ML092110147)
U.S. Nuclear Regulatory Commission (J. Wiebe)	Tennessee Department of Transportation (Commissioners Office)	September 10, 2009 (ML093080084)
Tennessee Historical Commission (E.P. McIntyre)	U.S. Nuclear Regulatory Commission (J. Wiebe)	September 22, 2009 (ML093510985)
Eastern Band of Cherokee Indians (T. Howe)	U.S. Nuclear Regulatory Commission	September 29, 2009 (ML0928605910)
U.S. Fish and Wildlife Services (M. Jennings)	U.S. Nuclear Regulatory Commission (J. Wiebe)	October 9, 2009 (ML0929301820)
Tennessee Valley Authority (M. Bajestani)	U.S. Nuclear Regulatory Commission	October 22, 2009 (ML093510833)
U.S. Nuclear Regulatory Commission (J. Wiebe)	Tennessee Valley Authority (A. Bhatnagar)	December 3, 2009 (ML093030148/ ML093290073)
Tennessee Valley Authority (M. Bajestani)	U.S. Nuclear Regulatory Commission	December 23, 2009 (ML100210358)
Tennessee Valley Authority (M. Bajestani)	U.S. Nuclear Regulatory Commission	February 25, 2010 (ML100630116)
Tennessee Historical Commission (E. Patrick McIntyre, Jr)	U.S. Nuclear Regulatory Commission (J. Wiebe)	March 5, 2010 (ML100770290)
Tennessee Valley Authority (M. Bajestani)	U.S. Nuclear Regulatory Commission	April 9, 2010 (ML101130392)
Tennessee Valley Authority (M. Bajestani)	U.S. Nuclear Regulatory Commission	May 12, 2010 (ML101340589)
Tennessee Valley Authority (M. Bajestani)	U.S. Nuclear Regulatory Commission	July 2, 2010 (ML101930470)
Tennessee Valley Authority (R.M. Krich)	U.S. Nuclear Regulatory Commission	July 6, 2010 (ML101890069)
Tennessee Valley Authority (E.E. Freeman)	U.S. Nuclear Regulatory Commission	January 4, 2011 (ML110060510)
Tennessee Valley Authority (M. Gillman)	U.S. Nuclear Regulatory Commission	February 7, 2011 (ML110400384)
Tennessee Valley Authority (E.E. Freeman)	U.S. Nuclear Regulatory Commission	March 24, 2011 (ML110871475)
Tennessee Valley Authority (E.E. Freeman)	U.S. Nuclear Regulatory Commission	March 28, 2011 (ML110890472)

Author	Recipient	Date of Letter/Email
Tennessee Valley Authority (D. Stinson)	U.S. Nuclear Regulatory Commission	May 19, 2011 (ML11143A083)
Tennessee Valley Authority (D. Stinson)	U.S. Nuclear Regulatory Commission	May 20, 2011 (ML11146A044)
Tennessee Valley Authority (D. Stinson)	U.S. Nuclear Regulatory Commission	May 26, 2011 (ML11152A160)
Tennessee Valley Authority (R.M. Krich)	U.S. Nuclear Regulatory Commission	July 28, 2011 (ML11215A098)
U.S. Nuclear Regulatory Commission (S. Campbell)	Absentee Shawnee Tribe of Oklahoma (K. Kaniatobe)	November 1, 2011 (ML11301A320)
U.S. Nuclear Regulatory Commission (S. Campbell)	Tennessee Valley Authority (M. Skaggs)	November 2, 2011 (ML11299A153)
U.S. Nuclear Regulatory Commission (S. Campbell)	U.S. Environmental Protection Agency	November 2, 2011 (ML11299A184)
U.S. Nuclear Regulatory Commission (S. Campbell)	Tennessee Historical Commission (E.P. McIntyre, Jr)	November 2, 2011 (ML11304A040)
U.S. Nuclear Regulatory Commission (S. Campbell)	Advisory Council on Historic Preservation (R. Nelson)	November 2, 2011 (ML11305A245)
U.S. Nuclear Regulatory Commission (S. Campbell)	U.S. Fish and Wildlife Service (M. Jennings)	November 2, 2011 (ML11304A083)
U.S. Nuclear Regulatory Commission (S. Campbell)	Tennessee Department of Agriculture (K. Givens)	November 2, 2011 (ML11305A191)
U.S. Nuclear Regulatory Commission (S. Campbell)	Monroe County Economic Development (S. Burris)	November 2, 2011 (ML11304A171)
Choctaw Nation of Oklahoma (I. Thompson)	U.S. Nuclear Regulatory Commission (P Milano)	November 21, 2011 (ML12053A441)
Tennessee Valley Authority (E. Freeman)	U.S. Nuclear Regulatory Commission	November 21, 2011 (ML11329A001)
U.S. Environmental Protection Agency (H. Mueller)	U.S. Nuclear Regulatory Commission (S. Campbell)	December 15, 2011 (ML12004A168)
U.S. Department of the Interior (J. Stanley)	U.S. Nuclear Regulatory Commission (C. Bladey)	December 19, 2011 (ML12005A211)
TN Department of Transportation (A. Andrews)	U.S. Nuclear Regulatory Commission (S. Campbell)	December 20, 2011 (ML12018A397)
U.S. Fish and Wildlife Service (M. Jennings)	U.S. Nuclear Regulatory Commission (S. Campbell)	December 20, 2011 (ML12004A167)
Tennessee Valley Authority (J.W. Shea)	U.S. Nuclear Regulatory Commission	December 27, 2011 (ML12009A072)

Author	Recipient	Date of Letter/Email
Chickasaw Nation (J. Keel)	U.S. Nuclear Regulatory Commission (P Milano)	January 19, 2012 (ML12053A439)
U.S. Department of the Interior (J. Stanley)	U.S. Nuclear Regulatory Commission (J. Poole)	January 23, 2012 (ML12023A185)
Tennessee Valley Authority (B. Brickhouse)	U.S. Nuclear Regulatory Commission (C. Bladey)	January 27, 2012 (ML12040A052)
Tennessee Valley Authority (G. Arent)	U.S. Nuclear Regulatory Commission	March 12, 2012 (ML12073A363)
Tennessee Valley Authority (R. Hruby)	U.S. Nuclear Regulatory Commission	June 11, 2012 (ML12166A068)
Tennessee Valley Authority (R. Hruby)	U.S. Nuclear Regulatory Commission	December 12, 2012 (ML12346A009)

Appendix D

Scoping Comments and Responses

WATTS BAR NUCLEAR PLANT, UNIT 2 – COMPLETE LIST OF COMMENTS, SUGGESTIONS, AND STAFF RESPONSES CONDENSED FROM THE OCTOBER 6, 2009, PUBLIC SCOPING MEETING

On October 6, 2009, a Category 3 public meeting (two sessions) was held between the U.S. Nuclear Regulatory Commission (NRC) and interested public at the Magnuson Hotel, 1421 Murrays Chapel Road, Sweetwater, Tennessee 37874. The purpose of the meeting was to present an overview of the environmental review process for Watts Bar Unit 2 operating license application and to obtain public comments regarding the scope of the environmental review.

Scoping meeting attendees provided either written statements or oral comments that the NRC recorded and a certified court reporter transcribed. In addition, during the scoping period, the NRC received four letters and five emails providing comments on the proposed action. The staff considered all comments and suggestions received.

The meeting summary was issued on October 21, 2009, and is available electronically from the Publicly Available Records component of NRC's Agencywide Documents Access and Management System (ADAMS) under accession number ML092880764. ADAMS documents can be found at https://www.nrc.gov/reading-rm/adams.html.

The following selection of public comments has been broken down into two categories:

1) Public comments that are covered in the supplemental final environmental statement (SFES) (equivalent to an environmental impact statement [EIS])

2) Public comments concerning issues that are outside the scope of review

Table A-1 identifies the individuals providing comments in alphabetical order; their affiliation, if given; the ADAMS accession number that can be used to locate the correspondence; and the correspondence identification number (ID). Table A-2 identifies individual comments covered in the SFES and those comments outside the scope of review.

Table A-1. Individuals Providing Comments During the Comment Period

Commenter	Affiliation (if stated)	Comment Source and ADAMS Accession #	Correspondence ID
Burris, Shane	Monroe County	Meeting Transcript (ML092870331)	0003
Cobb, Jim	Tennessee House District 31	Meeting Transcript (ML092870331)	0003
Curran, Diane	Harmon, Curran, Spielberg & Eisenberg, LLP	Letter (ML093080581)	0010
Gottfried, Yolande		Letter (ML093090656)	0008
Harris, Ann		Meeting Transcript (ML092870331)	0003
Harris, Ann		Meeting Transcript (ML092870338)	0004
Howe, Tyler	Eastern Band of Cherokee Indians	Letter (ML092860591)	0006
Jennings, Mary	U.S. Fish and Wildlife Service	Letter (ML092930182)	0005
Jones, Ken	Meigs County	Meeting Transcript (ML092870338)	0004
Kurtz, Sandy		Meeting Transcript (ML092870338)	0004
Mastin, Mary		Meeting Transcript (ML092870331)	0003

McCluney, Ross	BREDL	Meeting Transcript (ML092870331)	0003
Paddock, Brian	Sierra Club, Tennessee Chapter	Meeting Transcript (ML092870331)	0003
Reynolds, Bill		Meeting Transcript (ML092870331)	0003
Reynolds, Bill		Meeting Transcript (ML092870338)	0004
Safer, Don		Email (ML093060311)	0013
Safer, Don		Meeting Transcript (ML092870331)	0003
Smith, Stewart		Meeting Transcript (ML092870338)	0004
Yager, Ken	Tennessee Senatorial District 12	Letter (ML093090655)	0007
Zeller, Lou	Blue Ridge Environmental Defense League	Letter (ML093080360)	0015
Zeller, Lou	Blue Ridge Environmental Defense League	Meeting Transcript (ML092870331)	0003

Table A-2
Category No. 1
Public Comments that are Covered in the Supplemental Final Environmental Statement (SFES) (Equivalent to an EIS)

Comment: The Organizations [Southern Alliance for Clean Energy, the Sierra Club, Blue Ridge Environmental Defense League, Tennessee Environmental Council, and We the People] respectfully submit that the EIS should consider, at a minimum, the environmental concerns raised in their hearing request to the NRC, which is now pending before the Atomic Safety and Licensing Board. (0010 [Curran, Diane])

Response: When preparing the SFES, the NRC staff will consider concerns expressed by commenter's that are within the scope of the environmental review.

Comment: Given all those concerns and the fact that things have certainly changed since 1978, when the first Environmental Impact Statement was done and those supplements in 1995, I think NRC should recommend to TVA that they start all over with a new, from ground zero, Environmental Impact Statement. (0004 [Kurtz, Sandy])

Comment: The National Environmental Policy Act requires that before undertaking a major federal action, an agency must take a "hard look" at the environmental consequences of the action (Baltimore Gas and Elec. Co. v. Natural Resources Defense Council. Inc., 462 U.S. 87, 97 (1983)). Where an agency has not yet taken the major federal action, it must consider "new and significant information" that bears on the environmental impacts of the proposed action [Marsh v. Oregon Natural Resources Council, 490 U.S. 360, 371-72(1989)]. Also, federal regulations require supplementation where the proposed action has not been completed, if: "(1) there are substantial changes in the proposed action that are relevant to environmental concerns; or (2) there are significant new circumstances or information relevant to environmental concerns and bearing on the proposed action or its impacts." [10 C.F.R. 51.92(a)] The environmental effects of the two side-by-side Watts Bar facilities raise the issues of segmentation and cumulative impacts. (0015 [Zeller, Lou])

Response: The commission expects the staff to take the requisite "hard look" at new information on the need for power and alternative sources of energy and has authorized the staff to supplement the SFES if there is new and significant information relevant to these matters. The NRC staff will prepare the SFES in accordance with NEPA and 10 CFR Part 51. The analysis will address the environmental effects of operating the proposed WBN Unit 2.

Comment: I am really concerned about the water quality in the Tennessee River, and I think that as TVA goes forward with this

Environmental Impact Statement, they are going to be required to take a hard look at the new information on water quality, discharges of heavy metals, and serious long-term consequences from the Kingston coal ash spill. (0003 [Mastin, Mary])

Comment: Please, as you go forward with the environmental work on this, consider the water quality and the new information -- I mean, not only is there -- are there sediments on the bottom where the Clinch River comes into TVA, coming down from Oak Ridge, there apparently is some other stuff from some old paper mill or lumbering operations; there has been a huge concern about doing that very carefully. (0003 [Mastin, Mary])

Response: *Operating a nuclear plant involves discharging some effluents to nearby water bodies. The Clean Water Act designated the U.S. Environmental Protection Agency (EPA) as the Federal agency responsible for regulating effluent discharges to the Nation's waters. Although the NRC does not regulate effluents, it is responsible under NEPA to assess and disclose the expected impacts of the proposed action on water quality throughout the plant's life. The staff will assess water quality issues related to operating the proposed WBN Unit 2. Chapter 4 of the SFES will present the NRC staff's assessment of the nonradiological impacts to water quality. Chapter 4 will also address any cumulative effects of the proposed action.*

Comment: There's a whole lot of assumptions about what's a normal condition in the river and what's a normal year, and I think if you've noticed, the last decade we've seen increasing changes, perhaps due to climate change, where the definition of what's normal needs to be re-examined. (0003 [Paddock, Brian])

Comment: Operating Watts Bar 1 nuclear plant requires 188.2 million gallons per day of water drawn from the river. Each day, of that amount, 14.3 million gallons is evaporated into the air, not returned to the river. Yet another reactor, a second reactor here, drawing out so much water causes me to ask how much can we draw out of the river on any given day in the same reservoir. (0004 [Kurtz, Sandy])

Comment: The Tennessee River is already overstressed and does not need additional warm water discharge and water lost from evaporative cooling. (0008 [Gottfried, Yolande])

Response: *Nuclear plants consume water due to the evaporation of some of the water used to cool plant components. The NRC staff will address the impact of consumptive water losses on the sustainability of local and regional water resources. Although the NRC does not regulate or manage water resources, it is responsible, under NEPA, to assess and disclose the impacts of the proposed action on water resources. Chapter 4 of the SFES will assess impacts on water resource sustainability related to operating the proposed WBN Unit 2*

Comment: The second point in the scope of the environmental assessment is that there's an interaction here, because the State of Tennessee has just released the draft NPDES, National Pollution Discharge Elimination System, permit for the Watts Bar nuclear plant. That seems to be talking just about Unit 1, but in fact the way TDEC has written the draft permit, it's not clear if you could turn the switch on Watts Bar 2 if it were ready and use that same permit. And there are a number of defects and concerns specifically with that permit. We're going to talking with TDEC about this, and the time for public comment has been extended, so that permit is probably not going to be coming down the road until early next year, at the best, but here are some of the difficulties. And we're assuming -- and I think TVA asserts this in their comments on the NPDES -- that the phase 2 regulations don't apply here; that the phase 2 content of this permit under Section 316 is remitted to TDEC in terms of its best professional judgment. That could change if EPA puts the phase 2 regulations back into effect following the most recent Supreme Court decision. But right now it's up to TDEC, and there are limitations in both the Clean Water Act and in the state regulations. One of the main problems is that most of the environmental information that TVA brought to TDEC for the renewal and extension of the NPDES for the nuclear plant basically was ten and twelve years old. (0003 [Paddock, Brian])

Response: The NRC staff will discuss current surface water quality in Chapter 2 of the SFES and impacts from operating the proposed WBN Unit 2 in Chapter 4. TVA has indicated in its application that the discharge from WBN Units 1 and 2 will meet discharge limits stated in the current National Pollutant Discharge Elimination System (NPDES) permit. The NRC staff will consider discharge limits in its evaluation of impacts on the Tennessee River.

Comment: There is, we think right now, a clear failure of TVA with respect to the NPDES, and we think if they were held to this in the EIS for the additional thermal impacts from Watts Bar 2, that they simply have not been able to show that they won't violate the water quality criteria. They don't provide data on the drift community, the spacial or temporal distribution of the plankton in the mixing zones. The mixing zones, by the way, according to the diagram, as I read it -- and I admittedly am no expert on this -- seem to be substantially larger. And by the way, the initial mixing zone in the renewed permit that's proposed actually goes border to border in the river. There is no way for aquatic life to go down the river without being in either what essentially is a dead zone immediately next to the discharges or on the cooler but active side of the river where they would have impacts. (0003 [Paddock, Brian])

Response: The NRC staff will consider water quality impacts from operating the proposed WBN Unit 2 on the Tennessee River, including the effects of thermal discharge on aquatic life. Chapter 4 of the SFES will present results of this analysis.

Comment: As was mentioned earlier, you now have operating six nuclear plants plus one thermal plant on the same river system, and you're now about to add a seventh, and the cumulative impacts of this amount of cooling water, cooling water loss from evaporation, thermal -- cumulative thermal effects and so forth, needs to be looked at. TVA has already experienced the situation where, during summer peaks, it had to derate downstream nuclear plants. Building another one toward the top of the river system, when it simply, as a consequence of the thermal discharge, will then have to shut down the plants lower on the river system during the hottest times of the peak loads, is not going to make any sense at all. So TVA may have run out of running room in terms of thermal discharges. Let's identify that now before we go ahead and license this plant. In fact, let's make sure that we do it in such a way that those of us who are ratepayers don't wind up for another while elephant that's never licensed to operate. (0003 [Paddock,

Brian]}

Response: Chapter 4 of the SFES will address consumptive use and water quality impacts on the Tennessee River, including the thermal impacts of discharge to the Tennessee River, from operating the proposed WBN Unit 2. Chapter 4 also will present cumulative impacts to the Tennessee River from operating WBN Units 1 and 2 and other facilities.

Comment: The Tennessee River is stressed already -- the quality of the river. It has fish that are not safe to eat. There is the impact of the Kingston toxic fly ash spill which must be taken into consideration when assessing water quality, because we all live downstream. (**0004** [Kurtz, Sandy])

Response: Chapter 4 of the SFES will address impacts on Tennessee River water quality from operating the proposed WBN Unit 2. Chapter 4 also will present cumulative impacts to the Tennessee River from operating WBN Units 1 and 2 and other facilities.

Comment: I am very afraid that we are killing the aquatic life in the Tennessee River and that the thermal discharges from Watts Bar 1, Watts Bar 2, then you go down to Nickajack or Sequoyah, and Nickajack, you start up there where Oak Ridge -- there are still sediments with radionuclides -- I don't know the technical language on this, but I know that TDEC and EPA and TVA have been very concerned about the dredging as they are trying to clean up the Kingston coal ash spill and not getting down to the bottom and stirring up all of this really terrible stuff that's there. (**0003** [Mastin, Mary])

Comment: I'm working with scientists who have talked to us about the discharges from selenium; you got arsenic and mercury; you got heavy metals; you've got fragile fish; you've got mollusks. You have got a whole downstream river system and people who are dependent on your doing this with a great amount of care. (**0003** [Mastin, Mary])

Response: The NRC staff will address the cumulative impact on the aquatic biota in the Tennessee River in Chapter 4 of the SFES. The staff will consider thermal discharges from facilities, including WBN Unit 1, Sequoyah Nuclear Plant, and Kingston Fossil Plant, as part of the cumulative impact analysis. The staff also will discuss water quality issues related to radionuclides and heavy metals that exist in river sediments as a result of past operations at Oak Ridge, and the Kingston coal ash spill and subsequent cleanup activities.

Comment: There are a lot of questions with respect to the mortality of mussels downstream, even though TVA has spent a good deal of effort over the years relocating mussels. I'm not sure when we started rebuilding natural populations in different places in order to allow this kind of project to go forward, but it seems to me that the impact on mussels and the impact of mussel relocation needs to be documented currently. (**0003** [Paddock, Brian])

Response: The NRC staff will assess the impacts of operating the proposed WBN Unit 2 on the aquatic biota in the Chickamauga Reservoir including any plans for future relocation of mussels and impacts from relocation. Chapter 4 of the SFES will address impacts on aquatic biota from operating the proposed WBN Unit 2.

Comment: The temperature of the water returned is hotter, not good for aquatic life, and in droughts it can't be cooled enough and so has to be shut down, just as has happened summer before last, I think it was. (0004 [Kurtz, Sandy])

Response: Chapter 4 of the SFES will address impacts on the aquatic biota in the Chickamauga Reservoir from thermal discharges from the proposed WBN Unit 2.

Comment: So the point I'd like to make in response to my enormous sympathy to the economic problems of the area, and the mention of jobs in solid-state and other areas, is that renewable energy is a really labor-intensive operation, so that your intensive worker group that comes in to build the nuclear power plant, usually from outside the region, most of those leave when the plant is built, and a moderately small task force remains. Whereas if you instead focused on attracting some of this new technology development and factories, you could build up this region enormously, building and making environmentally benign technology to provide what electricity is needed. (0003 [McCluney, Ross])

Comment: Our unemployment rate in Monroe County right now is over 16 percent, so we would like to see jobs from that plant as it is being constructed and then once it's completed. (0003 [Burris, Shane])

Comment: Also, if they run out of money, there are provisions in the technical specifications to shut the plants down and put them in a safe condition so the public is not threatened. That being said, I really admire Mr. Burris for the comments he made about the economic impact this will have on our area, but I can tell you that the Nuclear Regulatory Commission does not have compassion at the level that they're really concerned about jobs. (0003 [Cobb, Jim])

Comment: So anyway, the green economy is how we're going to get back, and part of that green economy is to learn how to reintegrate our rural areas, our smaller towns with our urban centers and create the -- you know, in Nashville people are nuts about local produce. There's a whole industry of local growers that is growing up all around Nashville, and people are making a living at it. It's hard work, it's honest work. You get your fingernails dirty, but it's just an old-fashioned way to do it. And, you know, getting back to more locally based economies with an eye toward creating jobs in our rural counties is definitely something that we need to do, but these nuclear plants don't create very many jobs after construction. (0003 [Safer, Don])

Comment: The project will generate thousands of jobs during construction period and 250 permanent jobs in a region characterized by double digit unemployment. (0007 [Yager, Ken])

Response: Chapter 4 of the SFES will address regional socioeconomic impacts of the proposed action, including impacts to the local economy, employment, transportation, aesthetics and recreation, housing, education, community infrastructure, and social services.

Comment: [A]s an economic developer in the state of Tennessee, most economic developers know that the United States and the state of Tennessee's manufacturing base runs on cheap power. And if your cap and trade bill passes in Congress, the electric bill will go up about 300 percent, and also that will end manufacturing in this country as we know it, and we will only be one mass distribution center. (0003 [Burris, Shane])

Comment: Our community is suffering economically, and it's important for future economic development and the future health of our community that we have reliable -- cheap, reliable power so that we can continue to bring industry in to this community. (0004 [Smith, Stewart])

Response: The price of electricity is outside the regulatory scope of licensing actions; however, the NRC staff will evaluate the regional socioeconomic impacts of the proposed action in Chapter 4 of the SFES, including impacts to the local economy, transportation, aesthetics and recreation, housing, education, community infrastructure, and social services.

Comment: The Tribal Historic Preservation Office of the Eastern Band of Cherokee Indians is in receipt of the notification to act as a consulting party for the above-referenced project information and would like to thank you for the opportunity to comment on this proposed Section 106 activity. The EBCI THPO accepts the invitation to act as a consulting party on the above referenced Section 106 undertaking(s) as mandated under 36 C.F.R. 800. (0006 [Howe, Tyler])

Comment: The project's location is within the aboriginal territory of the Cherokee People. Potential cultural resources important to the Cherokee people may be threatened due to adverse effects expected from the level of ground disturbance required for this project. (0006 [Howe, Tyler])

Comment: Please send all related archaeological, cultural resource and historical investigatory materials, completed by the applicant to this office for review so we can make proper comments that pertain to accomplishing our NHPA requirements. (0006 [Howe, Tyler])

Response: As outlined in 36 CFR 800.8, "Coordination with the National Environmental Policy Act of 1969," and Section 106 of the National Historic Preservation Act of 1966, as amended (NHPA), the NRC will fulfill the requirements of NEPA and NHPA by consulting with and requesting input from the Eastern Band of Cherokee Indians. Chapters 2 and 4 of the SFES will provide historic and cultural resources information. The NRC will consult with the Eastern Band of Cherokee Indians to identify cultural resources important to the Tribe to avoid or minimize any potential adverse effects to historic properties from this undertaking.

Comment: Talking about a community, I don't see you taking this up to Farragut and putting the nuclear plant in the middle of Farragut, where the houses all cost like $750,000 or 2 or 3 million. (**0004** [Harris, Ann])

Response: Chapter 4 of the SFES will specifically address potential impacts of the proposed action on low income and minority populations.

Comment: Nuclear plants do have radioactive leaks into the water, which they say is insignificant, but since radiation is cumulative, how much is too much for humans and other life to absorb without health impacts? (**0004** [Kurtz, Sandy])

Comment: The sources of the contamination include leaks from pipes and vales and other water-bearing components and airborne discharges from cooling towers. These radioactive discharges are difficult to quantify and may be underestimated. (**0015** [Zeller, Lou])

Response: The NRC staff will evaluate the release of radioactive materials into the environment from WBN Units 1 and 2. Chapter 4 of the SFES will address the cumulative impacts from releases of radioactive effluents from WBN Units 1 and 2.

Comment: I think as -- since this reactor was proposed in the '60s, designed in the -- or licensed in the '70s, we had a lot of opportunity to have all these nuclear plants that have been operating. And I haven't seen any public health studies about the communities that are downwind, you know, with the windrows of where the wind blows, and if it's true that nobody is getting sick, that their cancer rates and leukemia rates are not elevated, wonderful; I would love to see it. But I haven't seen it. I've looked for it. It's not easy to find. I think in this Environmental Impact Statement we need to have a clear study of Watts Bar 1; Sequoyah, the two units, and -- well, in particular those three, because they're the same design of reactor. (**0003** [Safer, Don])

Comment: I read of a study completed in Germany. Since 1991 in fact they have done several studies in Europe regarding the health of children who live within ten miles of nuclear facilities, primarily in England and Wales. And what they found out was that there was a statistically significant increased incident rate -- I want to say that right, because these are studies -- significant increased incident rate for leukemia's among children within the five-kilometer zones around the sites. That is, the closer -- and it seemed that the closer you got to the plant, the more -- the higher the incidence. This is of great concern and I think should be looked into before we add another reactor. (**0004** [Kurtz, Sandy])

Comment: I know that the lady before me made mention of a high incidence of leukemia within a close proximity of the plant. I'm somehow unaware of that. We have children in Meigs County -- I have a son that grew up in Meigs County, went to high school in

Meigs County, and I've never heard of a high incidence of leukemia; that's -- but I will investigate that to see if there are. (0004 [Jones, Ken])

Comment: I was born and raised in Meigs County, but I won't live there anymore. There's more to radiation exposure than cancer, and there's a lot of it. (0004 [Harris, Ann])

Response: These comments refer to health effects to populations around nuclear power plants. The NRC's primary mission is to protect the public health and safety and the environment from the effects of radiation from nuclear reactors, materials, and waste facilities. The NRC's regulatory limits for radiation protection are set to protect workers and the public from the harmful health effects of radiation on humans. The limits are based on the recommendations of standards-setting organizations. Radiation standards reflect extensive scientific study by national and international organizations. The NRC has reviewed a number of studies that looked at the incidence of cancers in the vicinity of nuclear power plants in the United States. The studies did not observe a correlation between the radiation dose from nuclear power facilities and cancer incidence in the general public. Some studies the NRC recognized include those conducted by the following organizations: the National Cancer Institute, the University of Pittsburgh, the Illinois Public Health Department, the Connecticut Academy of Sciences and Engineering, the American Cancer Society, and the Florida Bureau of Environmental Epidemiology. Chapter 4 of the SFES will evaluate the impacts to human health from radioactive emissions.

Comment: You don't have -- there's no water testing in this river of radionuclides by an outside source. That's according to TDEC's own mouth. That's not my opinion. They trust TVA. Well, we trusted TVA up at Kingston. There's tritium in the soil and the water, above legal limits. It's sitting there, and nobody's doing anything about it, you're just pumping more. And this idea that tritium won't hurt you -- why do we use it to make bombs go off faster than what they did when just a normal bomb? There's no wastewater program to stop the radionuclides going into the Chattanooga and others' drinking water. (0003 [Harris, Ann])

Comment: There is also -- there is radiation already in the river sediment, and another nuclear reactor will only add more. Nuclear plants put radionuclides in the water that no one tests for. (0004 [Kurtz, Sandy])

Comment: The NRC -- when you go to the website, look up the word tritium, and you go down through there, and you go and see what all the things are. There's a statement there -- it's very short; I think it's got -- I'll count them in a minute -- like a dozen words in the statement. The NRC does not believe that there's any safe level of exposure to radiation. (0004 [Harris, Ann])

Comment: We respectfully submit that the EIS should consider the issue of tritium releases into the Tennessee River by the proposed reactor. (0013 [Safer, Don])

Comment: Nuclear power plants generate tritium in the course of their operation and release it both to the atmosphere and to water bodies. Tritium releases have also occurred as a result of malfunctions. (0013 [Safer, Don])

Comment: Tritium, a radioactive form of hydrogen . . .combines with oxygen to make radioactive water. As radioactive water, tritium can cross the placenta, posing some risk of birth defects and early pregnancy failures. Ingestion of tritiated water also increases cancer risk. (**0013** [Safer, Don])

Comment: Tritium releases generally constitute the largest routine releases from nuclear power plants and as such have caused widespread contamination of water bodies at low-levels. (**0013** [Safer, Don])

Comment: All of this is particularly relevant to public health issues considering the widespread usage of the water from the Tennessee River especially as the municipal drinking water supply downstream in Chattanooga. (**0013** [Safer, Don])

Comment: The NRC must include in its SEIS the impacts of tritium emissions from both Watts Bar Unit 1 and Unit 2 upon the environment and public health. (**0015** [Zeller, Lou])

Comment: Tritium releases are the largest routine radioactive emissions from nuclear power plants. The chemical compound H2O with a radioactive H3 (Tritium) is virtually impossible to contain because nuclear plants are thermoelectric units which rely upon the heating of water to drive steam turbine-powered electric generators. (**0013** [Safer, Don])

Comment: Nuclear power plants contaminate the water bodies used for cooling water. Watts Bar Unit 2, like Unit 1, would be cooled by cooling towers drawing makeup water from Chickamauga Reservoir. The contamination of the area surrounding Watts Bar is as follows [Annie Makhijani and Arjun Makhijani, Science for Democratic Action Vol. 16, No. 1, August 2009 (Sources by plant from Annual Radiological Environmental Operating Reports for 2006. Sourcelink at http://www.nrc.gov/reactors/operating/ops-experience/tritium/plant-info.html)]:

	Drinking water	Surface Water
Picocuries per liter	606	588

These levels of tritium contamination of drinking water and the river are found 24 and 9.9 miles from the Watts Bar reactor, respectively. They are excessive and harmful to human health. (**0015** [Zeller, Lou])

Comment: That tritium emissions are released to the environment is well known and even acknowledged in NRC "lessons learned" documents. At minimum, the NRC must account for these releases in its EIS. Further, the agency should undertake a top to bottom review of its monitoring and control of tritium emissions. (**0015** [Zeller, Lou])

Response: *The NRC staff will review and evaluate the monitoring for radionuclides in the environs around the WBN plant and the Tennessee River. Chapters 2 and 5 of the SFES will address radiological monitoring of all pathways, including water. Chapter 5 also will discuss tritium monitoring at the WBN site. Chapter 4 will present results from the radiological monitoring and any potential* environmental impacts.

Comment: Tennessee Valley Authority is irradiating Tritium-Producing Burnable Absorber Rods (TPBARs) for the U.S. Department of Energy (DOE). The production of radioactive tritium for defense purposes is authorized by License Amendment No. 48 issued October 8, 2003. However, the tests conducted during the sixth cycle of irradiation revealed disturbingly high levels of tritium to the reactor coolant system outside of acceptable limits; in fact, the emissions were 9.6 times higher than predicted by TVA's analytical model. **(0015** [Zeller, Lou])

Comment: The questions which NRC must address are: (1) How were predictions by TVA and DOE nearly an order of magnitude too low? (2) What was the impact upon the local environment caused by the unexpected excess before it was discovered? (3) What are the implications for Watts Bar Unit 2? (4) What evidence do we have that TVA's predictive analysis is now reliable? **(0015** [Zeller, Lou])

Response: This comment is related to tritium production from WBN Unit 1 and is not within the scope of the environmental review for the proposed WBN Unit 2. However, the cumulative impacts from the releases (including tritium) from WBN Unit 1 will be considered and addressed in Chapter 4 of the SFES.

Comment: And the situation, as I understand it, in the environmental assessment that's being done right now is that indefinite on-site retention of spent fuel is proposed. So I hope you folks locally are prepared to take care of this stuff for at least a quarter of a million years, because with respect to spent fuel, it's pretty clear that Yucca Mountain is dead. I'm not sure exactly the state of the post mortem and rites, but it appears that the federal government is not going to invest more in the development of that site, and no other site has as yet been suggested even as a possible target. **(0003** [Paddock, Brian])

Comment: TVA, of course, has no right, even if Yucca Mountain were to open, to send the waste from Watts Bar 2, as I understand it, to that repository, even if it were to open, and it simply has, as far as I can see, no real plan other than just keep stacking it up locally. **(0003** [Paddock, Brian])

Comment: I'm going to start by going into the storage casks -- the spent-fuel storage casks that are being placed by the river right now. They're going to be placed there with greater frequency if this second plant goes on line. **(0003** [Safer, Don])

Comment: I think it's important to know that inside of those casks the radiation is far worse than what went in. The radionuclides in there, there's plutonium, which didn't even exist on the face of the earth until we started fooling with the atom 60, 70 years ago, and that's one of the most awful substances on the face of the earth. It is bomb-making material, but one atom of that that gets into your lungs, if it gets airborne, will give you lung cancer; it will kill you. It burns on contact with the air, spontaneously. It's sitting in there. **(0003** [Safer, Don])

Comment: It's not a whole big lot of plutonium in there; that's why reprocessing is such a nightmare, because to get enough plutonium to make it work, you've got to create a lot of other waste. But inside of there is just this cauldron of about 500 degrees -- it's too hot at the beginning, for the first five years, to put these fuels rods into these dry casks; they have to be put into the storage pools, which are overloaded currently and have had to be modified because of the lack of any real storage solution. And then after five years they go into these concrete-steel dry cask storage that are not hardened, and they are out -- I've seen them at Browns Ferry; they are just out in the open. **(0003** [Safer, Don])

Comment: [I]n those casks, that cauldron of 500-degree Fahrenheit radioactive material that's 1000 or 100,000 times more radioactive than the original fuel rods is doing who knows what. I mean, I asked -- I've forgotten your name, but I asked three gentlemen from the NRC earlier today, in private, or in a conversation at the open house, What's going on inside of those casks? Has anybody taken one of those apart after ten years? To my knowledge, nobody has, and what I've heard is that it's all sort of, you know, just kind of decomposing. Nothing stays the same. You put it in there, and its 500 degrees of boiling radioactive science experiment. And they were supposed to last for about or 30 years at first; now they're saying, well, they'll go for 50 and probably a hundred. Well, it's your community here that is the guinea pig on this, as well as the community at every other nuclear reactor site, because that's what's happening with all of these; there's no plan at all to move them away from your community, and these things, as Mr. Paddock said, they remain toxic for literally several hundred thousand years. **(0003** [Safer, Don])

Comment: [T]hey [nuclear plants] leave these legacy of these storage casks that our grandchildren, our great-grandchildren and those beyond that will not remember us will curse us for those storage casks.

Comment: [T]here is the storage of radioactive waste and the legacy it leaves for the future; there is no solution now, and we hear people say, We're going to figure it out. They've been working on it for a long time, and so far we actually seem to be going backwards. Yucca Mountain is closed and, in fact, if it were open, it would be immediately filled up, as I have heard, because we've already stored enough to fill it up. Where does our radioactive waste go? **(0003** [Safer, Don])

Comment: Somehow somebody's got to start stopping and looking, because you haven't dealt with the waste. **(0004** [Harris, Ann])

Comment: There is still no solution to the problem of storing nuclear waste. **(0006** [Gottfried, Yolande])

Response: The NRC evaluated the safety and environmental effects of long-term storage of spent fuel and, as set forth in the Waste Confidence Rule at 10 CFR 51.23 (available at http://www.nrc.gov/reading-rm/doc-collections/cfr/part051/part051-0023.html), the NRC generically determined that "if necessary, spent fuel generated in any reactor can be stored safely and without significant environmental impacts for at least 30 years beyond the licensed life for operation (which may include the term of a revised or renewed license) of that reactor at its spent fuel storage basin or at either onsite or offsite independent spent fuel installations. Further, the Commission believes there is reasonable assurance that at least one mined geologic repository will be available within the first quarter of the twenty-first century and sufficient repository capacity will be available within 30 years beyond the licensed life for operation of any reactor to dispose of the commercial high-level waste and spent fuel originating in any such reactor and generated up to that time." On October 9, 2008, the NRC published for public comment a proposal to amend its generic determination of no significant environmental impact for the temporary storage of spent fuel after cessation of reactor operation codified at 10 CFR 51.23(a) (73 FR 59547) and a related update and proposed revision of its 1990 Waste Confidence Decision (73 FR 59551). Chapter 4 of the SFES will address the impact of the uranium fuel cycle, including disposal of low level radioactive waste and spent fuel.

Comment: And some people have said that the electricity you get from the nuclear reactor is not really the primary component or the primary outcome; it's really all this nuclear waste, because the electricity you generate, we use it or we don't, and it's gone. (0003 [Safer, Don])

Response: According to 10 CFR 51.95(b), the SFES, which is a supplement to the FES-OL, will only cover matters regarding radioactive waste material (low-level, high-level, and transuranic wastes) that differ from the FES-OL or provide significant new information concerning issues discussed in the FES-OL. Chapter 4 of the SFES will discuss issues related to radioactive waste management.

Comment: [T]he Watts Bar Lake area already is highly polluted, particularly at the junction with the Clinch River and is already a designated Superfund site. And I have not had a chance to review the documents, but it's not clear to me that the -- what happened - - if there's any mobilization of those upstream legacy sediments from that Superfund site and moving down into the cooling-water intakes for this plant. The same thing is true with respect to the coal ash spill, because we've already seen the coal ash migrate during high-water events. They now they're going to get it out of there by -- worst of it out of there by next year, but they also say there won't even be the phase 2's plan for getting some of the rest of it cleaned up until next year. To the extent that those heavy metals are in solution, are in compounds and can travel freely with the flow of the river, you essentially have a different condition in the river at the point that you hit the cooling-water intakes, and we're not sure that the environmental assessment at this point has recognized that condition and has looked at the consequences of having heavy metals in solution in larger proportions at the point of intake and discharge from the cooling water. (0003 [Paddock, Brian])

Comment: These proposed [tritium] releases should be considered as an addition to the existing releases from Watts Bar Unit 1 which have been increased by the production of weapons grade tritium for the DOE. (0013 [Safer, Don])

Comment: The requirements of NEPA may not be avoided by segmentation of a project [River v. Richmond Metropolitan Authority, 481 F.2d 1280 (4th Cir. 1973)]. Segmentation arises when the comprehensive environmental impact of a project is not given full consideration or that analysis of the impact is done after permitting agency decisions are made and the project is underway [Daniel R. Mandelker, NEPA Law and Litigation, 9-25 (2nd ed. 2004)]. The principal criteria for the determination segmentation are whether the parts of a project are interdependent, the original intent and whether the parts may be considered alone. Watts Bar Units 1 and 2 are co-located facilities. They share certain structures, systems and components. Cumulative actions are those which have significantly greater impacts when viewed with other actions or which have increasing effect caused by successive additions. Council of Environmental Quality Regulations Implementing NEPA [Sec. 1508.7 Cumulative impact. "Cumulative impact" is the impact on the environment which results from the incremental impact of the action when added to other past, present, and reasonably foreseeable future actions regardless of what agency (Federal or non-Federal) or person undertakes such other actions. Cumulative impacts can result from individually minor but collectively significant actions taking place over a period of time provided that reasonably foreseeable future actions are to be considered in a cumulative impact analysis. The consecutive licensing of Watts Bar Units 1 and 2 in close proximity are actions which are plainly foreseeable. Therefore, NRC must account for the combined impact of Watts Bar Units 1 and 2 in its EIS. (0015 [Zeller, Lou])

Response: The NRC staff determines cumulative impacts by evaluating results from the proposed action in combination with other past, present and reasonably foreseeable actions, regardless of who takes the actions. The NRC staff will evaluate cumulative impacts associated with operating the proposed WBN Unit 2 for each affected resource. Chapter 4 of the SFES will present the results of cumulative impact analyses.

Comment: So my concern is that there are lots of moves afoot to reduce our needs for electricity in the Tennessee Valley and around the country that aren't really addressed in TVA's Environmental Impact Statements, that I've been able to find. In particular, I'll refer to sections relating to alternatives, alternatives to building the plant. And sometimes TVA will put a little bit in about that, in other cases, so I searched the most recent Environmental Impact Statement prior to this meeting, and what I found was a statement that referred back to that 1995 -- December 1995 earlier Environmental Impact Statement for finding something about alternatives. (0003 [McCluney, Ross])

Comment: We don't know -- because I couldn't find that document -- whether those alternatives were just alternatives to the design of the plant, alternatives to mitigate environment impact, or whether it actually included alternative power sources or other options for reducing the need for the plant in the first place. So I believe TVA is fairly deficient in that area. Even if the 1995 report addresses the subject, a whole lot has happened since then, in 14 years. There's been an enormous amount of research, development, and promulgation of energy-efficient technology and renewable energy choices. It doesn't take a particularly astute observer to know about a lot of this. If you watch TV, and especially if you go to the science channels -- Discovery, National Geographic, and these channels -- if you read the paper, read magazines, you'll see about this, because everybody's excited about these relatively pollution-free or somewhat benign alternatives -- energy alternatives. (0003 [McCluney, Ross])

Comment: Millions and even billions of private money have been spent to explore, develop and actually commercialize an enormous variety of technologies we still don't know too much about unless you really dig in. A good -- some good searches on the internet will reveal a lot of this technology, a lot more about it, and yet we see nothing about this in TVA's reports. So the question is, Do they fail to include it because they've already decided, years ago, that solar can't work here, or whatever decision they make, and so because they made that decision -- and if we trace it back, we may have to go back to the original -- I fear we have to go back to the original Environmental Impact Statement in 1978. So I glanced through this document to see if I could find a reference to that, and there was nothing there. So I fear that the really viable alternatives in renewable energy and energy efficiency have not been addressed and therefore the decision could be one based on inadequate information that will endanger the public. (0003 [McCluney, Ross])

Comment: But even if the demand is lower, that doesn't mean they won't have to build new plants, because hopefully they'll be taking out of operation all those dirty coal plants, and so they'll need to replace some of those, and I admit that. But I'd hate to see it with nuclear, when abundant natural energy is available from the sun and from other sources, outside this region, with long distance transport of energy as well as within this region, and yet TVA is silent on this. So what I urge the Nuclear Regulatory Commission to do is insist that, before they give any permit to this Unit 2, that TVA do a truly comprehensive study of these other alternatives: improved energy efficiency and renewable energy development. (0003 [McCluney, Ross])

Comment: They [TVA] can put the solar systems out and lease the rooftops of customers in a whole new mode of power plant production which is called distributed energy. The beauty of distributed energy is they're relatively small; they're distributed over the region. They're not terror-susceptible, because you want to take out the power in the region? How many rooftops do you have to go and knock out in order to have an event? So distributed power has an inherently higher security factor to it. And the utility can participate; in fact, it already is, in very tiny, little minuscule power programs, where the homeowner pays to put the solar power on their roof, and then the utility pays them a double price for the electricity that's generated. So I think if they could look at that model more look at these new technologies, including battery storage -- battery storage is amazing; I thought it was the unsolvable problem, because solar power, we know, is intermittent, and therefore we need a way to store electricity or some other form that can be turned into electricity and then produce it where it's needed. TVA has a facility for that near my home in Chattanooga; it's pump storage on the top of a mountain, and then they pull the water down when the need the power at peak periods. So there are options available, and so I urge NRC to insist that TVA do this truly comprehensive study. If they do that, I suspect that what TVA will discover is they can withdraw their application for this new plant. (0003 [McCluney, Ross])

Comment: But one of the things I think TVA should be held to respond to in its environmental assessment is how poor its energy efficiency and conservation programs are. And I say that with respect to the staff who I've sat with a number of times and discussed with them the activities that they're rolling out, including the home energy audits and retrofits and so forth, and with respect to the State of Tennessee, which is going to I think not only get on board with solar generation but is going to join the national effort to invigorate the purchase of Energy Star appliances. Unfortunately, TVA, in its approach to energy efficiency and conservation, has made a number of missteps. If you'll remember the strategic plan, the first thing it did was to fail to have a target even for efficiency and conservation. After a good deal of public debate and lobbying, it put in, I believe, a 1400-megawatt cumulative demand reduction target, and as it has carried that out, by limiting its instructions to its consultants, the reports of which have not been released to the public on energy efficiency and conservation and the limited results that have probably come if you tell them only to look at a very narrow slice of the issue, is that you now have programs that really go to peak shaving only. There has been no effort really to engage with reducing baseload demand, and clearly the Watts Bar 2 plant is about baseload demand, not just about peaks. And it seems to me that as part of the environmental assessment, TVA should be made to explain why it does not expect the baseload

objective point of view about nuclear energy and went through to the completion of entirely comprehensive studies and assessments and found the opposite to that claim to be true. The folks who have done these studies are high experts in the fields of energy production technologies and the economics of operating these technologies. They know what they're talking about, and their studies have been thorough. The Institute of Energy and Environmental Research is a primary and star example, and this book that they've produced contains excellent documentation of the massive data and analysis that supports the view that alternative sources to both coal-burning and nuclear power can meet our future energy needs. The scope of NRC's Environmental Impact Statement for Watts Bar 2 should therefore include full attention to and genuine consideration of what's in this report, and don't expect it to be an easy read; it's highly technical and deep; but also in addition to this report, the other comprehensive studies that have been done. (**0004** [Reynolds, Bill])

Comment: In particular, in looking at these other studies that started out objective and neutral about nuclear energy, they should look at -- in the EIS process, they ought to look first at the real-world potential for renewals and implementation of more efficient end-use energy practices and conservation to displace the need for a Watts Bar 2. That would be component of a responsible and honest Environmental Impact Statement about the proposed licensing Watts Bar 2. (**0004** [Reynolds, Bill])

Comment: Secondly, in particular this EIS should fully assess the comparative financial cost of Watts Bar 2 -- capital cost and operating cost over the life of the plant -- in contrast to those same costs from meeting future energy needs while protecting environmental health and climate stability through applications of renewable resources and proved efficiencies in end-use energy use and conservation. (**0004** [Reynolds, Bill])

Comment: The money would be better spent on less dangerous alternative energy technologies and energy conservation. (**0008** [Gottfried, Yolande])

Response: The commission expects the staff to take the requisite "hard look" at new information on the need for power and alternative sources of energy and has authorized the staff to supplement the SFES if there is new and significant information relevant to these matters. Alternative energy sources, including energy-efficiency programs, conservation, and renewable energy sources, will be considered and discussed in the SFES.

Comment: Information available to the Service does not indicate that wetlands exist in the vicinity of the proposed project. However, our wetland determination has been made in the absence of a field inspection and does not constitute a wetland delineation for the purposes of Section 404 of the Clean Water Act. The Corps of Engineers and Tennessee Department of Environment and Conservation should be contacted if other evidence, particularly that obtained during an on-site inspection, indicates the potential presence of wetlands. (**0005** [Jennings, Mary])

Response: The applicant is responsible for obtaining a Section 404 permit, and the U.S. Army Corps of Engineers is responsible for ensuring the applicant's compliance with its permit. Although Chapters 2 and 4 of the SFES will describe onsite habitats, including wetlands, this level of wetland information does not constitute a wetland delineation. If a Section 404 permit is needed, the U.S. Army Corps of Engineers will require a wetland delineation.

Category No. 2
Public Comments Concerning Issues that are Outside the Scope of Review

Comment: They don't want anybody there. I mean, this is quite obvious that the public -- this is another way to shut out the public, and it's a constant thing that we have going here. I mean, you're talking about computer usage. Does anybody see any big overwhelming public libraries over there in Spring City that people can go and pull up on -- the Federal Register? I mean, I get notices because I have hounded you people for years to stay on the mailing list, but not everybody knows to do that, or people suddenly find out things. (0004 [Harris, Ann])

Comment: And this visit by the ACRS in the Federal Register -- do you all not all work together? Is this another group of people that's got their own little fiefdoms hanging around through the agency? (0004 [Harris, Ann])

Response: The NRC's mission is to regulate the safe uses of radioactive materials for civilian purposes to ensure the protection of public health and safety and the environment. As part of this mission, the NRC is responsible for reviewing and issuing licenses for nuclear power facilities. The Advisory Committee on Reactor Safeguards (ACRS) is an advisory committee mandated by the Atomic Energy Act of 1954, as amended, under the Federal Advisory Committee Act (FACA). The ACRS is independent of the NRC staff and reports directly to the Commission, which appoints its members. The provisions of the FACA govern ACRS operational practices. The ACRS comprises recognized technical experts in their fields. It is structured so that experts representing many technical perspectives can provide independent advice, which can be factored into the Commission's decision-making process. Most Committee meetings are open to the public, and any member of the public may request an opportunity to make an oral statement during a Committee meeting.

Comment: We've paid billions of dollars out through DOE at these nuclear facilities to people that are really dying. We have two in our family that's already died from cancer that worked in Oak Ridge. One of them did not die from -- a third one did not die from cancer; he died from Parkinson's disease, and that was a miserable time to watch. (0004 [Harris, Ann])

Response: The commenter is referring to the National Institute for Occupational Safety and Health's Dose Reconstruction Project for Department of Energy Sites. The NIOSH program is not related to any NRC-licensed activities. This comment will not be addressed in the SFES.

Comment: And the final note is that the decommissioning funds that TVA already has set aside for its existing nuclear operations were badly depleted by the change in the economy and the stock market decline. TVA is already trying to figure out ways to steal money from within its operating budget and perhaps pass through charges to ratepayers to rebuild that decommissioning fund, along with the retirement funds for its employee retirees, and the whole issue of an adequate decommissioning fund and how that's to be accomplished and whether it's really adequate in an age when you don't have nearly the options for the disposal of high-level radioactive materials which come when you disassemble a plant -- unless they're planning to just, you know, build a mountain over the thing, which I guess is the other option. **(0003** [Paddock, Brian])

Comment: But I would again ask that decommissioning -- both its costs and its practicability -- be listed as one of the environmental concerns that has to be addressed. **(0003** [Paddock, Brian])

Comment: And they're in DC now, asking for more funds. That doesn't even address the issue of decommissioning funds, which they had a major start on back in 1995, but somehow those funds got -- nobody could ever tell me what they spent them on. So at that point they had $257 million. The last time I asked, they had 42 million, so you -- I'll let you adjust your own mind as to where that money went. **(0003** [Harris, Ann])

Response: These comments concern decommissioning. Requirements for providing reasonable assurance that funds will be available for the decommissioning are provided in 10 CFR 50.75.

Comment: I'm also concerned about the high cost and the delayed return on that investment of a nuclear power plant. It's required to go through a lot of work like this meeting in preparation, a lot of analysis, and even when you get close to construction, it takes quite a while to get the plant operating and then tested and presumed safe enough to turn it on and finally start generating revenue. Well, in this economic time it's rather risky, and I'm sure -- I believe not a very good idea to invest so much money in something that may not be needed. **(0003** [McCluney, Rossi])

Comment: I see this is quite a problem to accomplish, in other words, a gargantuan challenge, at the very least. And environmental protection plan that could be fail-safe for eons to come would obviously run into costs over much time adding up to multi-trillions of dollars, I would imagine. Part of the gargantuan challenge, then, is creating such a plan that it provides and requires a funding system that will never fail. It will cost lots of dollars. If the funding system fails, the regulation enforcement will not be done, and it will present an unacceptable risk to the public. The Environmental Impact Statement must contain assessment of how these funds will be guaranteed. To me it is obvious those funds will have to come out of the pockets of either the ratepayers who buy the power or the taxpayers who bail out when the funds aren't there, or both, which is the kind of situation we have now, those of us who are ratepayers, in particular, with -- dealing with the cleanup of the toxic ash spill. **(0003** [Reynolds, Bill])

Comment: Couple of things that I want to address up front that Brian talked about earlier: TVA's debt that they admit to today is at

$29.5 billion. That's not my assessment anymore; that's what they admit to, but it's more like 42 billion whenever you take all that other rinky-dink stuff they don't count in; it's called creative bookkeeping. (**0003** [Harris, Ann])

Comment: Now they're asking us to believe -- or at least you to believe; they don't want to ask me -- that they can do Unit 2 at Watts Bar for less than $4 billion or thereabouts. Well, they started out telling people that they -- that Watts Bar 1 was $7 billion. That is not true. When you add in the interest, the amortized part of Unit 1 that you -- or Unit 2 that you already paid for, it comes up to closer to $12-1/2 billion. So now you're going to ask to be paid for probably another 6 to $8 billion on this one. (**0003** [Harris, Ann])

Response: The commission has authorized the staff to supplement the SFES if there is new and significant information relevant to these matters. In the SFES, the NRC staff will consider the cost of power produced by the proposed licensing action and the overall benefits and costs of operating the proposed WBN Unit 2. However, general issues related to the applicant's financial viability are outside of the NRC's regulatory scope, and the SFES will not consider them. The NRC has requirements for licensees at 10 CFR 50.75 to provide reasonable assurance that funds will be available for the decommissioning process.

Comment: We fully support licensing Watts Bar Number 2. (**0003** [Burris, Shane])

Comment: They [the NRC] are concerned about the health and safety of the public, the environmental impact, the physical security of the plants, and I firmly stand behind the continued construction and moving forward with Unit 2. (**0003** [Cobb, Jim])

Comment: And my recommendation to you folks from NRC is that you give serious consideration to issuing license for Watts Bar Unit 2. (**0004** [Jones, Ken])

Comment: As a member of this community or a member of the community that this plant serves, I would just like to speak out favorably for licensing of this plant. (**0004** [Smith, Stewart])

Comment: I fully support the decision of the Tennessee Valley Authority to complete construction of Unit 2 at the Watts Bar Nuclear Reactor site and urge favorable consideration from the NRC. (**0007** [Yager, Ken])

Comment: TVA's decision to complete construction of Unit 2 results from detailed studies of not only cost and energy needs, but environmental impacts as well. These studies satisfy me that the project is feasible and environmentally responsible. (**0007** [Yager, Ken])

Response: These comments provide general information in support of the application. They do not provide any specific information related to the environmental effects of the proposed action and will not be evaluated in the SFES.

Comment: I think that internationally scientists have, for my mind, proven that carbon emissions do have an effect on the environment, and I think that nuclear energy should play an important role in providing the energy that this country and this world needs, particularly this country: clean energy that does not contribute to global warming. (**0004** [Smith, Stewart])

Comment: I know that we have in this country had an incident that was certainly a serious incident. I'm getting on up there, a middle-age guy, and I can barely remember when that happened, and with the technology and as far as technology has come, I feel like this -- that we need to follow up with nuclear energy. (**0004** [Smith, Stewart])

Response: *These comments provide general information in support of nuclear power. They do not provide any specific information related to the environmental effects of the proposed action and will not be evaluated in the SFES.*

Comment: I heard concerns about, you know, we need to keep a scorecard that accepts nothing less than 100 percent, and I agree with that. The fact is that the Nuclear Regulatory Commission and Tennessee Valley Authority have a standard that the average person's 100 percent is probably the TVA and NRC's 50 percent. So I think that they go above and beyond the call of duty to make sure that we have safe power. (**0003** [Cobb, Jim])

Comment: I have lived with it for 35 years. I believe that TVA has proven to us that they can operate a nuclear plant in a safe, environmentally friendly manner. (**0004** [Jones, Ken])

Comment: I'd just like to say that the history of the Tennessee Valley Authority in operating nuclear plants has been very successful. (**0004** [Smith, Stewart])

Response: *These comments express support for the applicant. They do not provide any specific information related to the environmental effects of the proposed action and will not be evaluated in the SFES.*

Comment: I start from the national policy of the Sierra Club, which is that nuclear power plants should not be expanded as a source of energy in this country until we've solved the waste-disposal problem. (**0003** [Paddock, Brian])

Comment: This reactor should not be built. (**0003** [Zeller, Lou])

Comment: I have compiled a list of reasons, that I have just put together, as to why there should not be a second Watts Bar reactor. (**0004** [Kurtz, Sandy])

Comment: I am a concerned citizen living in eastern Tennessee and I wish to register my opposition to building (or continuing to build) the second reactor at Watts Bar. (0008 [Gottfried, Yolande])

Response: These comments provide general information in opposition to the proposed action. They do not provide any specific information related to the environmental effects of the proposed action and will not be evaluated in the SFES.

Comment: In addition to my general concerns about nuclear power -- I won't list all the concerns and fears; they're in the media. They've been examined quite a bit, and there's a lot of controversy about most of it, but I think the dangers are real; the potential environmental impact in the event of accidental releases of materials, either fuels or waste, are severe and consequential. What we're counting on is the probability, hopefully, of that happening being low, but as the number of these power plants and materials being transported across the country increase, the probability may change that something can happen, and if it does, it could spell serious consequences. (0003 [McCluney, Ross])

Comment: This spells danger to people in Rhea County, eastern Tennessee, if and when one of these reactors was to be breached. Combined with the fundamental problems of nuclear power, this presents an unacceptable risk in this case. (0003 [Zeller, Lou])

Response: These comments provide general information in opposition to nuclear power. They do not provide any specific information related to the environmental effects of the proposed action and will not be evaluated in the SFES.

Comment: TVA overall has a very mixed and, I think, unbalanced, poor environmental record, and I would invite the Commission to look at the inspector general's report on Kingston, which found a culture in TVA of dispersed responsibility, lack of accountability, lack of internal communication -- it was always somebody else's job. (0003 [Paddock, Brian])

Comment: I went to work for TVA at Watts Bar Nuclear Plant in nuclear construction in January 1982. They told me I'd be there nine months. It was nine years before I got a paycheck that did not have overtime on it. And I left under -- for me it was quite a -- I don't want to way victory, because I didn't really win anything; what I did is I turned some magnificently strong lights into some really dark areas of TVA's management, their money, their funding, how they spend that money, and how they abuse not only ratepayers, but they abuse each other, they abuse the public, they abuse their future, and they abuse my children and my grandchildren's future. (0003 [Harris, Ann])

Comment: Browns Ferry Nuclear Plant is listed by Region 2 as the worst nuclear plant program in America. Now, the same person that was over Browns Ferry's fiasco is heading up the Unit 2 fiasco at Watts Bar. The amount of money that was spent at Browns Ferry was two times the original designated amount, and longer term, so if -- TVA's habits have not changed in the past 25 years, the way I -- according to what TVA puts out. (0003 [Harris, Ann])

Comment: I mean, there were leaks; there were wet spots. There were studies that $26 million could have saved that whole billion-dollar nightmare. One of the ten worst environmental disasters on the planet is what that was called by Newsweek, and it could have been saved with $26 million worth of investment, and TVA would not spend it because of their slavish devotion to the bottom line and keeping our electric rates low, which I appreciate, but it's given everybody the wrong message. (0003 [Safer, Don])

Comment: There's other security guards at TVA that none of them knew anything about each other until they came to me; one from Browns Ferry, two from Sequoyah, one from Watts Bar, and then this woman out of corporate. This is the beginning of the same pattern that TVA went through back in the late '80s and the '90s, and I don't see why that we have to go over that same road and travel that same absolute harassing, demeaning, humiliating practice again, because the only people that come out on top of this is the media, and the only way that we can get anything done is through the media.NRC doesn't want to listen; TVA won't listen; the Inspector General won't listen, and the only people that we've got to go is to the media and the Congress, and we're there. (0004 [Harris, Ann])

Response: These comments express opposition to the proposed action. They do not provide any specific information related to the environmental effects of the proposed action and will not be evaluated in the SFES.

Comment: I'm told by inside sources that are working with the engineers that we have engineers on site that don't know the difference between a code plant and a noncode plant. Maybe the NRC can describe to the engineers that are working on Unit 2 at Watts Bar what the difference is and how they need to -- how they can see that what they're doing is not working. Browns Ferry is a noncode plant. Watts Bar Unit 2 is a code plant. And for those of you that don't know and didn't work at the plant, you'll just have to look it up and trust me on that one. I find that the evacuation plan -- and this is just kind of silly. I'm appalled that the NRC even lets this get put in print. But in the evacuation plan, that they're going to take the people that live north of the plant, in Spring City and ten miles on both sides of the river, and they're going to move them up the valley 20 miles downwind; that means north of -- the prevailing winds all move north in this valley. You can't -- it's just common sense -- and if you live here, you would know that and wouldn't question it. But to take people that would be evacuated from Watts Bar Nuclear Plant or the surrounding community and move them 20 miles up the valley to put them in storage in a gymnasium at the junior college -- I mean, I live there, in the connecting community. This is just beyond the pale. I mean, I just -- I don't know if the NRC -- if they just really and truly don't care anymore or if they're just too ignorant to ask anybody besides themselves, who don't trust each other. (0003 [Harris, Ann])

Comment: My mother lives in a direct line of eight miles from Watts Bar Nuclear Plant. She's 86 years old, and she's in severe bad health. I take care of her. In fact, somebody's hired today so I could be here with y'all. I know that you're going to enjoy what I have to say, but this is the truth. My mother gets a calendar; it's this size (indicating). She didn't know what it was, because she couldn't read it. And then we put all of the announcements on Knoxville and Chattanooga radios. What's the problem with putting it out on the local radios? My mother doesn't listen to Chattanooga and Knoxville; she can't even get them. She listens to Athens; she listens to Dayton; she listens to Crossville. What is it with you guys? My mother cannot read this calendar, and I go into it, and I find something that is so disgusting y'all all ought to get up and and walk out; I think you ought to be fired now, because in this calendar it says. Take this calendar and keep it with you wherever you go, so that whenever the accident happens, you'll know which direction to go in. And part of the direction is to come back toward the area that will be so bad that it'll be blocked off. What is it with you people? Don't y'all read what you write? Don't you ever look at it? I mean, it's just really disgusting. This is what you're going to my family. Think about -- there's other -- I'm not -- my mother's not the only elderly woman in these communities; she's not the only one. There's little children. I've got great-grandchildren that will be affected by this, sitting in close proximity to Watts Bar. (0003 [Harris, Ann])

Response: These comments relate to the adequacy of emergency plans, which is a safety issue that is outside the scope of the environmental review. As part of its site safety review, the NRC staff will determine, after consultation with the Department of Homeland Security and the Federal Emergency Management Agency, whether the emergency plans submitted by the applicant are acceptable.

Comment: I admit that TVA will need electricity, not necessarily because it expects a growth in demand -- I really don't think because of all this technology is getting out there that the demand will be as high as they think it's going to be; I think the lower growth in their Environmental Impact Statement, the one that's slightly negative, may be closer to the truth. (0003 [McCluney, Ross])

Comment: You know, the electric power that it will generate is very necessary. There's something that most people in this room may not know. They're going to build a company, Beikler, in Cleveland, Tennessee, that will build solar panels; they will also make semiconductors, but mostly solar panels. That build out, that plant will require a quarter to a third of a nuclear power plant to run its full operation. (0003 [Burris, Shane])

Comment: The second thing is basically the -- and this goes to the question of whether or not a license should be granted at all under NEPA standards, but also to the environment assessment, is options and alternatives, as Dr. McCluney addressed. Basically, you have a situation where, according to the reports to the Tennessee Valley board of directors, power production and sales have dropped approximately 9 percent during the current economic downturn, the end of which one can debate if it's begun to happen, let alone any true date for that. In the past TVA, in its power projection demands, including those I assume that were used when the board decided to go ahead and restart construction on Watts Bar Unit 2, was that there would be an annual 2 percent increase in demand. That in fact hasn't happened; the reverse has happened. And if in fact we were to have effective conservation and efficiency programs, it would never happen. We would go into a flat or declining demand usage, and we would have reduced energy intensity on a per capita basis in the TVA service area. (0003 [Paddock, Brian])

Comment: [D]emand for electricity is down. (0004 [Kurtz, Sandy])

Comment: [S]outheast Tennessee probably is one of the fastest growing areas from a standpoint of population in this state. In the last five or six years, we have seen a tremendous spurt of growth. And certainly when we experience those things, then we're going to see a higher demand for energy. (**0004** [Jones, Ken])

Comment: completion of Unit 2 makes good sense, because it uses an existing asset to meet the growing power needs of the Tennessee Valley. (**0007** [Yager, Ken])

Comment: There is no guarantee that the demand for power would justify the cost of this plant by the time it is completed. (**0008** [Gottfried, Yolande])

Response: In accordance with 10 CFR 51.95(b), unless otherwise determined by the Commission, this SFES will not include a discussion of need for power, or of alternative energy sources, or of alternative sites, or of any aspect of the storage of spent fuel for the nuclear power plant within the scope of the generic determination in § 51.23(a) and in accordance with § 51.23(b), and will only be prepared in connection with the first licensing action authorizing full-power operation. Therefore, this issue is outside the scope of the environmental review and will not be analyzed in the SFES.

Comment: One percent slackness on enforcement is a failing grade. Why? -- Because of what it can do to human beings and their lives and their health. Necessary ramifications, lesson learned, is the assertion that and Environmental Impact Statement that omits responsible, honest accounting for perpetual vigilance through the eons to come, continuously and consistently, is not worth the paper it's written on. So I'm here encouraging NRC to make sure they get all that covered, all that protection of human health and life in perpetuity, as long as the waste will last. (0003 [Reynolds, Bill])

Comment: I went there for an NRC hearing about the unscheduled shutdowns of that unit that that they brought back on line, the five of them in the first five or six months. It caused a big, huge slap on the wrist by the NRC. I will have to support some of what Ann said about the NRC seems to be the enabler of the nuclear industry and not the watchdog, and that's not any news for people that have been following this issue for quite a while. (0003 [Safer, Don])

Response: The NRC's mission is to regulate the safe uses of radioactive materials for civilian purposes to ensure the protection of public health and safety and the environment. The NRC has established an extensive regulatory process to ensure the integrity of each application review. The NRC can deny an application for an operating license based on the findings of its review. These comments do not provide specific information related to the environmental review and will not be addressed in the SFES.

Comment: This bears saying in a scoping session for the environmental impact assessment of a new nuclear power plant here, because the most noble and honorable Union of Concerned Scientists, who are not antinuclear, by the way, but they do totally responsible scientific evaluation and assessment of the nuclear power industry and, upon close scrutiny of the Nuclear Regulatory Commission's track record and their oversight of nuclear power plant operation, concluded as follows: Nuclear power is riskier than it should and could be. The United States has strong regulations on the books, but the Nuclear Regulatory Commission does not enforce them consistently. I agree with the implication in this statement that emphasizes the consistency. TVA has done a lot of good things; we all know that. We appreciate the great service they've done, but -- and it's not all their fault, because the regulations were not in place regarding the coal ash spill. Regulations are, according to the UCS, in place for strong management of nuclear power, so consistency is what's needed, unfailing consistency. NRC cannot be given a passing grade on their regulation enforcement for anything less than a perfect 100. (0003 [Reynolds, Bill])

Comment: How do you think this makes me feel, to know that I'm paying your salaries, and you're not doing your job. You're just accepting whatever TVA hands you, and TVA will hand you a bunch of garbage, because they will lie. Got it? I don't even want to have to say it anymore. You can't trust TVA. You can't trust TVA. How long do you have to have that said to you? And now you can't trust the NRC, because the NRC, they are so close in bed with TVA, that you're beginning to look a bit foolish, even from other people, not just me. Somehow or another this Environmental Impact Statement has to address these issues that concern and deal with people's lives on a day-to-day basis, and if these jobs are the best that TVA can provide, somebody else needs to be running TVA besides somebody that's running a bunch of serfdoms. (0003 [Harris, Ann])

Comment: I'm telling you, Region 2, we're asking for Congressional hearings on you and your inability to deal with TVA. This is a repeat of the 1985 and '86 hearings, and you can look for these to be just as disgusting whenever we uncover that pile of crap. (0003 [Harris, Ann])

Comment: We're not going to back down off of this, because the persecution of this -- she's a little, ol' grandmother; she's a clerk. She had a 18-year career in personnel, and nobody ever -- she never made a mistake. She had wonderful -- but the bottom line is that there's two women involved that come through the revolving door from the NRC, and they both lost their jobs and were removed from TVA, but then they went back to work for the NRC in in-house security. Now, what does that say about you, NRC? I can't trust you to do what you need to do, because you've still got the mentality that the workers don't know what we're doing, because management is always right. And what you found out after -- what was it? -- From 1984 to 1996 -- how many years is that? -- 12 years? You couldn't get it right, and TVA couldn't get it right, because everybody was talking about somebody; they wasn't talking to anybody, and nobody -- neither one of you were listening, and then the NRC -- I don't know what it's going to take. (0003 [Harris, Ann])

Response: *These comments are outside the scope of this review and do not provide specific information related to the environmental effects of the proposed action; therefore, they will not be evaluated further.*

Comment: I daresay I've learned a lot of valid lessons in my studies and private individual studies through the years, and I think I just recently, within in the past year, less than a year, have learned a most important new lesson that I think a lot of folks, including TVA itself, probably has learned as a result of the horrible disaster of the Kingston ash spill, not far from here, that you all probably are very well informed with the great disaster, and I'm not going to go into details about it. I bring this up at this time because I think it's a lesson learned that should be known and paid attention to in the practice of producing nuclear power plants and managing nuclear power plants and so on. (0003 [Reynolds, Bill])

Comment: I want to define a lesson learned that I think we should all apply, particularly to the scoping of building a new nuclear power plant here. And here's my definition: Regulations, monitoring inspection regimens, and compliance enforcement must absolutely be maintained and sustained with absolute unwavering consistency in perpetuity, as long as the waste remains. And we -- those who are informed about nuclear power waste products, some of those waste products remain lethal to human life and health for multiple centuries. There must never be a single occurrence of slacking in maintaining and sustaining protection of our supremely precious air, land, and water from exposure to the poisons contained in the waste produced by electrical power generation. Nothing akin to the Kingston coal ash spill should ever happen with nuclear power plants, whose waste is even more toxic than coal ash. (0003 [Reynolds, Bill])

Comment: And you cannot really think that you're going to have a safe 40- to 60-year operation of a nuclear plant in a culture where plant operations suffer from those same defects. Now, that was with respect to a fossil plant, where, if something goes wrong, ordinarily you think it's not going to be a big deal. Of course, that was a miscalculation, because when you lose 5 million tons of coal ash, it is a big deal. In fact, it's probably one of the biggest environmental disasters on the North American continent in our lifetimes. But please do look at the inspector general's report on the culture in TVA and decide what you have to do in terms of building that into the evaluation of environmental impacts. (0003 [Paddock, Brian])

Comment: Watts Bar Unit 2, as its sister reactor, Number 1, would utilize an ice-condenser containment structure -- many people have referred to this as an eggshell-type containment -- in order to reduce costs of construction, concrete and steel, in the construction of the containment vessel, that large domed structure. Ice-condenser units employ baskets of ice. During an event inside of a nuclear reactor, excess heat and pressure are created. Ice-condenser reactors are designed to reduce that heat and pressure by using baskets of ice. There are relatively few of these reactors in operation, and they are fraught with fundamental engineering flaws and also real-world difficulties in keeping baskets of ice free, operating over a period of decades, which they are required to do. The ice-condenser system should not be constructed in the 21st century; it should not have been constructed at all. (0003 [Zeller, Lou])

Comment: I am told by workers -- this is not engineers; this is workers, from the inside -- that the 21 million that you paid Bechtel to go in and see if Unit 2 could be brought up to speed, they spent their $21 million, walked around, and said, Yeah, we can do it; y'all have a good time. Then, guess what? Bechtel turned around and said, Okay, we're going to start letting them decide what all needs to be done. Bechtel's still looking at what needs to be done; they're still looking at it, because they're finding such massive amounts of rust and corrosion and equipment that cannot be used, won't be used, and cannot be replaced with what is there, because those people left and seen better days somewhere else that got the money, that took it and run. (0003 [Harris, Ann])

Comment: The cost-cutting measures designed to make construction cheaper result in some of the most dangerous reactors on the planet. A Sandia study which is memorialized in Nuclear Regulatory's own guidance documents, NUREG/CR-6427, in April 2000, states that ice-condenser plants are at least two orders of magnitude more vulnerable to early containment failure than other types of pressurized water reactors. Two orders of magnitude: ten times ten, 100 times more vulnerable to a catastrophic disaster. Hydrogen buildup during an event inside of a nuclear reactor is one of the reasons for this vulnerability. Measures over the years, which have been added to or retrofitted to existing ice-condenser reactors have addressed part of the problem. Buildup of hydrogen is why the pressure gets so high and can cause a rupture in the containment structure. Backfitting of hydrogen igniters over the years have not addressed the full problem. Ice condenser reactors are still vulnerable to hydrogen ignition during a reactor event which would otherwise be contained inside a more robust containment structure. (0003 [Zeiler, Lou])

Comment: So that's what going on inside those storage casks, which are going to be more and more along the river. They are not designed to be flooded. I don't know this particular site, I haven't seen it. I know at Browns Ferry they're not that high off of the river, and if they're flooded, then the cooling that is just a convection cooling with vents gets clogged with debris and what-not, and who knows what can happen. (0003 [Safer, Don])

Comment: Getting into that reactor design, that design dates from the 1960s. I was in high school when that thing was first proposed. I'm retired now. A lot of things have changed. You know, a lot of people in this room are not that much different in age from me; many are younger. But, my gosh, that design comes from the middle '60s; that was when the Mustang -- the first iteration of the Mustang was the hottest car going. You wouldn't buy the Mustang if it was in the showroom -- the 1965 -- well, you might buy it as an antique, but it's not going to perform up to environmental standards or whatever; the point being that this design was put together was an idea of cost containment and not safety. When it was originally designed and approved, there was -- Chernobyl had not happened. They thought an event like Chernobyl, an event like Three-Mile Island was not even possible; it was not in the design criteria for the original design, so that there -- and that's why they've had to go back with this hydrogen, you know, ignition system and how you take care of all that hydrogen. This was the cheapest reactor TVA could build at the time. It's a clear indication of the same culture that put that ash into the river. TVA was dumping that ash into that pile for 50 years. They had plenty of indications that the ash pile was suspect. (0003 [Safer, Don])

Comment: Back to that ice-condenser design, who can imagine putting 3 million pounds of ice in a nuclear reactor so that you can make the containment structure half as thick? My gosh, that's a fabulous idea. I applaud whoever came up with it. It's a wonderful idea. It's just like Rube Goldberg, though; it's stupid. You know, I mean, you have all that ice, which has problems with subsidence. I went on line, you know, last few days, and somebody patented an idea of what do you do with the ice that's compacted in there? The ice, from what I read, it's one-foot wide cylinders that are 50-feet tall, and they're wrapped with these steel containment things that are sort of straps. And so they can't get in there to replace the ice very easily, and somebody invented some sort of a -- I didn't look at the design, but some sort of a contraption to replace the ice, because they were having problems with the ice just melting away, which it does naturally, and not having the million pounds they needed to survive an incident, which is really a core meltdown, and to keep that containment structure, however fragile it is, from melting down. (0003 [Safer, Don])

Comment: For example, the most complete and recent probabilistic risk assessment suggests core melt frequencies in the range of 1 in 1000 per reactor year to 1 in 10,000 per reactor. A typical value is 3 in 10,000. I'm reading from David Lochbaum's monograph which quotes a Nuclear Regulatory Commission statement to US Congress, and that's what I am citing here. This is the NRC to the

Congress: Were this the industry average, then in a population of 100 reactors, which we have today, over a period of 20 years, the crude cumulative probability of a severe reactor accident would be 45 percent. That is for all reactors combined, including the more robust designs. The ice-condenser reactor can withstand half the pressure of the more robust old designs, not talking about the new AP-1000 and other designs which have not yet been built under CFR Part 52. (0003 [Zeller, Lou])

Comment: [T]his reactor plan relies on an outdated ice condenser plan that brings with it far more risk than is necessary. (0004 [Kurtz, Sandy])

Comment: That's not reliable power if you have to shut down the nuclear plants because of droughts and hot weather, an issue associated perhaps with climate change. (0004 [Kurtz, Sandy])

Comment: Most nuclear accidents happen due to human error. In the light of the Kingston fly ash spill, do you believe that TVA can avoid human error? And do you believe that TVA is choosing to use this old nuclear reactor design because it's the best technology available or because it's cheaper? (0004 [Kurtz, Sandy])

Comment: This reactor would use old technology, the ice condenser reactor, which is considered to have design flaws already. (0008 [Gottfried, Yolande])

Response: The issues raised in these comments are safety issues, and as such, are outside the scope of the environmental review and will not be addressed in the SFES. TVA provided a safety assessment for the proposed licensing action as part of its application. The NRC is developing a Safety Evaluation Report that analyzes all aspects of reactor and operational safety.

Comment: And in this letter it talks about this woman who worked at corporate security for TVA. She was drummed out because she asked too many questions, and she wanted to go by the rules. And the bottom line is that after a two year period, the young lady and TVA came to a mutually agreeable settlement, and then the NRC's Region 2 -- I don't know who they are; we keep getting all these different names of who they are, what they represent and what their agenda is. The bottom line is the NRC is going after this woman because they said that she was unauthorized to use documents when she was protesting her termination as retaliation against the issues that she had raised. TVA agreed, and they redacted the documents. Nobody was identified outside; no documents were taken off the jobsite. The bottom line is that the NRC's Office of Investigation, they're still pursuing this woman for criminal charges under federal -- they say federal laws; they can't tell us what they're looking for. I suspect that it's more of a fishing expedition than it is anything because somebody needs to keep a job, or they're doing something that they don't know what they're doing, or they're just totally incompetent and needed someplace to hide themselves. We went to the NRC's Office of Inspector General to try to stop it, and they told us that as long as there was an allegation against this woman by somebody at TVA, that they would pursue the issue, and they would not do any kind of investigation. Then, whenever we questioned that, TVA's Inspector General, they just didn't do anything. Of course, that's not unusual; that's their record of decision-making. And now we've been forced to file legal documents with the Commission over this issue. (0004 [Harris, Ann])

Comment: But the other thing is if I can't trust you to keep the security at these nuclear facilities and it's not even up and running, why should I trust you to do right whenever it's up and running? (**0004** [Harris, Ann])

Response: *Comments related to security and terrorism are not within the scope of the environmental review. The NRC is devoting substantial time and attention to terrorism-related matters, including coordination with the Department of Homeland Security. While these are legitimate matters of concern, they will continue to be addressed through the ongoing regulatory process as a current and generic regulatory issue that affects all nuclear facilities and many of the activities conducted at nuclear facilities. The Commission has affirmed that the National Environmental Policy Act (NEPA) does not require the NRC to consider the environmental consequences of hypothetical terrorist attacks on NRC-licensed facilities.*

Comment: I would like to see more development in recycling of our nuclear waste so that we can use that to the best of its ability. (**0004** [Smith, Stewart])

Response: *The recycling of nuclear waste is a national policy issue that is outside the scope of the environmental review of WBN Unit 2.*

Appendix E

Draft Supplemental Final Environmental Statement Comments and Responses

Appendix E

Draft Supplemental Final Environmental Statement Comments and Responses

E.1 Introduction

This supplemental final environmental statement (SFES) has been prepared in response to reinstatement of an application submitted to the U.S. Nuclear Regulatory Commission (NRC) by the Tennessee Valley Authority (TVA) for an operating license (OL). The proposed action requested in the TVA application is for the NRC to issue a facility operating license for a second light-water reactor located adjacent to an operating reactor within the existing Watts Bar Nuclear (WBN) station. This SFES updates NRC environmental statements published in 1978 and 1995 in response to the original application. It includes the NRC staff's analysis that considers and weighs the environmental impacts of operating an additional nuclear unit at the WBN site including an analysis of energy alternatives, and mitigation measures available for reducing or avoiding adverse impacts.

As part of the NRC review of the application, the NRC solicited comments from the public on the draft of this SFES, which was published in October 2011 (NUREG-0498 Supplement 2, NRC 2011a). The U.S. Environmental Protection Agency Notice of Filing in the *Federal Register* (76 FR 70130) indicated a 75-day comment period, commencing on November 10, 2011, to allow members of the public to comment on the results of the NRC staff's review. This was amended in the *Federal Register* on November 18, 2011 to a 45-day comment period (76 FR 71559). The NRC issued a Notice of Availability (76 FR 70169) of the draft SFES in the Federal Register that specified a 45-day comment period. On December 8, 2011, a public meeting was held in Sweetwater, Tennessee. At the meeting, the NRC staff described the results of the NRC environmental review, answered questions related to the review, and provided members of the public with information to assist them in formulating their comments. Based on comments received at the public meeting, the comment period was extended by the NRC to January 24, 2012 (76 FR 80409).

As part of the process to solicit public comments on the draft SFES, the NRC staff:

- Made the draft SFES available in the NRC's Public Document Room in Rockville, Maryland
- Made the draft SFES available on the Federal Rulemaking website http://www.regulations.gov (Docket ID NRC-2008-0369)

- Placed a copy of the draft SFES on the NRC website at http://www.nrc.gov/reading-rm/doc-collections/nuregs/staff/

- Provided a copy of the draft SFES to any member of the public who requested one

- Sent copies of the draft SFES to certain Federal, State, Tribal, and local agencies

- Published a notice of availability of the draft SFES in the Federal Register (76 FR 70169)

- Filed the draft SFES with the U.S. Environmental Protection Agency (EPA)

- Announced and held an afternoon and evening public meeting on December 8, 2011, in Sweetwater, Tennessee, to describe the results of the environmental review, answer any related questions, and take public comments.

Approximately ten members of the public attended these meetings and five attendees provided oral comments. A certified court reporter recorded these oral comments and prepared written transcripts of the meetings. The transcripts of the public meetings are part of the public record for the proposed project and were used to establish correspondence between comments contained in this volume of the SFES to oral comments received at the public meeting. In addition to the comments received at the public meeting, the NRC received a total of nine letters and submissions with comments including a submission by TVA. The comment period closed on January 24, 2012; however, the NRC did, to the degree permitted by the schedule, consider comments submitted after the comment period ended.

The portion of the public meeting transcripts containing questions and comments from the public are provided in Section E.4 of this appendix. The complete transcripts of the public meetings are available by accessing NRC's Agencywide Documents Access and Management System (ADAMS) at http://www.nrc.gov/reading-rm/adams.html, and searching for Accession Number ML113630081. Persons who do not have access to ADAMS or who encounter problems in accessing the documents located in ADAMS should contact the NRC's Public Document Room reference staff at 1-800-397-4209 or 301-415-4737, or by e-mail at pdr@nrc.gov. The ADAMS accession numbers for comment correspondence are provided in Table E-1. Copies of the comment correspondence are provided following the meeting transcripts in Section E.4 of this appendix. Comments and transcripts are also available on http://www.regulations.gov; a website for information on the development of Federal regulations and other related documents issued by the United States government.

E.2 Disposition of Comments

After the comment period, the NRC staff considered and categorized all comments received. To identify each individual comment, the NRC staff reviewed the public meeting transcripts and each letter or electronic comment received related to the draft SFES. As part of the review, the NRC staff identified statements that they believed were related to the proposed action and recorded the statements as comments. Each comment was assigned to a specific subject area, and similar comments were grouped together. Finally, responses were prepared for each comment or group of comments. Table E-1 provides a list of commenters identified by name, affiliation (if given), comment number, and the source of the comment.

Table E-1. Individuals Providing Comments on the Draft Supplemental Final Environmental Statement for Watts Bar Unit 2

Commenter	Affiliation (if stated)	Comment Source and ADAMS Accession #	Correspondence ID
Anonymous	Self	www.regulations.gov comment (ML12012A113)	0006
Andrews, Ann	State of Tennessee, Department of Transportation	Letter (ML12018A397)	0007
Brickhouse, Brenda	Tennessee Valley Authority	Letter (ML12040A052)	0010
Budnick, Donna	Self	www.regulations.gov comment (ML12005A082)	0001
Curran, Diane	Southern Alliance for Clean Energy	Letter (ML12030A100)	0008, 0015, 0016, 0017
Ferris, Kathleen	Self	Meeting Transcript (ML113630069)	0003
Harris, Ann	Self	Meeting Transcripts (ML113630069, ML113630077)	0003, 0004
Hogue, Gregory	United States Department of the Interior	Letter (ML12023A185)	0009
Keel, Jefferson	The Chickasaw Nation	Letter (ML12053A439)	0013
Kurtz, Sandy	Self	Meeting Transcript (ML113630069)	0003
Mueller, Heinz	EPA	Letter (ML12004A168)	0005
Riden, David	Self	Meeting Transcript (ML113630077)	0004
Safer, Don	Self	Meeting Transcript (ML113630069)	0003
Stanley, Joyce	U.S. Department of the Interior	Letter (ML12005A211)	0002
Thompson, Ian	Choctaw Nation of Oklahoma	Letter (ML12053A441)	0014

Sections E.2.1 through E.2.22 presents comments and NRC staff responses to them, grouped by similar issues, as shown in Table E-2.

Many comments specifically addressed the scope of the environmental review, analyses, and issues contained in the draft SFES, including comments about potential impacts to specific environmental resources, the agency review process, and the public comment period. Responses to each of these comments are provided in Sections E.2.1 through E.2.17. When the comments resulted in a change in the text of this SFES, the corresponding response refers the reader to the appropriate section of the report where the change was made. Revisions to the text from the draft SFES are indicated by vertical lines beside the text in this SFES.

Table E-2. Comment Categories in Order as Presented in this Appendix

E.2.1	Comments Concerning Process - COL
E.2.2	Comments Concerning Process - NEPA
E.2.3	Comments Concerning Site Layout and Design
E.2.4	Comments Concerning Geology
E.2.5	Comments Concerning Hydrology - Surface Water
E.2.6	Comments Concerning Hydrology - Groundwater
E.2.7	Comments Concerning Ecology - Terrestrial
E.2.8	Comments Concerning Ecology - Aquatic
E.2.9	Comments Concerning Socioeconomics
E.2.10	Comments Concerning Historic and Cultural Resources
E.2.11	Comments Concerning Health - Radiological
E.2.12	Comments Concerning Accidents - Severe
E.2.13	Comments Concerning the Uranium Fuel Cycle
E.2.14	Comments Concerning Decommissioning
E.2.15	Comments Concerning Related Federal Projects
E.2.16	Comments Concerning Cumulative Impacts
E.2.17	Comments Concerning Alternatives - Energy
E.2.18	General Comments in Opposition to the Licensing Action
E.2.19	Comments Concerning Issues Outside Scope - NRC Oversight
E.2.20	Comments Concerning Issues Outside Scope - Safety
E.2.21	Comments Concerning Issues Outside Scope - Security and Terrorism
E.2.22	General Editorial Comments

Other comments addressed topics and issues that are not part of the environmental review for this proposed action. These comments included questions about the NRC's safety review, general statements of support or opposition to nuclear power, observations regarding national nuclear waste management policies, comments on security and terrorism, comments on the NRC regulatory process in general, and comments on NRC regulations. These comments are included in Sections E.2.18 and E.2.21, but the responses to such comments are not as detailed because they addressed issues that do not directly relate to the environmental effects of this proposed action and are thus outside the scope of the National Environmental Policy Act (NEPA) review of this proposed action. Section E.2.22 contains general editorial comments and NRC staff responses.

E.2.1　Comments Concerning Process – COL

Comment: We would ask for an extension of 45 days so that people have an opportunity to comment on this outside of the holiday period. And I don't know that that can be granted today, but I think that's a formal request, as formal as I can get right here. (**0003-2-2** [Safer, Don])

Comment: So I'm making a formal request that we have an extension for 45 days. (**0003-2-7** [Harris, Ann])

Response: As a result of these comments, the draft SFES comment period was extended from 45 to 75 days (November 10, 2011 through January 24, 2012) as indicated in the Federal Register notice posted on December 23, 2011 (76 FR 80409).

Comment: In 2.6, the radiological environment, it references a report, Annual Radiological Environmental Operating Report, RAMP, and also the Annual Radioactive Affluent Release Report. I believe those are from TVA, but -- and I know you all are not TVA. But I'm just wondering how to get a hold of this document. (**0003-2-3** [Safer, Don])

Response: The TVA Annual Radiological Environmental Operating Reports and Annual Radioactive Effluent Release Reports for WBN Unit 1, which are cited in Section 2.6 of this SFES, are publicly available in the ADAMS Public Electronic Reading Room. These documents are accessible from the NRC website at http://www.nrc.gov/reading-rm/ADAMS.html by searching the Accession Number provided with each reference in Section 2.10 (References) of this SFES.

Comment: So I don't know where you [NRC] got your information. You may have gotten it from different agencies. You said federal, state, and local. Well, some of these with information in here that whenever I went, I got different information. So I'm having a hard time dealing with your numbers and the information that you're giving as opposed to what I'm getting from the same agencies.....So they're making me do FOIA requests over your documents that you requested to put into this. (**0003-2-4** [Harris, Ann])

Response: The NRC is an independent regulatory agency that is charged with the responsibility of overseeing the commercial nuclear power industry. The NRC staff independently reviews the applicant's submittals and related documents. The NRC staff also reviews and obtains information from onsite audits, meetings with Federal, State, and local officials, and various agency and institutional sources. After carefully considering all this information, the NRC staff prepares an independent assessment of environmental impact. The sources of information used in this document are provided in reference sections at the end of each chapter. If the reference source does not hold a copyright (for example on books or technical journals), the document is uploaded to the ADAMS Public Electronic Reading Room. These documents are accessible from the NRC website at http://www.nrc.gov/reading-

rm/ADAMS.html by searching the Accession Number provided with each reference. A search in ADAMS of the Accession Number will provide a link to the document. Books and technical journals can be obtained from the publisher. The URL is provided in references where information on the internet has a copyright.

Comment: Earlier today we talked about some of the documents that you used to make your judgment in here and some of them refer to 40-year-old documents. Now I realize some things haven't changed, but a lot more has changed than has not.

And I'm wondering on these documents where you used TVA's documents when you did use them and did you just accept TVA's documents without going back and checking to verify in those old documents and did you go and look for new information concerning those same documents because I'm not finding consistency between what you've put in and some things that I personally know about? And I'll put those in my comments. But I'd like to know how you made those determinations. (**0004-1-1** [Harris, Ann])

Response: The NRC staff considers the type of information that is available and determines whether it is still appropriate for use as is, in lieu of, or in combination with current available information that is more indicative of the affect the nuclear plants will have on the environment. For example, the site's geologic characterization has not changed in the last 40 years, so the older studies are applicable and useful for the evaluation process. For aquatic ecology, it is important to use the past studies to determine if the aquatic ecology has changed since the conditions that existed before the operation of WBN Unit 1. The applicant submits their documents under oath and affirmation that they are correct. However, the NRC has a process to review the documents and request clarification or further information if the documents are unclear or appear to contain contradictions. In addition, the NRC verifies the information provided by the applicant and compares it to other technical documents and expert studies. In some cases both sets of information are used.

E.2.2 Comments Concerning Process – NEPA

Comment: The Department of the Interior (Department) has reviewed the Draft Supplemental Environmental Impact Statement (SEIS) [Draft Supplemental Final Environmental Statement (SFES)] Related to the Operation of the Watts Bar Nuclear Plant. We have no comments at this time. (**0002-1** [Stanley, Joyce])

Response: This comment states that the U.S. Department of the Interior has no comments on this SFES.

Comment: At this time, the Tennessee Department of Transportation is unaware of any conflicts with your proposed project. Please feel free to contact me in the future if other questions arise. (**0007-1** [Andrews, Ann])

Response: This comment states that the Tennessee Department of Transportation has no conflicts with this project. No changes have been made to this SFES in response to this comment.

Comment: EPA finds that this document appropriately includes an analysis that evaluates the environmental impacts of the proposed action of relicensing [issuing an operating license for] WBN Unit 2. (**0005-1** [Mueller, Heinz])

Response: This comment states the U.S. Environmental Protection Agency finds the analysis of the environmental impact of the proposed action appropriate. No changes have been made to this SFES in response to this comment.

E.2.3   Comments Concerning Site Layout and Design

Comment: Page/Line 3-13/ 21: In the *that* statement "The WBN site is located on a 2.7-m (9-ft)- wide navigable channel...." "wide" should be replaced with "deep." (**0010-9** [Brickhouse, Brenda])

Response: Section 3.2.4 of this SFES was modified as noted in this comment.

Comment: The Draft FSEIS however, does not mention the condition of the WBN Unit 2 facility. EPA recommends more discussion on the condition of the WBN Unit 2 physical condition relative to relicensing. NRC should discuss any historical maintenance activities that will demonstrate the condition and structural integrity of Unit 2. The identified additional information (data, analyses, and/or discussions) should be included (or referenced as appropriate) in the Final SEIS. (**0005-2** [Mueller, Heinz])

Comment: We are also, request additional clarifying information on the on-going structural safety analysis and repairs, upgrades and/or retrofits to Watts Bar Unit 2, be mentioned in the FSEIS. (**0005-3** [Mueller, Heinz])

Response: Most of the equipment at WBN Unit 2 was fabricated and installed during the original construction period along with the equipment for WBN Unit 1. Because layup activities were terminated for a period of time after 2001, it is necessary to ensure the equipment is still capable of meeting its required design specifications. Thus, TVA proposed to perform inspections or evaluations, refurbishment or replacements, and system testing to ensure the plant meets its original licensing, design and equipment vendor specifications. The NRC staff concluded in a letter dated July 2, 2010 (NRC 2010) that the TVA program provides for reasonable assurance that the potential degradation effects would be adequately reviewed. The NRC staff found that the program, when properly implemented, would adequately manage the identification of potential degradation effects and refurbishment activities. NRC follow-up inspections will be used to determine if requirements are being adequately implemented.

The NRC staff modified Section 3.1 of this SFES to provide a brief discussion of the NRC staff's review of the TVA construction and refurbishment program for WBN Unit 2 and to note that the NRC staff would conduct inspections to ensure implementation of the program. Structural safety analysis is outside the scope of the environmental review, but is addressed in the NRC's safety review, as documented in a Final Safety Evaluation Report (FSER). An operating license cannot be granted until both environmental and safety reviews are complete.

Comment: The reason [hardened onsite spent fuel storage] that's not happening is strictly cost. And when you're talking about cost, this whole reactor is nuclear power on the cheap. And I don't know why we're accepting the cheapest possible nuclear power plant. TVA tried to build a new AP 1000, two of them at Bellefonte. They found out they was going to be so much more expensive than finishing this reactor and the Bellefonte Unit 1 that they backed off from it. Well, excuse me, but this is not the place to cut costs. If they want to build these things, they have to be state-of-the-art. This is far from state-of-the-art. (**0003-7-3** [Safer, Don])

Response: This comment raises concerns related to the safety of the method of spent fuel storage (in the spent fuel pool rather than in onsite storage casks) and the completion of the existing facility rather than building a completely new facility. The environmental review in this SFES assesses the effects of operating WBN Unit 2 on the environment, based on the design proposed by the applicant. The separate NRC safety review considers the ability of the proposed design to safely operate. No changes have been made to this SFES in response to this comment.

Comment: This ice condenser design is really a joke in the industry. And I mean I talked to the operators at Sequoyah and they just kind of grinned when I asked them about – the ice condenser design means there's three million pounds of ice, literally three million pounds of frozen water, that's in the reactor within the containment structure. And should they have a loss of coolant, all of that hot gas is supposed to go through that ice room to lessen the pressure. And so they've made the containment less sturdy than the other reactors around the country and around the world. Nobody else is building any ice condenser designs ever again. They were built back in the `70s. Sequoyah, ice condenser designs; Watts Bar 1 is an ice condenser design. There's no justification for finishing this thing. (**0003-7-4** [Safer, Don])

Response: This SFES for the WBN Unit 2 operating license considers the environmental impacts of the design and operating parameters as proposed by the applicant. The environmental risk of design basis accidents (e.g., loss of coolant) and severe accidents (e.g., core melt) are addressed in Sections 6.1 and 6.2 of this SFES, respectively. Structural safety analysis is outside the scope of the environmental review, but is addressed in NRC's parallel safety review which is documented in a Final Safety Evaluation Report (FSER). No changes have been made to this SFES in response to this comment.

E.2.4   Comments Concerning Geology

Comment: My question has to do with the geology, the underground structures that this plant has been built upon. And my question is whether this is karst, k-a-r-s-t. Don't ask me what that stands for. But it's limestone. And I'm wondering whether this is being built and has been built on limestone topography? (**0003-2-1** [Ferris, Kathleen])

Response: The geology of the WBN site and vicinity is discussed in Section 2.2.1.2 of this SFES. The site is underlain by the Conasauga Shale, which is approximately 16 percent limestone and 84 percent shale. Severe karstic features typical of rock with a high percentage of calcium carbonate "are not found anywhere within the Conasauga Formation" (TVA 2009). The carbonate rich limestones of the Knox Group are found approximately one mile to the southeast of the WBN site and exhibit karst characteristics. However, the Knox Group is not present beneath the site nor is the site underlain by karst formations. No changes have been made to this SFES in response to this comment.

E.2.5   Comments Concerning Hydrology – Surface Water

Comment: let's go over one issue. You talk about the tritium in the water. And I know nobody don't want to hear about it. And you're here sick and tired of hearing me talk about it. I'm sick and tired of having to deal with it. But the other thing is, you've not dealt with the tritium. You call it a spill. Three years of over the limit and then you didn't even do anything to TVA about it to begin with. That is still sitting out there. Don't tell me the tritium is gone because I know better. (**0003-2-5** [Harris, Ann])

Response: Tritium in the groundwater at the WBN site is discussed in Section 2.2.3.2 of this SFES. The NRC staff characterizes the concentration of tritium observed in groundwater, describes the origin of the tritium as a leak from WBN Unit 1, indicates that plant modifications have been made to stop the leak, and presents information on the monitoring program in place to track tritium concentrations in groundwater. Tritium concentrations in groundwater during 2010 are reported in this SFES and the concentration has dropped to approximately one tenth of the EPA drinking water standard, in part due to natural decay of tritium (tritium has a half life of 12 years) and likely also due to dilution.

Comment: And already huge corporations are buying up water supplies all over the world, which means that before long anybody who can't afford to buy water won't have clean water to drink or may not have water at all because there are water wars going on. We've already had it over the Tennessee River here where Georgia and North Carolina want their share of our water, right? Global warming and climate change, which I see you have noted in your study, are going to affect the supplies of water and threaten -- think about it -- land masses are shrinking; populations are growing. The demand for water will ever be greater, particularly if we are able to continue in what we think of as an advanced civilization. The single largest use of fresh water in the United States is thermal nuclear -- no, I'm sorry – the thermal energy, either by nuclear or coal. And I have -- the study that I referred to has a pie chart, shows that 41 percent of the water, the largest usage is for these forms of energy production. (**0003-6-2** [Ferris, Kathleen])

Response: *The current use of Tennessee River water in the vicinity of the WBN site is presented in Section 2.2.2 of this SFES. The impact of operation of WBN Unit 2 on water supply is presented in Section 4.2.2.1 and shows that operating WBN Unit 2 will consume approximately 0.1 percent of the Tennessee River flow past the plant, resulting in a small impact on surface water use in the region.*

Comment: One of the things that the Union of Concerned Scientists have pointed out is that in the Southeast United States we have a particularly severe problem of water and energy production. That drought and heat have caused many – and we all know this -- many closings, shut-downs of nuclear reactors because the water is too hot or there's not enough of it. Same thing has happened in – and the drought is threatening the nuclear industry in Europe now. (**0003-6-4** [Ferris, Kathleen])

Response: *The cumulative impact of past, present, and reasonably foreseeable actions including the operation of WBN Unit 2 are discussed in Section 4.14.4.2 of this SFES. The NRC staff concluded that past, present, and reasonably foreseeable actions in the region have adversely affected the thermal conditions in the Tennessee River; however, it also concluded that the temperature increase attributable to operation of WBN Units 1 and 2 are predicted to be negligible compared to the temperature increase attributable to air temperature and solar heating. The NRC staff also reviewed the impact of past, present, and reasonable foreseeable actions, including climate change, in the region on water availability. Water consumption is expected to increase on the Tennessee River due to power generation, population growth, industrial development, and irrigation. It is unclear if climate change will result in an increase or a decrease in runoff in the region. The NRC staff concluded that although increases in consumptive use may be detectable, and climate change could result in a change to runoff, that these changes would be unlikely to noticeably alter the resource.*

Comment: [T]his is right where TVA is building all these plants, six, going on seven, on the Tennessee River. And TVA wants to put four more at Watts Bar. Now that Tennessee River provides drinking water for the cities of Knoxville, Chattanooga, Huntsville, all the communities in between. The TVA's plan is to become, as Mr. Kilgore said, the foremost producer of nuclear energy in the country. And that means this Watts Bar 2. It also means the plant at Bellefonte, the first one, and then three more. And I propose that this is a threat to our drinking water. It's not what your study says. (**0003-6-5** [Ferris, Kathleen])

Response: *The NRC staff discusses the cumulative impact on surface water of past, present, and reasonably foreseeable actions, including the operation of WBN Unit 2, in Section 4.14.4.2 of this SFES. The NRC staff concluded that past, present, and reasonably foreseeable actions in the region have adversely affected the chemical and thermal conditions in the Tennessee River; however, it also concluded that the effects of operating WBN Unit 2 are predicted to be negligible compared to the thermal and chemical effects of other actions. The assessment does not consider the construction and operation of additional reactors at the WBN site (beyond WBN Units 1 and 2) because the NRC staff does not consider additional reactors to be reasonably foreseeable until an application is prepared and submitted. Should applications for additional reactors be submitted to the NRC, the impact of those actions will be considered during the licensing process.*

Comment: And then recently I was a contractor at Fort Calhoun on the Missouri River. NRC was concerned about the data that the utility had used for Fort Calhoun as far as the flood projections were for the Missouri River. Fort Calhoun had used data from the Corps of Engineers. And NRC has a process to calculate each utility to calculate that without utilizing the data from the Corps of Engineers. And the reason why I was there was to look back over their information that they were going to present to NRC. And they did extensive updates. Don't need to go into that.

But the bottom line is the information provided by the Corps of Engineers was faulty. And they made great improvements at Fort Calhoun and they're still working on it. If you watch the news, if they hadn't prepared for it, they'd be in a lot of trouble, lot worse trouble. And I attribute NRC pushing them to correct what they had there and it made a bad situation a lot better.

And my question is has TVA depended on the Corps of Engineers data for anything related to the [flood projections for the] Tennessee River at Watts Bar? And if they have, will NRC then go back to the Tennessee Valley Authority and ask them the same prudent questions they asked the utility owner on the Missouri River to do it in accordance with the federal regulations and not depend on the Corps of Engineers? (**0004-1-2** [Riden, David])

Response: *Flood analysis is outside the scope of the environmental review, but is addressed in the NRC safety review documented in a Final Safety Evaluation Report (FSER).*

E.2.6   Comments Concerning Hydrology – Groundwater

Comment: Karst topography is limestone. It's got cracks and crevices everywhere. If it gets into -- if radiation gets -- or pollution gets into that, you have got an effect on the ground water. (**0003-6-6** [Ferris, Kathleen])

Response: *The geology of the WBN site and vicinity is discussed in Section 2.2.1.2 of this SFES. The site is underlain by the Conasauga Shale which is approximately 16 percent limestone and 84 percent shale. Severe karstic features typical of rock with a high percentage of calcium carbonate "are not found anywhere within the Conasauga Formation" (TVA 2009). The carbonate rich limestones of the Knox Group are found approximately one mile to the southeast of the WBN site and exhibit karst characteristics. However, the Knox Group is not present beneath the site nor is the site underlain by karst formations. No changes have been made to this SFES in response to this comment.*

E.2.7 Comments Concerning Ecology – Terrestrial

Comment: I'm deeply resentful that there is a fan crane 2,500 acre island down here that is so contaminated that the geese are even -- they're not even coming in there anymore. The cranes don't want to go there. You can't entice them; you can't put enough food on them to entice them in is what I'm seeing. (**0003-5-4** [Harris, Ann])

Response: The NRC staff is not familiar with a fan crane. However, the sandhill crane is found in the vicinity of the WBN site. It is the most abundant of the world's crane species (USGS 2006). They are widespread, and most populations are stable or increasing in size. Sandhill cranes that occur in large numbers in Tennessee are likely migratory, and breed in the Great Lakes States and winter in central Florida. Loss and degradation of habitat is the single greatest threat to this species. Operation of WBN Unit 2 is not expected to contribute to the loss or degradation of sandhill crane habitat and is not expected to noticeably affect regional or local sandhill crane populations. No changes have been made to this SFES in response to this comment.

Comment: Page/Line 2-17/18: Consider replacing or supplementing the statement, "During summer, gray bats are known to roost in two caves within 8 km (5 mi) from the WBN site," with the following more specific information from the FSEIS, Section 3.4.3, page 60, "Small numbers (less than 500) of gray bats continue to roost in a cave approximately 3.3 miles from the project." (**0010-4** [Brickhouse, Brenda])

Response: Text was added to Section 2.3.1.2 to further clarify gray bat seasonal occurrence in the vicinity.

Comment: Page/Line 4-15 I 39-40: TVA recommends the following revision "While TVA does not conduct studies of avian mortality, no noticeable events of avian mortality associated with the existing transmission system have been recorded by TVA." (**0010-7** [Brickhouse, Brenda])

Response: The comment provides clarification of monitoring and reporting activities. Section 4.3.1.1 has been modified to clarify TVA avian monitoring and reporting along the existing transmission system as suggested.

E.2.8 Comments Concerning Ecology – Aquatic

Comment: [Two plants in the same place makes twice as much risk] for ongoing aquatic danger to the aquatic species (**0003-4-5** [Kurtz, Sandy])

Comment: [Southern Alliance for Clean Energy's Statement of Disputed Material Facts] Description of the Proposed Project General Information

18. The present proceeding pertains to the OL for WBN Unit 2. The added operation of WBN Unit 2 may result in minimal increased demands on that aquatic environment both for cooling water intake and cooling water discharge. Disputed as to the term "minimal." As discussed in Dr. Young's Declaration throughout, the already-stressed Tennessee River aquatic environment will be further stressed by additional CCW intake and discharge and increased SCCW discharge to accommodate the operation of both WB1 and WB2 cooling towers and the increased cumulative cooling tower blowdown discharge to the Tennessee River as a result of WB2 operation. The combined operation of two units will have substantial impacts on the Tennessee River. (**0016-1-2** [Curran, Diane])

Response: These comments concern the additional stress to the aquatic environment from operation of WBN Unit 2 and the cumulative effects from the operation of WBN Unit 1. The NRC staff discusses the additional quantity of water that would be removed from the Tennessee River to operate WBN Unit 2 (such that both units can operate simultaneously) in Sections 3.2.2.1 and 4.3.2.1. The NRC staff also discusses the increase in the thermal discharge from the supplemental condenser cooling water (SCCW) discharge and the blowdown discharge from the condenser cooling system through the diffusers in Sections 3.2.2.4 and provides the increment from the addition of WBN Unit 2 in Table 3-4. Section 4.2.2.2 contains a description of the thermal discharge from the outfalls and includes a figure showing the relative locations of the mixing zones for Outfalls 101 and 113. Section 4.14.6 of the SFES addresses the cumulative effects on the aquatic biota of past, present and reasonably foreseeable future projects in the Tennessee River in combination with the operation of the co-located units.

Comment: Since the preparation of the DSEIS, the laurel dace (*Chrosomus saylori*) was listed as endangered (76 FR 48722 4874 1) on September 8, 2011, and is known to occur with the project assessment area. The sheepnose mussel (*Plethobasus cyphyus*) is proposed for listing as endangered (76 FR 3392 3420) and occurs in the project assessment area. (**0009-2** [Hogue, Gregory])

Response: The NRC staff described sheepnose mussel in Section 2.3.2.2 as a species that the U.S. Fish and Wildlife Service proposed for listing on January 19, 2011 (76 FR 3392). The NRC staff updated the discussion of the laurel dace as a candidate species to indicate it was listed as endangered on September 8, 2011 (76 FR 48722). The NRC staff revised Section 2.3.2.2 to address this comment.

Comment: The [U.S. Department of Interior's] Tennessee Ecological Services Field Office has completed section 7 consultation with the NRC on the project and has no other substantive comments on the DSEIS to offer at this time. (**0009-3** [Hogue, Gregory])

Response: The NRC staff made no changes to the SFES as a result of this comment.

Comment: In Contention 7 (which was admitted by the Atomic Safety and Licensing Board ("ASLB") in Tennessee Valley Authority (Watts Bar nuclear Plant, Unit 2), LBP-09-26, 70 NRC 939, 981-90 (2009)), SACE has challenged the adequacy of TVA's FSEIS for WBN2 to address the impacts of WBN2 on aquatic organisms.... Although TVA conducted additional environmental studies that were intended to address our concerns, they are not sufficient to support TVA's claim that the aquatic environmental impacts of WBN2 are insignificant. (**0008-1** [Curran, Diane])

Comment: TVA [in "Motion for Summary Disposition of Contention 7" (Nov. 21, 2011)] claims that it has conducted studies that resolve the three major deficiencies identified in Contention 7. As discussed in Dr. Young's attached Declaration, this is not correct. With respect to the inadequacy of TVA's previous data and analyses, TVA has made some progress by collecting new data on entrainment, impingement, freshwater mussels, and thermal impacts during 2010. But TVA has only started to catch up with its failure to collect the appropriate data that would be reasonably sufficient to evaluate impacts on aquatic resources by collecting only one year of data for entrainment, impingement, freshwater mussels, and thermal impacts over the preceding years. TVA still has not collected an amount of data that is reasonably necessary to evaluate the effects of WBN1 on aquatic organisms in the Tennessee River, and therefore it does not have enough information to extrapolate the impacts of WBN2. (**0015-3** [Curran, Diane])

Comment: [From Southern Alliance for Clean Energy's Statement of Disputed Material Facts] Description of TVA's Aquatics Studies

31. As noted in ¶ 10 above, TVA conducted a number of aquatics studies in direct response to the assertions made by SACE and its expert, Dr. Young, in Contention 7. Those studies, which are described in more detail below, collectively provide data on fish and mussel populations in the WBN vicinity, and the entrainment, impingement, and hydrothermal impacts on those species that result from operation of WBN Unit 1. In addition, TVA conducted some of the studies to resolve alleged errors in TVA's original studies identified by SACE and Dr. Young. Undisputed that TVA conducted the studies described in pars. (A) through (G) below. Disputed that the studies resolve Dr. Young's concerns, as discussed throughout his Declaration. (**0016-3-1** [Curran, Diane])

Comment: [From Southern Alliance for Clean Energy's Statement of Disputed Material Facts] Description of TVA's Aquatics Studies Mollusk Survey of the Tennessee River Near [WBN] (Rhea County, Tennessee) (Oct. 28, 2010, Revised Nov. 24, 2010) ("Mollusk Survey"), and Discussion of the Results of the 2010 Mollusk Survey of the Tennessee River Near [WBN] (Rhea County, Tennessee) (Mar. 2011) ("Discussion of Mollusk Survey")

50. Because WBN Unit 1 was in operation in 2010 and had been in operation for more than a decade, this survey inherently reflects the impact of the operation of WBN Unit 1 on the mussel community in the WBN vicinity. Disputed as to a one year survey capturing the population trend of a mussel community. It was reasonable for TVA to have contracted for a multi-year study when it was decided to apply for the operating license. (**0016-3-14** [Curran, Diane])

Comment: SUMMARY OF MY [Dr. Shawn Paul Young] PROFESSIONAL OPINION REGARDING TVA'S ASSERTIONS.

2. With respect to the inadequacy of TVA's previous data and analyses, TVA has made some progress by collecting new data on entrainment, impingement, freshwater mussels, and thermal impacts during 2010. But TVA has only started to catch up with its failure to collect the appropriate data that would be reasonably sufficient to evaluate impacts on aquatic resources by collecting only one year of data for entrainment, impingement, freshwater mussels, and thermal impacts over the preceding years. TVA still has not collected an amount of data that is reasonably necessary to evaluate the effects of WBN1 on aquatic organisms in the Tennessee River, and therefore it does not have enough information to extrapolate the impacts of WBN2. See pars. III-A.5, III-B.3-4, and III-C.1-2 below. (**0017-1-1** [Curran, Diane])

Response: The NRC staff understands these comments to dispute the sufficiency of the environmental studies that TVA performed to collect data to evaluate the effects of WBN Unit 1 and thereby extrapolate to estimate the impacts of operating WBN Unit 2.

Table 5-1 in the SFES provides a list of aquatic studies conducted between 1973 and 2011 and reviewed by the NRC staff. The NRC staff discusses the results of the studies in further detail in Sections 2.3.2.1 and 4.3.2. TVA submitted additional studies after the draft SFES was published, including two fish sampling studies above and below the Watts Bar Dam, a year-long impingement study at the IPS, and additional months of a previously reported entrainment sampling for the IPS and SCCW. The NRC staff reviewed the additional studies and incorporated a discussion of the study results into the SFES text in Sections 4.3.2 and 5.5.2 and Table 5-1 (along with citations). The studies reviewed by the NRC staff include

- *mussel surveys prior to the operation of WBN Unit 1 for 11 distinct years between 1975 and 1994*

- *mussel surveys for 1996, 1997, and 2010 following the operation of WBN Unit 1*

- *preoperational fish sampling studies from 1976 to 1995*

- *14 years of fish sampling studies from 1996 to 2010 following the start of WBN Unit 1*

- *two years of SCCW impingement studies (August 2005 to August 2007) during operation of WBN Unit 1*

- *two IPS impingement studies (March 1996 to September 1997 and March 2010 to March 2011) during operation of WBN Unit 1*

- *two entrainment studies (1975 for the intake of the Watts Bar Fossil Plant, and March 2010 to March 2011 for the SCCW during operation of WBN Unit 1)*

- *ichthyoplankton study of SCCW in 2000, during operation of SCCW for WBN Unit 1*

- *two entrainment studies at the IPS (1996 and 1997 during operation of WBN Unit 1 and March 2010 through March 2011 during operation of WBN Unit 1)*

- *ichthyoplankton studies (six years of sampling [1976-1979 and 1982-1985]) on Chickamauga Reservoir near the site prior to the operation of WBN Unit 1.*

The NRC staff believes that available information is sufficient to meet the intent of NEPA to perform the assessment of the impacts of operation of WBN Unit 2 and disclose the impacts of the proposed action on the aquatic resources in the Tennessee River as discussed in Section 4.3.2.

Comment: By TVA's own admission, the Tennessee River "*is the most diverse temperate freshwater ecosystem in the world.*" Programmatic EIS for Reservoir Operations Study, §4.7.1 Neither TVA nor the NRC Staff has grappled with the significance of the impacts of WBN2 to aquatic organisms, and thus they have given no serious consideration to mitigation measures that could protect the fragile and extraordinarily important ecosystem of the Tennessee River. (**0008-4** [Curran, Diane])

Comment: By falsely painting a rosy picture of aquatic health in the river, TVA understates the significance of the impacts of WBN1 and WBN2, and thus minimizes the benefits that could be achieved by implementing alternatives that would reduce the impacts of the cooling system on organisms in the river. (**0015-9** [Curran, Diane])

Response: The NRC staff understands these comments to suggest that TVA and the NRC do not understand the significance of the effect of WBN Units 1 and/or 2 on the aquatic biota and, as a result, have not adequately addressed mitigation or alternatives that would reduce the effect of operating WBN Unit 2.

In Section 4.3.2 of the SFES, the NRC staff discusses the significance of the impacts of the operation of WBN Unit 2 on aquatic organisms. Section 4.14.6 of the SFES considers other past, present, and reasonably foreseeable future actions (cumulative actions) that could affect aquatic ecology of the WBN site. The NRC staff found the direct and indirect impacts on aquatic biota due to operation of WBN Unit 2 would be SMALL (so minor that they would neither destabilize nor noticeably alter any important attributes of the aquatic resources) and that the cumulative impacts caused by the aggregate of past, present, and reasonably foreseeable future actions would be LARGE.

The NRC staff did not discuss mitigation for the SCCW intake because the average monthly flow through the SCCW intake for the operation of WBN Units 1 and 2 will be slightly less than the flow through the SCCW while operating Unit 1 only, and within the uncertainty in the estimate for flow while operating either one or two units. The NRC staff did not discuss mitigation for the IPS intake because the IPS is a closed-cycle system and the rates of impingement and entrainment are low. The NRC staff did not discuss mitigation for discharges from the outfalls because TVA has a current National Pollutant Discharge Elimination System (NPDES) permit for both units (TDEC 2011; TVA 2011a).

Comment: [T]here are environmental concerns with additional information requested in the FSEIS. Specifically, as outlined in EPA's comment letter dated May 14, 2007, referenced subject, TVA's Draft Supplement Environmental Impact Statement Watts Bar Nuclear Plant Unit 2. [excerpt from EPA's May 14, 2007 letter:] "protecting the environment involves ... continuing measures to limit bioentrainment and other impacts to aquatic species from surface water withdrawals and discharges, and compliance with the NPDES Permit." (**0005-5** [Mueller, Heinz])

Response: In 2007, the EPA expressed concern about continuing measures to limit bioentrainment and other impacts to aquatic species from surface water withdrawals and discharges, as well as compliance with the NPDES permit. In preparing the SFES, the NRC staff reviewed TVA's environmental impact statement as well as supporting information and responses to requests for additional information. The EPA and its delegated States, not the NRC, regulate entrainment and impingement as well as the effects of surface water discharges under the Clean Water Act through NPDES permits. The NRC discloses such impacts in the environmental impact statements it prepares under NEPA, but does not regulate the impacts. As discussed in Section 3.3.2, TVA has received an updated NPDES permit for operation of Units 1 and 2 and had previously been in compliance with the existing NPDES permit for Unit 1.

Comment: TVA [in "Motion for Summary Disposition of Contention 7" (Nov. 21, 2011)] claims that it has conducted studies that resolve the three major deficiencies identified in Contention 7. As discussed in Dr. Young's attached Declaration, this is not correct......The combined operation of WBN1 and WBN2, by itself, may cause changes in how Watts Bar Dam is operated. TVA and the NRC Staff both acknowledge that in order to stay within thermal discharge limits stated in the NPDES that requests for additional discharge from Watts Bar Dam may be needed. Thus, operating WBN alone would change reservoir operations in the middle-Tennessee Basin that would be supported by water releases or hydrological adjustments in upper- Tennessee River Basin. The effects of more alterations to the hydrological cycle of the basin on aquatic organisms, especially the already declining native fish and freshwater mussel species, must be addressed. Given the extensive portfolio of energy and industrial facilities that the Tennessee River supports and that the management agencies must maintain adequate water for all these facilities, this is an extremely important omission. (**0015-11** [Curran, Diane])

Comment: SUMMARY OF MY [Dr. Shawn Paul Young] PROFESSIONAL OPINION REGARDING TVA'S ASSERTIONS.

7. Finally, TVA still does not address the cumulative impacts of WBN2 in conjunction with the impacts of the numerous water impoundments on the Tennessee River, or with other industrial facilities such as the ten fossil fuel-burning plants, the six nuclear reactors that are already in operation, and the five additional reactors for which TVA has sought operating licenses. The combined operation of WBN1 and WBN2, by itself, may cause changes in how Watts Bar Dam is operated. TVA and the NRC Staff both acknowledge that in order to stay within thermal discharge limits stated in the NPDES that requests for additional discharge from Watts Bar Dam may be needed. Thus, operating WBN alone would change reservoir operations in the middle- Tennessee Basin that would be supported by water releases or hydrological adjustments in upper-Tennessee River Basin. The effects of more alterations to the hydrological cycle of the basin on aquatic organisms, especially the already declining native fish and freshwater mussel species, must be addressed. Given the extensive portfolio of energy and industrial facilities that the Tennessee River supports and that the management agencies must maintain adequate water for all these facilities, this is an extremely important omission. (**0017-1-7** [Curran, Diane])

Comment: STATEMENT OF [Dr. Shawn Paul Young] PROFESSIONAL OPINION REGARDING ADEQUACY OF TVA'S RECENT BIOLOGICAL STUDIES TO ADDRESS THE ENVIRONMENTAL IMPACTS OF THE PROPOSED WATTS BAR 2 NUCLEAR POWER PLANT ON AQUATIC ORGANISMS

Failure to Discuss Cumulative Impacts

1. TVA has not addressed the cumulative impacts on the Tennessee River Basin from combined operation of WBN Units 1 and 2. The combined operation will increase cooling water needs and increase thermal and chemical discharge. These consequences of adding yet another energy production facility will have adverse impacts on the whole system with large impacts to the upper-basin tributaries that also support highly diverse and unique fish and mussel species. TVA manages the Tennessee River as one hydrosystem; thus, changes in water consumption or changes in flow to accommodate energy and industrial facilities in one area will affect the rest of the system. Further, the quantity of water available at Watts Bar Dam and then released into Chickamauga Reservoir determines the management of the rest of the hydrosystem, especially water releases from the upper basin. Therefore, if WBN Plant requires flow in order to operate at maximum efficiency and to remain within NPDES permit limits, the entire upper basin or at least the aquatic ecology of 10 different tributaries with a high number of fish and mussels will be affected. This is

supported by the following excerpts from TVA's discussion of water management policy on its website (http://www.tva.gov/river/lakeinfo/systemwide.htm):

• "In May 2004, the TVA Board of Directors approved a new policy for operating the Tennessee River and reservoir system. This policy shifts the focus of TVA reservoir operations from achieving specific summer pool elevations on TVA-managed reservoirs to managing the flow of water through the river system. The new policy specifies flow requirements for individual reservoirs and for the system as a whole."

• "System-wide flow requirements ensure that enough water flows through the river system to meet downstream needs."

• "When water must be released to meet downstream flow requirements, a fair share of water is drawn from each reservoir. System-wide flows are measured at Chickamauga Dam, located near Chattanooga, Tenn., because this location provides the best indication of the flow for the upper half of the Tennessee River system."

• "If the total volume of water flowing into Chickamauga Reservoir is less than needed to meet system-wide flow requirements, additional water must be released from upstream reservoirs, resulting in some drawdown of these projects. How much water is released depends on the time period and the total volume of water in storage in 10 tributary reservoirs: Blue Ridge, Chatuge, Cherokee, Douglas, Fontana, Nottely, Hiwassee, Norris, South Holston and Watauga." (**0017-4-11** [Curran, Diane])

Response: *These comments state that the cumulative impacts on the Tennessee River Basin from combined operation of WBN Units 1 and 2 have not been addressed and indicate that the operation of one or both WBN units would change the operation of Watts Bar Dam, altering the reservoir operations and water releases from the upper Tennessee River Basin and affecting the biota in the upper basin.*

The NRC staff discusses the additional quantity of water that would be removed from the Tennessee River to operate WBN Unit 2 in Sections 3.2.2.1 and 4.3.2.1, such that both units can operate simultaneously. Further, the NRC staff discusses the increase in the thermal discharge from the SCCW discharge and the blowdown discharge from the condenser cooling system through the diffusers in Section 3.2.2.4, and provides the increment from the addition of WBN Unit 2 in Table 3-4. Section 4.14.6 of the SFES addresses the cumulative effects of past, present, and reasonably foreseeable future projects in the Tennessee River in combination with the operation of the co-located units.

In Sections 2.2.1.1 and 4.14.4.1, the NRC staff discusses the operation of the reservoirs in the Tennessee River and references the Reservoir Operations Study conducted in 2004 (TVA 2004). The Reservoir Operations Study is a programmatic environmental impact statement. As such, it is appropriate for the NRC staff to tier from the information in this study rather than reproducing the analysis in the SFES. The Reservoir Operations Study included operation of WBN Unit 1 but not the operation of WBN Unit 2.

In Section 4.14.6, the NRC staff discusses the cumulative impacts of operation of WBN Unit 2 on the aquatic biota considering the past, present, and reasonably foreseeable future projects in the defined geographical region. In addition, the NRC staff discusses the coordination between the operation of the Watts Bar Dam and WBN Units 1 and 2 in Sections 3.2.2.4 and 4.2.2.2, to keep the discharges from the WBN units within the temperature limits of the NPDES permit.

As a result of this comment, the NRC staff has included the following statement in Section 4.14.6:

> *Increasing the volume of water released from Watts Bar Dam is one of five options TVA can use to ensure that the thermal discharge from operation of WBN Units 1 and 2 remains within the NPDES limits as discussed in Section 4.2.2.2. If this option is chosen, the water released from Watts Bar Dam could have a slight and indiscernible effect on the water levels in Tennessee River reservoirs and tributaries upstream and downstream of Watts Bar Dam and a slight and indiscernible effect on the biota in those reservoirs and tributaries.*

Comment: In Contention 7 (which was admitted by the Atomic Safety and Licensing Board ("ASLB") in Tennessee Valley Authority (Watts Bar nuclear Plant, Unit 2), LBP-09-26, 70 NRC 939, 981-90 (2009)), SACE has challenged the adequacy of TVA's FSEIS for WBN2 to address the impacts of WBN2 on aquatic organisms....The DEIS has not resolved the issues raised in Contention 7 because it merely adopts the analysis and conclusions of TVA's FSEIS with respect to aquatic impacts. Our continuing concerns about the inadequacy of TVA's environmental analysis are documented in Contention 7. (**0008-15** [Curran, Diane])

Response: *The NRC staff understands that the commenter is concerned that the NRC staff's draft SFES adopted the analysis and conclusions of TVA's FSEIS.*

The SFES presents the NRC staff's independent review that considers and weighs the environmental impacts of the proposed action for operation of WBN Unit 2. After receipt of the TVA EIS, the NRC staff visited the site; met with TVA staff, public officials, and the public; reviewed comments received during the scoping process; requested additional documents from TVA; and located and reviewed peer reviewed articles and other documents not published by TVA related to the site and environment. The NRC staff also contacted Federal, State, Tribal, regional, and local agencies to solicit comments and information. Following this review, the staff requested that TVA provide additional information, documents, and data. The NRC staff's approach to quantifying the impact of operation of WBN Unit 2 was different from TVA's in several key areas (e.g., the NRC staff did not determine impacts based on the Reservoir Fish Assemblage Index [RFAI] and the Benthic Macroinvertebrate Index). The NRC staff documented its independent review of aquatic ecology in Sections 2.3.2, 4.3.2, 4.14.6, and 5.5.2 of the SFES.

Comment: TVA's finding [in 2007 FSEIS] that WBN Unit 2 will have no significant impacts on aquatic life in the Tennessee River is inadequately supported in the following respects:

1. TVA's conclusion that cumulative impacts will be insignificant is based on the faulty premise that the aquatic ecosystem that will be affected by WBN Unit 2 is currently in a good state of health. In fact, data in TVA's own environmental studies, as well as available literature, show that the health of the Tennessee River ecosystem, including Lake Chickamauga where WBN Units 1 and 2 are located, is damaged, fragile, and quite vulnerable to the additional impacts that would be posed by WBN Unit 2's cooling water system. Young Declaration at ¶ III.A.1.

The Tennessee River is an extraordinarily diverse and unique ecosystem that supports over 200 fish species, including twenty species that are found only in the Tennessee River...... Yet the ecosystem also harbors the highest number of imperiled species of any large river basin in North America..... TVA incorrectly portrays the ecosystem as healthy, when its health and diversity are actually in steep decline...... TVA asserts, for example, that the freshwater mussel communities are in "excellent" health because their population is "constant." But, in fact, the mussel population is only constant because it is not reproducing, which is a sign of poor health. ... By characterizing the health of fish and benthic organisms as "good" or "excellent," TVA rationalizes its failure to take a hard look at the reasons why these species are declining. While dams may be the primary cause of these ill effects, they are not the only contributor. (**0008-13** [Curran, Diane])

Comment: [Contention 7: Inadequate Consideration of Aquatic Impacts]
TVA claims that the cumulative impacts of WBN Unit 2 on aquatic ecology will be insignificant (FSEIS Table S-1 at page S-2, and Table 2-1 at page. 30). TVA's conclusion is not reasonable or adequately supported, and therefore it fails to satisfy 10 C.F.R. § 51.53(b) and NEPA. TVA's discussion of aquatic impacts is deficient in three key respects. First; TVA [in 2007 Final Supplemental Environmental Impact Statement] mischaracterizes the current health of the ecosystem as good, and therefore fails to evaluate the impacts of WBN2 in light of the fragility of the host environment. (**0008-6** [Curran, Diane])

Response: These comments assert that TVA incorrectly characterized the "aquatic ecosystem" as being in "good health" and the cumulative impacts of operation as insignificant. These comments refer to the TVA EIS (TVA 2008), rather than the analysis provided by the NRC staff in the draft SFES.

In this SFES, the NRC staff describes the changes in the ecosystem since the early 1900s. In Section 2.3.2, the NRC staff describes the effect of impoundment of the river and the effects of the introduction and success of non-native and invasive aquatic fish, invertebrate, and plant species that "have clearly changed the environment of the Tennessee River aquatic communities." In Section 2.3.2.1, the NRC staff discusses the significant decline in freshwater mussel and fish species. In Section 4.14.6, the NRC staff further indicates that "...[t]he aquatic resources are not stable in the sense of persisting as they were in the past or are today." In Section 9.6, the NRC staff concludes that the cumulative impact for aquatic ecology would be LARGE because of other activities that have affected the environment. The NRC defines LARGE as "environmental effects that are clearly noticeable and are sufficient to destabilize important attributes of the resource." The NRC staff added the words "aquatic ecology" to Section 9.6 to clarify that this impact level relates specifically to aquatic ecology.

Comment: In Contention 7 (which was admitted by the Atomic Safety and Licensing Board ("ASLB") in Tennessee Valley Authority (Watts Bar nuclear Plant, Unit 2), LBP-09-26, 70 NRC 939, 981-90 (2009)), SACE has challenged the adequacy of TVA's FSEIS for WBN2 to address the impacts of WBN2 on aquatic organisms.... TVA also distorts the aquatic impacts of WBN2 by characterizing the baseline condition of the Tennessee River as a reservoir rather than a free-flowing river that has been adversely affected by dams and industrialization. (**0008-2** Curran, Diane])

Comment: To accept TVA's assertion [in "Motion for Summary Disposition of Contention 7" (Nov. 21, 2011)] that for purpose of an EIS affecting this unique ecosystem, current deteriorated condition could be considered appropriate for purposes of evaluating impacts and alternatives would be equivalent to pounding nails into its coffin. If the narrow species diversity of a reservoir is considered the baseline for the WBN2 environmental analysis, then any hope of mitigation measures to sustain or restore the vestiges of diversity that remain will be effectively extinguished by the environmental analysis whose purpose is to protect the environment. (**0015-1** [Curran, Diane])

Comment: If TVA and the NRC are allowed to ignore the true baseline condition of the river in the EIS for Watts Bar, then not only is any opportunity for mitigation of the effects of WBN2 lost, but future decisions will be affected by the bad assumptions of these EISs. That outcome is not consistent with the purposes of NEPA. (**0015-13** [Curran, Diane])

Comment: For instance, as Dr. Young discusses in his Declaration in Section F, TVA operates the dams and the power plants on the Tennessee River as a single system. This system includes ten different tributaries with a high number of fish and mussel species. By failing to use a baseline that takes into account the fragile health of these tributaries, TVA effectively writes off any mitigation measures that could aid their survival and consigns them to oblivion. (**0015-2** [Curran, Diane])

Comment: TVA [in "Motion for Summary Disposition of Contention 7" (Nov. 21, 2011)] claims that it has conducted studies that resolve the three major deficiencies identified in Contention 7. As discussed in Dr. Young's attached Declaration, this is not correct......Further, despite alarming evidence of significant decline in the diversity and numbers of aquatic organisms in the Tennessee River in the vicinity of WBN, TVA continues to assert that the aquatic health of the river is good. The only way that TVA can present such a clean bill of health is to mischaracterize the baseline condition of the Tennessee River as a large reservoir where one would expect to see a limited number of species of aquatic organisms. In reality, the Tennessee River is a fragile and rapidly deteriorating riverine ecosystem with remnants of the greatest species diversity of any river in the United States. (**0015-8** [Curran, Diane])

Comment: SUMMARY OF MY [Dr. Shawn Paul Young] PROFESSIONAL OPINION REGARDING TVA'S ASSERTIONS.

6. Further, despite alarming evidence of significant decline in the diversity and numbers of aquatic organisms in the Tennessee River in the vicinity of WBN, TVA continues to assert that the aquatic health of the river is good. The only way that TVA can present such a clean bill of health is to mischaracterize the baseline condition of the Tennessee River as a large reservoir where one would expect to see a limited number of species of aquatic organisms. In reality, the Tennessee River is a fragile and rapidly deteriorating riverine ecosystem with remnants of the greatest species diversity of any river in the United States. By falsely painting a rosy picture of aquatic health in the river, TVA understates the significance of the impacts of WBN1 and WBN2, and thus minimizes the benefits that could be achieved by implementing alternatives that would reduce the impacts of the cooling system on organisms in the river. (**0017-1-6** [Curran, Diane])

Response: The comments address the baseline used in the analysis of environmental impacts. The NRC staff agrees with the commenter that the baseline for cumulative analysis must be the Tennessee River before impoundment, construction, and operation of power-producing facilities and the introduction of non-native aquatic biota. In the SFES, the NRC staff describes the Tennessee River, and indicates that the character of the river was altered by a series of impoundments constructed from the late 1930s to the 1960s. In Section 2.3.2.1, the NRC staff indicates "impoundments have altered the dynamics of river flow" at the location of the site. The NRC staff lists factors that have accompanied the placement of dams (e.g., changes in spring floods, lack of previous expansive rocky or gravel shoal areas that once existed, changes in water depth and temperature, reductions in the amount of dissolved oxygen, and increased sedimentation). Further, the NRC staff notes that resource managers (and others) have introduced species, including nuisance species, into the river system, thus affecting the native aquatic biota. In addition, the NRC staff notes that chemical contaminants from upriver facilities may affect aquatic biota.

In Section 2.3.2.1, the NRC staff indicates, "[T]he assemblage of organisms in the river changed in response to the impoundments" and discusses the decline in the mussel abundance since the 1940s and changes to the fish population since the late 1930s. In Section 4.14.6, the NRC staff also discusses impoundment of the river, other power-producing facilities, overfishing, agriculture-related activities, and other anthropogenic processes that have degraded the ecosystem.

Although the baseline for cumulative impacts occurs prior to impoundment of the river, the direct and indirect impact analysis in Chapter 4 of the SFES looks at the incremental effect of operating WBN Unit 2 (i.e., the pathways of entrainment, impingement and the effects of thermal discharge) on the species currently inhabiting the Tennessee River near the WBN site. The NRC staff's analysis in Chapter 4 does not consider how plant operations would have affected species that had previously lived in the river, but are no longer found in the river. The NRC staff's interpretation of NEPA is to use existing conditions as a baseline for incremental (direct and indirect) effects of the possible future operation of WBN Unit 2.

Comment: [From Southern Alliance for Clean Energy's Statement of Disputed Material Facts] Description of TVA's Aquatics Studies
Analysis of Fish Species Occurrences in Chickamauga Reservoir – A Comparison of Historic and Recent Data (Oct. 2010) ("Fish Species Occurrences Study")

41. In analyzing the collective historical fish survey data for the Chickamauga Reservoir, this study takes into consideration the variations in survey methods employed over the past 60 years. Variations in survey methodology preclude direct comparisons between historical and recent surveys. This study also compared the results of fish sampling efforts in various Tennessee River reservoirs subject to similar conditions to understand widespread patterns and behavior of species in reservoir environments.

    Disputed. While the study may acknowledge the variations in survey methods employed over the years, it does not cure the mistakes of the past, and instead perpetuates them. TVA either has an "extensive" fish species survey/study for historical comparison, which shows significant decline of fish species overtime, including since operation of Unit 1, or TVA has an unreliable, outdated, and inadequate means to properly evaluate impacts from WBN. The different sampling methods do not detract from the fact that there has been a decline in fish species pre- and post-WBN operation, which is evidence that the health of the fish community is poor See Dr. Young's Declaration at pars. III-E.1-20. (**0016-3-8** [Curran, Diane])

Comment: [From Southern Alliance for Clean Energy's Statement of Disputed Material Facts] Description of TVA's Aquatics Studies. Analysis of Fish Species Occurrences in Chickamauga Reservoir – A Comparison of Historic and Recent Data (Oct. 2010) ("Fish Species Occurrences Study")

45. Finally, the study found that changes in fish survey methods account for some of the changes in findings of species occurrence and abundance. Certain survey methods, such as hoop nets, trap nets, and cove rotenone sampling, that were effective for targeting certain species, are no longer in use.

Undisputed in that this is a conclusion of the study. Disputed as being used as rationale for the decline of the fish community. Even with TVA's many changes in methods, a clear pattern of declining indigenous fish species and their abundance pre- and post-WBN operation is clear. See Young Declaration, ¶¶ III-E.1-20 and III-D.4-7, III-D.10, and III-A.6-9. (**0016-3-12** [Curran, Diane])

Comment: STATEMENT OF [Dr. Shawn Paul Young] PROFESSIONAL OPINION REGARDING ADEQUACY OF TVA'S RECENT BIOLOGICAL STUDIES TO ADDRESS THE ENVIRONMENTAL IMPACTS OF THE PROPOSED WATTS BAR 2 NUCLEAR POWER PLANT ON AQUATIC ORGANISMS. E. RFAI Study and Fish Species Occurrences Study [Comparison of Fish Species Occurrence and Trends in Reservoir Fish Assemblage Index Results in Chickamauga Reservoir Before and After WBN Unit 1 Operation (June 2010) ("RFAI Study") and Analysis of Fish Species Occurrences in Chickamauga Reservoir – A Comparison of Historic and Recent Data (Oct. 2010) ("Fish Species Occurrences Study")

3. Second, TVA's summation of data in the Fish Species Occurrence study is biased, and TVA attempts to portray sampling gear changes as the reason for the decline of fish species near WBN and Chickamauga Reservoir in general to mask the reality that the fish community has experienced significant decline pre- and post-WBN operation from cumulative man-made impacts to the aquatic ecosystem. (**0017-4-3** [Curran, Diane])'

Comment: [From Southern Alliance for Clean Energy's Statement of Disputed Material Facts] Description of TVA's Aquatics Studies. Analysis of Fish Species Occurrences in Chickamauga Reservoir – A Comparison of Historic and Recent Data (Oct. 2010) ("Fish Species Occurrences Study")

44. The study found that another reason for the change in species diversity and abundance is that most species that have not been collected in recent times have historically never been caught frequently or in large numbers in Chickamauga Reservoir.

     Undisputed that this is a conclusion of the study. Disputed as a rationale for the decline of indigenous species present and decline of indigenous species abundance. The fact that species have not been caught in the reservoir is a meaningful indication of the decline of indigenous fish species. See Young Declaration, ¶¶ III-E.1-20. (**0016-3-11** [Curran, Diane])

Response: *The Southern Alliance for Clean Energy's (SACE's) comments address statements in the TVA report "Analysis of Fish Species Occurrences in Chickamauga Reservoir – A Comparison of Historic and Recent Data" (Simmons 2010). The NRC staff reviewed Simmons (2010) while developing the SFES. The NRC staff agrees with SACE that there have been changes to the fish population from "cumulative man-made impacts." In Section 2.3.2.1 of the SFES, the NRC staff states "The fish populations in the Tennessee River have changed considerably as a result of human-initiated activities…" and "[a]s with the mussel community,* *the fish community appears to be changing in response to historical changes in land use, river regulation, and other human activities." In Section 4.14.6, the NRC staff further indicates that "aquatic communities can change slowly in response to stress: they have been changing for a long time, are changing now, and will probably continue to change for the foreseeable future. The aquatic resources are not stable in the sense of persisting as they were in the past or are today." Further, based on TVA's historical data, the SFES highlights the emerald shiner as a species that has "declined substantially in numerical importance – most obviously downstream of the Watts Bar Dam in the period from 1976 to 1997." Potential reasons for the decline cited by the NRC staff include water quality and competition with an introduced fish species.*

The NRC staff acknowledges that changes in survey equipment introduce a confounding affect into the study design that may account for some of the changes in findings of species occurrence and abundance. The NRC staff addresses historical changes in species composition in the SFES in Section 2.3.2.1, changes in survey methodologies in Section 5.5.2, and cumulative impacts in Section 4.14.6.

Comment: STATEMENT OF [Dr. Shawn Paul Young] PROFESSIONAL OPINION REGARDING ADEQUACY OF TVA'S RECENT BIOLOGICAL STUDIES TO ADDRESS THE ENVIRONMENTAL IMPACTS OF THE PROPOSED WATTS BAR 2 NUCLEAR POWER PLANT ON AQUATIC ORGANISMS. Mollusk Survey, Discussion of Mollusk Survey, and Revised Aquatics Study [Mollusk Survey of the Tennessee River Near [WBN] (Rhea County, Tennessee) (Oct. 28, 2010, Revised Nov. 24, 2010) ("Mollusk Survey") and Aquatic Environmental Conditions in the Vicinity of Watts Bar Nuclear Plant During Two Years of Operation, 1996-1997 (June 1998, Revised June 2010) ("Revised Aquatics Study")]

8. In paragraph 74 of their affidavit, TVA's experts assert that I erroneously extrapolated TVA's characterization of the Reservoir Benthic Macroinvertebrate Index (RBMI") for the benthic macroinvertebrate community in the WBN vicinity, to the freshwater mussel community specifically. They are incorrect. My opinion is based on a passage in TVA's FSEIS on page 55 which states: Another aspect of the Vital Signs Monitoring Program is the benthic index, which assesses the quality of benthic communities in the reservoirs (including upstream inflow areas such as that around WBN). The tailwaters of Watts Bar Dam support a variety of benthic organisms including several large mussel beds. One of these beds has been documented along the right-descending shoreline immediately downstream from the mouth of Yellow Creek. To protect these beds, the state has established a mussel sanctuary extending 10 miles from TRM 520 to TRM 529.

9. Since the institution of the Vital Signs Monitoring Program, the quality of the benthic community in the vicinity of the WBN site has remained relatively constant. The riverine tailwater reach downstream of Watts Bar Dam and WBN rated "good" in 2001 and the rating has increased to "excellent" in 2003-2005 (Appendix C, Tables C-4 and C-5)(emphasis added). This paragraph specifically discusses freshwater mussels as part of the benthic

community evaluated under TVA's Vital Signs Monitoring program. Mussels are benthic macroinvertebrates, and are represented in Metric 2 – "Long-lived Organisms" of the Reservoir Benthic Index (Table 6. Biological Monitoring of the Tennessee River near Watts Bar Nuclear Discharge, 2008). Therefore I did not misinterpret the passage stated in the FSEIS in expressing my opinion that when only four out of 64 (i.e., 6% of) freshwater mussel species once found in the vicinity of WBN remain reproductively viable, in no way can any aspect of the aquatic community be rated in "excellent" health. (**0017-3-4** [Curran, Diane])

Response: This comment reflects a disagreement between TVA and Dr. Shawn Paul Young regarding the interpretation of a statement in the TVA EIS. It also refers to the Reservoir Benthic Macroinvertebrate Index for the benthic macroinvertebrate community (specifically freshwater mussels) and TVA's conclusions regarding the health of the aquatic community. The NRC staff did not use the Reservoir Benthic Macroinvertebrate Index or rely on TVA's conclusions regarding the health of the reservoir. The NRC staff did, however, use the same data that TVA both obtains and applies in calculating its indices (Simmons and Baxter 2009; Simmons et al. 2010; Simmons 2011). The NRC staff also considered and used other relevant data. The NRC staff presents its conclusions in terms of SMALL, MODERATE, or LARGE levels of impact as defined in Section 1.2 of the SFES. These impact levels are not equivalent to the TVA scale of reservoir health.

The NRC staff concluded that the level of impact for the aquatic environment from operation of WBN Unit 2 was SMALL after performing a review of the site and vicinity, which included the studies referenced in the comments. The NRC staff's review is in Sections 2.4.1, 4.3.2 and 5.5.2. The NRC staff also concluded that the level of impact for cumulative effects on aquatic ecology would be LARGE, based on its analysis of past, present, and reasonably foreseeable future activities as discussed in Section 4.14.6.

Comment: STATEMENT OF [Dr. Shawn Paul Young] PROFESSIONAL OPINION REGARDING ADEQUACY OF TVA'S RECENT BIOLOGICAL STUDIES TO ADDRESS THE ENVIRONMENTAL IMPACTS OF THE PROPOSED WATTS BAR 2 NUCLEAR POWER PLANT ON AQUATIC ORGANISMS. Mollusk Survey, Discussion of Mollusk Survey, and Revised Aquatics Study [Mollusk Survey of the Tennessee River Near [WBN] (Rhea County, Tennessee) (Oct. 28, 2010, Revised Nov. 24, 2010) ("Mollusk Survey") and Aquatic Environmental Conditions in the Vicinity of Watts Bar Nuclear Plant During Two Years of Operation, 1996-1997 (June 1998, Revised June 2010) ("Revised Aquatics Study")]

10. Another factor which indicates that the health of macroinvertebrates in general is declining is the dominance of only four species including the Asiatic clam, a non-native, invasive species. As shown in the "Revised Aquatics Study" at page 34, during operational monitoring in 1996-1997, only four of 104 aquatic invertebrate species found made up 87.5%. Further, the average density of aquatic macroinvertebrates per square meter actually declined by more than 50% from 1997 to 2008 in the vicinity of WBN. In 1997, 424

organisms per square meter were reported (Appendix C. Aquatic Ecological Health Determinations for TVA Reservoirs – 1997). In 2008, only 187 organisms per square meter were reported (Table 8. Biological Monitoring of the Tennessee River near Watts Bar Nuclear Discharge, 2008). In 2007 and 2008, even TVA's Reservoir Benthic Index (RBI) score used to monitor the macroinvertebrate community fell to the "fair" category. (**0017-3-5** [Curran, Diane])

Response: The NRC staff understands this comment to pertain to the measurement of the health of the macroinvertebrate population based on the average density of macroinvertebrates per square meter between 1997 and 2008. The NRC staff is cautious about comparing data obtained between 1996 and 1997, and data obtained between 2008 and 2010 for benthic macroinvertebrates because of two confounding factors. First, the sampling methods were different. The initial sets of measurements (1982 to 1997) were conducted using a Hess sampler. Samples from 1999 to the present were obtained using a Ponar sampler. The NRC staff added text to discuss the differences between the two sampling methods in Sections 2.3.2.1 and 5.5.2.3.

Second, TVA installed an aeration system in the reservoir upstream of Watts Bar Dam in early summer 1996 to reduce stratification near the dam. This increased the dissolved oxygen levels in the water released through the dam. The installation of the aeration system upstream of the WBN site is a confounding influence that makes it difficult to observe patterns specifically relating to WBN Unit 1 operations.

Because of the differences in sampling techniques and the effect of the aeration system, the NRC staff did not make comparisons between measurements in the two periods but did include a discussion of the organisms present in the benthic habitat. The NRC staff updated Table 2-9 to provide the most recently available three years of data (Simmons and Baxter 2009; Simmons et al. 2010; Simmons 2011) for locations at TRM 527.4 in Chickamauga Reservoir near the site and TRM 533.3 above the dam in Watts Bar Reservoir. Simmons et al. (2010) and Simmons (2011) contain new information that was unavailable at the time the draft SFES was written.

Comment: [From Southern Alliance for Clean Energy's Statement of Disputed Material Facts] Description of TVA's Aquatics Studies

33. First, this study provides a detailed explanation of TVA's RFAI methodology. TVA created the RFAI methodology based on industry standards for biological indices, including those approved by TDEC and the U.S. Environmental Protection Agency ("EPA"), for use in its Vital Signs monitoring program. TVA has conducted fish sampling in the Chickamauga Reservoir every year since 1993, in support of this program.

Undisputed as to the conduct of the RFAI study every year since 1993. Disputed as to the consistency, accuracy, and usefulness of the study to portray aquatic health in the Tennessee River near WBN1. (**0016-3-2** [Curran, Diane])

34. RFAI methodology uses twelve fish community metrics from four general categories: Species Richness and Composition; Trophic Composition; Abundance; and Fish Health. For each metric, scores are given on a scale from 1 to 5, with a score of 5 indicating optimum health. The resulting scores range from 12-60, broken down as follows: 12-21 ("Very Poor"), 22-31 ("Poor"), 32-40 ("Fair"), 41-50 ("Good"), or 51-60 ("Excellent"). RFAI scores have an intrinsic variability of ±3 points.

Undisputed as to the description of the RFAI methodology. Disputed as to the consistency, accuracy, and usefulness of the RFAI methodology to portray aquatic health in the Tennessee River near WBN1. (**0016-3-3** [Curran, Diane])

35. RFAI methodology addresses all five attributes or characteristics of a Balanced Indigenous Population ("BIP"), which is required by the Clean Water Act. If an RFAI score reaches 70% of the highest attainable score of 60 (i.e., 42), or if fewer than half of the RFAI metrics receive a low (1) or moderate (3) score, then normal community structure and function are considered to be present, indicating that BIP is maintained.

Undisputed that this is a description of TVA's methodology for compliance with the BIP requirement. Disputed as to the fact that RFAI methodology only addresses four not five attributes, and to the consistency, accuracy, and usefulness of this methodology to portray aquatic health in the Tennessee River near WBN1. (**0016-3-4** [Curran, Diane])

36. Second, this study evaluates the health of the aquatic environment in the WBN vicinity based on recent fish surveys and the RFAI methodology. The study found that RFAI scores from the site downstream of the WBN intake and thermal discharge have averaged 44 from 1996 to 2008 (i.e., during operation of WBN Unit 1), indicating that the aquatic health of that area is "good" even during WBN operation.

Undisputed that this is a description of TVA's RFAI study, results, and conclusions. Disputed as to the consistency, accuracy, and usefulness of this methodology to portray aquatic health in the Tennessee River near WBN1 and the concluding scores to properly correlate with the true health of the fish community. (**0016-3-5** [Curran, Diane])

37. Third, this study compares the health of that environment as reflected in RFAI scores from before and after WBN operation. Scores from every sample year (1993-2008) were at least 42, i.e., 70% of the highest attainable score of 60. As a result, the study concluded that both before and after WBN operation, BIP has been maintained. Undisputed that this is a description TVA's RFAI study, results, and conclusions.

Disputed as to the consistency, accuracy, and usefulness of this methodology to portray aquatic health in the Tennessee River near WBN1.

38. SACE has not challenged the methodology or findings of this study with this Board. Undisputed that SACE has not challenged the most recent iteration of the RFAI study before the Board. Contention 7, however, criticizes the methodology and results of previous RFAI studies, which have not changed in any significant respect. *See* Young Declaration, ¶ E-III.1. (**0016-3-6** [Curran, Diane])

Comment: ISTATEMENT OF [Dr. Shawn Paul Young] PROFESSIONAL OPINION REGARDING ADEQUACY OF TVA'S RECENT BIOLOGICAL STUDIES TO ADDRESS THE ENVIRONMENTAL IMPACTS OF THE PROPOSED WATTS BAR 2 NUCLEAR POWER PLANT ON AQUATIC ORGANISMS. RFAI Study and Fish Species Occurrences Study [Comparison of Fish Species Occurrence and Trends in Reservoir Fish Assemblage Index Results in Chickamauga Reservoir Before and After WBN Unit 1 Operation (June 2010) ("RFAI Study") and Analysis of Fish Species Occurrences in Chickamauga Reservoir – A Comparison of Historic and Recent Data (Oct. 2010) ("Fish Species Occurrences Study")]

1. TVA uses Reservoir Fish Assemblage Index ("RFAI") "scores" to provide general ratings of the fish community within TVA reservoirs. As discussed by TVA's experts in par. 55, TVA uses the RFAI to determine whether a "Balanced Indigenous Population" is being maintained as required by the EPA under the Clean Water Act. As discussed in Contention 7 and my supporting declaration, I believe TVA's RFAI scores are biased and misleading, and do not properly reflect the true state of the Tennessee River's aquatic resources. TVA's RFAI Study and Fish Species Occurrence Study do not resolve my concerns. (**0017-4-1** [Curran, Diane])

2. In the Fish Species Occurrence Study, TVA analyzed and scored new and historical fish survey data to determine the current presence of fish species, and compared the presence of species before and after operation of WBN Unit 1. TVA claims that a comparison of scores between 1993 and 2008 shows that both before and after operation of WBN1, TVA has maintained a "balanced indigenous population" ("BPI")....In the RFAI Study, TVA also concludes that "long-term data trends suggest that the ecological health of the fish community in Chickamauga Reservoir inflow has been maintained." See page 13 of Attachment 9. Furthermore, TVA states that: "The species composition of the fish assemblage of Chickamauga Reservoir has changed somewhat, but not markedly, over the decades of sampling by TVA."

3. In my professional opinion, the RFAI and Fish Species Occurrence studies does not present a reliable or reasonably accurate picture of the health of aquatic organisms near WBN1, for several reasons. First, TVA's method for conducting RFAI studies has changed over the years, making the scores difficult to compare. And the history of the RFAI program indicates

that the older scores are unreliable because the methodology for deriving those scores was questioned by EPA and others. In an EPA guidance document, for example, EPA includes improvement of the RFAI in a list of "Research Needs:"

Research Needs – TVA has been actively developing assessment tools for its reservoirs for several years. The move to a multimetric approach for reservoir fish began in 1990. Successive steps in this development process have brought continued improvement to the RFAI. Potential improvements in the fish indices include using a simple random sampling design rather than a fixed station design to enhance statistical validity with little increase in variability. Use of the index in reservoirs or other river systems is necessary to test its performance under a wider range of conditions than is available in the Tennessee river. Correlation with known human-induced impacts remains a critical need before general acceptance of the fish index as a reliable method to address reservoir environmental quality.

EPA 841-B-98-007 - Lake and Reservoir Bioassessment and Biocriteria: Technical Guidance Document, Appendix D: Biological Assemblages, Section D.5 Fish, pp. 176-177 (Undated). (http://water.epa.gov/type/lakes/assessmonitor/bioassessment/upload/lakereservoirbioasses s-biocrit-app-d.pdf) (emphasis added). (**0017-4-2** [Curran, Diane])

4. The scientific community has also criticized the RFAI's inability to correlate with environmental degradation or accurately reflect true patterns in environmental health within and among reservoirs:

    More recently, a second TVA reservoir version of the IBI [Index of Biotic Integrity] has been developed, termed the Reservoir Fish Assemblage Index (RFAI, Jennings, Karr, and Fore, personal communication). The RFAI has a somewhat different set of 12 metrics (Table 4), with the changes in metrics designed to improve sensitivity to environmental degradation and to increase adaptability to different types of reservoirs. However, results from applications of both the original TVA version and the newer RFAI have often not accurately reflected what are believed to be the true patterns in environmental health within and among reservoirs, and additional modifications will probably be necessary to develop better versions of the IBI for impoundments (Jennings, personal communication).

    Davis, W. S., and T. S. Simon, Biological Assessment and Criteria: Tools for Water Resource Planning and Decision Making, pp. 260-261 (Lewis Publishers: 1995) (emphasis added).

5. However, even the biased RFAI scores declined post-operation, thus undermining TVA's claim that the RFAI scores show that the "good health" of aquatic organisms near WBN1 has not declined. (**0017-4-4** [Curran, Diane])

6. Some of the problems with TVA's RFAI methodology can be seen in the 12 metrics described in Paragraph 52 of TVA Joint Affidavit for assessing four general categories of fish health characteristics: Species Richness and Composition, Trophic Composition, Abundance, and Fish Health. For each metric, scores are given on a scale from 1 to 5, with a score of 5 indicating optimum health.

7. TVA's RFAI scores are predominantly biased by inappropriate assessments of the first category "Species Richness and Composition" and its 8 metrics (i - viii), and the lack of appropriate metrics within the third category "Abundance" (metric xi).

8. Species Richness and Composition – Metric (i) is described as:

    i. Total number of indigenous species: Greater numbers of indigenous species are considered representative of healthier aquatic ecosystems. As conditions degrade, numbers of species at an area decline.

    Metric (i) is misleading because it reports only the mere presence of a species, and does not account for its actual abundance, reproductive viability, and future existence within the fish community under evaluation. There is no metric to account for this within the "Abundance" category. A threatened or endangered species would register positively under this metric even though its future existence is doubtful. Several indigenous species were present in only one or two years within a decade sampling period. Again, there is no metric to account for these important trends of indigenous fish decline within the "Abundance" category. Further, the percent of native species is biased by hatchery stockings of species that may otherwise have disappeared from Chickamauga Reservoir.

9. Appendix 1 of Attachment 9 to TVA's Motion illustrates my point. Appendix 1 shows that only one Largescale stoneroller was captured in 2004 and 2008 and zero were captured in all other years from 1999-2009. Yet, these two individuals that were collected during a 10-year sampling period represent species presence in Tables 2 and 3. Similarly, River redhorse (two individuals) and Smallmouth redhorse (one individual), which are Catostomids or suckers, show population trends near WBN similar to the Largescale Stoneroller. Thus, while one or two individual fish could not reasonably be characterized as a healthy or even viable population, the RFAI considers its presence as a positive attribute. Further, several intolerant species were found during 2009 in the following numbers: Chestnut Lamprey (0), Steelcolor shiner (4), Emerald Shiner (1), Black redhorse (5), Golden redhorse (3), Northern Hogsucker (0). In comparison, several tolerant species were found during 2009 in the following numbers: Bluegill (471), Gizzard shad (131), and Largemouth bass (61). Nevertheless, in 2009, TVA gave this metric a score of 5 (see Attachment 9, p. 144, Appendix 2-A). In my view, given the extremely low abundance of indigenous fish species and the high abundance of tolerant species, this metric should receive a score of 1, or an equivalent metric should be incorporated into the "Abundance" category to properly represent the extremely low abundance of numerous indigenous species. (**0017-4-5** [Curran, Diane])

10. Metric (ii) in the category of "Species Richness and Composition" is described as:

> ii. Number of centrarchid species: Sunfish species (excluding black basses) are invertivores and a high diversity of this group is indicative of reduced siltation and suitable sediment quality in littoral areas.

> Metric (ii) yields misleading results because it uses only one of several families of fishes that are commonly used to assess the status of a fish community, and because Centrarchids are not representative of the most vulnerable indigenous fish species. TVA neglected to use other families more representative of the Tennessee River such as Percidae (which includes darters), Catostomidae (i.e.,suckers), and Cyprinidae (i.e., minnows). These families were highly diverse and plentiful historically; are intolerant to human disturbance and pollution; and all have suffered severe decline in the Tennessee River. TVA gave this metric a 5, the highest score. The only attribute this reflects is that Centrarchids, which thrive in reservoirs, are well-represented. If one of the other three families were used, this metric would be scored a 1.

11. Metric (iii) in the category of "Species Richness and Composition" is described as:

> iii. Number of benthic invertivore species: Due to the special dietary requirements of this species group and the limitations of their food source in degraded environments, numbers of benthic invertivore species increase with better environmental quality.

> As with metric (i), metric (iii) evaluates only the presence of a species, and does not account for its actual abundance, reproductive viability, and future existence in the environment under evaluation. Again, there is no similar metric in the "Abundance" category to measure the actual numbers of a species. If those factors were taken into account, TVA could not have given this metric a score of 3. Given the steep decline of benthic invertivores as described in par. 9, the score should be 1.

12. Metric (iv) in the category of "Species Richness and Composition" is described as:

> iv. Number of intolerant species: This group is made up of species that are particularly intolerant of physical, chemical, and thermal habitat degradation. Higher numbers of intolerant species suggest the presence of fewer environmental stressors. The higher number of these species would be a positive indicator

> Metric (iv) should account for status of suckers, minnows, and darters as well as locally endangered or extirpated species such as sturgeon and paddlefish because these fish are intolerant and in decline. As with metrics (i) and (iii), metric (iv) evaluates only the presence of a species, and does not account for its actual abundance, reproductive viability, and future existence in the environment under evaluation. Again, there is no

similar metric in the "Abundance" category to measure the actual numbers of a species. If those factors were taken into account, TVA could not have given this metric a score of 5. This metric suffers from the same bias as Metric (i). TVA gave this metric a score of 5, but it should have received a score of 1. (**0017-4-6** [Curran, Diane])

13. Metric (v) and Metric (vi) in the category of "Species Richness and Composition" are described as:

    v. Percentage of tolerant individuals (excluding Young-of-Year): This metric signifies poorer water quality with increasing proportions of individuals tolerant of degraded conditions.

    vi. Percent dominance by one species: Ecological quality is considered reduced if one species inordinately dominates the resident fish community.

    Metric (v) should identify a fish species community that is dominated by species tolerant of disturbance and poor water quality. Metric (vi) should identify a fish species community that is unbalanced and dominated by only one or few species. These are negative attributes whose scores should be inversely proportional to the degree they exist. TVA's RFAI sampling shows a high percentage of tolerant species such as bluegills. See par. 19 below. Further, the fish community is currently dominated by bluegills (See par. 19); thus, the score should be a 1.

    TVA, however, gave Metric (v) a score of 3. TVA correctly gave Metric (vi) a score of 1, which is evidence that the fish community no longer supports a balanced indigenous population.

14. Metric (vii) in the category of "Species Richness and Composition" is described as:

    vii. Percentage of non-indigenous species: This metric is based on the assumption that non-indigenous species reduce the quality of resident fish communities.

    Like metrics (v) and (vi), this is a negative attribute, whose score should be inversely proportional to the degree it exists. Metric #7 should identify a fish species community that has a significant number of non-indigenous species, i.e. species that are not indigenous to the Tennessee River whether intentionally or unintentionally stocked. TVA sampling shows several non-indigenous species present; and, that the percent of native species is biased by hatchery stockings of species that may otherwise have disappeared from Chickamauga Reservoir. TVA properly scored this metric with a 1, again indicating that the fish community no longer supports a balanced indigenous population. (**0017-4-7** [Curran, Diane])

15. Metric (viii) in the category of "Species Richness and Composition" is described as:

> viii. Number of top carnivore species: Higher diversity of piscivores is indicative of the availability of diverse and plentiful forage species and the presence of suitable habitat.

> Metric (viii) should identify a fish species community that is in proper balance with an adequate carnivore population, or fish that eat other fish and serve as the upper food chain predators. However, this metric may also be biased by hatchery stockings that are used to support a sport fishery. Often hatchery supplementation is used to artificially support a fish population for recreational purposes when the aquatic system no longer supports natural reproduction. Recreational fisheries often target these predatory fish species such as striped bass, sauger, and walleye, all of which are stocked by the State of Tennessee into Chickamauga Reservoir because of lack of natural reproduction to support fishing. The lack of reproduction is due to the alterations of the Tennessee River and the resulting poor ecological health. While TVA scored this metric at 5, the score should be a 3.

16. The category "Abundance" is as equally important as "Species Richness and Composition"; yet, "Abundance" is only represented by one metric (metric xi) as compared to "Species Richness and Composition" which is represented by eight metrics. This is a major omission that leads to the inappropriately high RFAI scores that overstates the health of the fish community. Metric xi is described as:

> xi. Average number per run (number of individuals): This metric is based upon the assumption that high quality fish assemblages support large numbers of individuals.

> Metric (xi) is highly biased by the ever-increasing numbers of bluegills and other species that thrive in a man-made environment and now dominate the fish community. The increase of bluegills masks the low number of other native species in decline. TVA, scoring this metric based upon the definition, gave it a 5. However, if this category incorporated similar metrics as "Species Richness and Composition" based upon actual abundance, or number of individuals captured, all of the metrics designed to monitor indigenous fish species would receive RFAI scores of 1, the lowest possible. (**0017-4-8** [Curran, Diane])

17. Paragraph 53 of the Joint Affidavit describes the method for evaluating total RFAI scores as follows:

> Because there are 12 metrics, RFAI scores range from 12 to 60. The aquatic community health is indicated by the following ranges of scores: 12-21 ("Very Poor"), 22-31 ("Poor"), 32-40 ("Fair"), 41-50 ("Good"), or 51-60 ("Excellent").

TVA's final 2009 RFAI score for the area near WBN Plant was a 44 in the "Good" category. Correcting for the bias of the RFAI would lead to a score of 28, or a "Poor" rating of the health of the fish community. I believe the "poor" rating, which is a significantly different picture of the fish community in the vicinity of WBN than that of TVA's analyses, more accurately represents the status of the fish community of the Tennessee River in the vicinity of WBN Plant.

18. The score that I estimated is also consistent with other data which show a decrease in the level of diversity and the size of existing populations since WBN1 began operating. For instance, a comparison of the NRC's 1978 Final Environmental Impact Statement (FEIS) for WBN Units 1 & 2 (Table C-21) and the NRC's 2008 Final Supplemental Environmental Impact Statement (FSEIS) for WBN Unit 2 (Table 3.3.1) shows that the Chickamauga Reservoir experienced a 24% decline of freshwater fish species between 1970-73 and 1991-1996. Further, Vital Signs and Biological Monitoring reports from 1994 list 36 fish species that were captured in Upper Chickamauga Reservoir, and reports from 1999-2009 show the number of species declined to between 24 and 31 for a given year, another 14% decline. (**0017-4-9** [Curran, Diane])

Response: *These comments relate to the TVA's use of the RFAI to describe the impact of the operation of WBN Unit 1 and to evaluate the potential impact from operation of WBN Unit 2. These comments state that TVA's use of the RFAI is biased, inappropriate, or misused.*

The NRC staff agrees that the use of the RFAI to describe the potential environmental impacts to fish from operation of WBN Unit 2 is inappropriate. The NRC staff did not base the impact determination in the SFES on the RFAI methodology or scores. The NRC staff reviewed the TVA analysis and the data TVA obtained in support of its Vital Signs Monitoring Program. The NRC staff also considered and used other relevant data beyond that used for the Vital Signs Monitoring Program.

The NRC staff's analysis assesses impacts using NRC-defined SMALL, MODERATE, and LARGE impact levels (Section 1.2), which are not equivalent to the TVA scale of reservoir health. The NRC staff found that the direct and indirect impact level from operation of WBN Unit 2 on the aquatic environment that would be SMALL, as discussed in Section 4.3.2.7 and Section 9.5.3. The NRC staff also concluded that the cumulative impacts on aquatic ecology would be LARGE, based on the NRC staff's analysis of past, present, and reasonably foreseeable future activities as discussed in Section 4.14.6.

Comment: [From Southern Alliance for Clean Energy's Statement of Disputed Material Facts]
Description of TVA's Aquatics Studies
Analysis of Fish Species Occurrences in Chickamauga Reservoir – A Comparison of Historic and Recent Data (Oct. 2010) ("Fish Species Occurrences Study")

42. This study found that species occurrence and abundance in the Chickamauga Reservoir has changed from 1947 to 2009. Many of these changes took place before operation of WBN Unit 1 began. Undisputed to the extent that TVA asserts that many of the changes in species occurrence and abundance in the Chickamauga Reservoir took place before the operation of WBN1 began. Disputed to the extent that TVA implies that changes after WBN1 operation began are insignificant. See Young Declaration, ¶¶ III-E.1-20. (**0016-3-9** [Curran, Diane])

Comment: [From Southern Alliance for Clean Energy's Statement of Disputed Material Facts] Description of TVA's Aquatics Studies
Analysis of Fish Species Occurrences in Chickamauga Reservoir – A Comparison of Historic and Recent Data (Oct. 2010) ("Fish Species Occurrences Study")

43. One major cause of this change is impoundment of the Tennessee River, which began in the 1930s and has altered habitats required for various life stages of aquatic species. Some of the species not found in recent surveys require unimpounded, free flowing riverine environments.

  Undisputed to the extent that impoundment of the Tennessee River is a major cause of the decline in species occurrence and abundance. Disputed to the extent that TVA implies that changes after WBN1 operation began are insignificant. See Young Declaration, ¶¶ III-E.1-20 and III-D.4-7, III-D.10, and III-A.6-9. (**0016-3-10** [Curran, Diane])

Comment: [From Southern Alliance for Clean Energy's Statement of Disputed Material Facts] Description of TVA's Aquatics Studies
Analysis of Fish Species Occurrences in Chickamauga Reservoir – A Comparison of Historic and Recent Data (Oct. 2010) ("Fish Species Occurrences Study")

46. As a result, this study concluded that there is no basis to support a finding that operation of WBN Unit 1 caused the observed changes in fish species and occurrence in the Chickamauga Reservoir.

  Undisputed as to the study's stated conclusion. Disputed as to whether the conclusion is accurate that there is no basis to support a finding that operation of WBN1 caused the observed changes in fish species and occurrence. See Young Declaration, ¶¶ III-A.1-14, III-B.1-5, III-C.1-12, and III-E.1-20. (**0016-3-13** [Curran, Diane])

Comment: STATEMENT OF [Dr. Shawn Paul Young] PROFESSIONAL OPINION REGARDING ADEQUACY OF TVA'S RECENT BIOLOGICAL STUDIES TO ADDRESS THE ENVIRONMENTAL IMPACTS OF THE PROPOSED WATTS BAR 2 NUCLEAR POWER PLANT ON AQUATIC ORGANISMS

RFAI Study and Fish Species Occurrences Study [Comparison of Fish Species Occurrence and Trends in Reservoir Fish Assemblage Index Results in Chickamauga Reservoir Before and After WBN Unit 1 Operation (June 2010) ("RFAI Study") and Analysis of Fish Species Occurrences in Chickamauga Reservoir – A Comparison of Historic and Recent Data (Oct. 2010) ("Fish Species Occurrences Study")]

19. Evidence that the fish community near WBN is greatly unbalanced may be found by analyzing TVA electrofishing data in Aquatic Ecological Health Determinations for TVA Reservoirs –1994, Table 8, Page 352, and within Biological Monitoring of the Tennessee River Near Watts Bar Nuclear Plant Discharge, 2008, Table 3, Page 18. These data show that in 1994, bluegill – a species that thrives in man-made habitats and are thus popular for stocking in small ponds across the United States – comprised only 27% of all fish in TVA's sampling in Upper Chickamauga Reservoir. However, during 2008 sampling, bluegill comprised 63% of all fish captured in Upper Chickamauga Reservoir at areas downstream of Watts Bar Nuclear Plant Discharge. Upon further examination, Centrarchids in general (the family of fishes that is comprised of bluegill, sunfishes, and black-basses) make up 78% of all fish near WBN. A fish community that is made up of 78% bluegill, sunfishes, and black-basses is more indicative of a farm pond than the most biologically diverse freshwater ecosystem in North America. Further, by adding gizzard shad, another species that may thrive in reservoirs, the percent increases to 91%. This results in a very low abundance, whether stated in terms of percent composition and actual numbers, of other native riverine fish species that should be found in the Tennessee River near WBN. When this is compared to 1994 when Centrarchids comprised only 58% and gizzard shad 10%, there is evidence that the fish community is extremely unbalanced, and the percent of indigenous riverine species has continued to decline since WBN1 became operational.

20. Thus, these data show that the fish community has undergone significant negative changes since WBN1 became operational and the current health of the fish community is poor. The data certainly do not support the existence of a Balanced Indigenous Population or "BIP." (**0017-4-10** [Curran, Diane])

Response: The commenter questions whether changes to the reservoir following the start of operation of WBN Unit 1 are insignificant. The commenter states that the current health of the fish community is poor and that significant negative changes have occurred in the fish community since WBN1 became operational.

The NRC staff agrees that the largest percentage of fish that occur in the sampling from the Chickamauga Reservoir in the vicinity of the Watts Bar site are fish that thrive in reservoirs, rather than the indigenous riverine species. The NRC staff addressed this issue in Section 4.14.6 of the SFES by stating that "aquatic communities can change slowly in response to stress: they have been changing for a long time, are changing now, and will probably continue to change for the foreseeable future. The aquatic resources are not stable in the sense of persisting as they were in the past or are today." The NRC staff considered past, present, and reasonably foreseeable future projects (including operation of WBN Unit 1) in Section 4.14.6 and found the impacts from cumulative operations to be LARGE (clearly noticeable and sufficient to destabilize important attributes of the resource). Further, the NRC staff stated in Section 2.3.2.1 that "…the largest drop in species abundance occurred between the surveys taken from 1975 to 1989… and those from 1990 to 1995…" WBN Unit 1 did not begin operation until 1996.

Comment: [From Southern Alliance for Clean Energy's Statement of Disputed Material Facts]
Description of TVA's Aquatics Studies
Mollusk Survey of the Tennessee River Near [WBN] (Rhea County, Tennessee) (Oct. 28, 2010, Revised Nov. 24, 2010) ("Mollusk Survey"), and Discussion of the Results of the 2010 Mollusk Survey of the Tennessee River Near [WBN] (Rhea County, Tennessee) (Mar. 2011) ("Discussion of Mollusk Survey")

52. These studies agree that the Chickamauga Reservoir in the WBN vicinity is not the ideal habitat for mussels. Still, the 2010 survey found that the mussel community in the WBN vicinity is in substantially similar condition as it was near the end of the previous operational monitoring period (1996 to 1997), in both species composition and the number of mussels collected. In addition, the 2010 survey collected juveniles of at least five mussel species, evidencing reproduction of mollusks in the WBN vicinity.

    Undisputed as to the agreement a reservoir may not be ideal habitat for mussels. Disputed as to what results the consultant produced versus what conclusions TVA drew from that data. Disputed as to the mussel community in the WBN vicinity being in substantially similar condition as it was near the end of the previous operational monitoring period and the significance of the collection of five juvenile mussel species. See Young Declaration, ¶¶ III-D.1-7. (**0016-3-15** [Curran, Diane])

Comment: [From Southern Alliance for Clean Energy's Statement of Disputed Material Facts]
Description of TVA's Aquatics Studies
Mollusk Survey of the Tennessee River Near [WBN] (Rhea County, Tennessee) (Oct. 28, 2010, Revised Nov. 24, 2010) ("Mollusk Survey"), and Discussion of the Results of the 2010 Mollusk Survey of the Tennessee River Near [WBN] (Rhea County, Tennessee) (Mar. 2011) ("Discussion of Mollusk Survey")

53. As a result, this study concluded that there is no basis to support a finding that the relatively low densities of mussels in the WBN vicinity are the result of operation of WBN Unit 1.

    Undisputed that this is the conclusion stated. Disputed as to the accuracy and reasonableness of the conclusion. See Young Declaration, ¶ III-D.4-7.
    (**0016-3-16** [Curran, Diane])

Comment: STATEMENT OF [Dr. Shawn Paul Young] PROFESSIONAL OPINION REGARDING ADEQUACY OF TVA'S RECENT BIOLOGICAL STUDIES TO ADDRESS THE ENVIRONMENTAL IMPACTS OF THE PROPOSED WATTS BAR 2 NUCLEAR POWER PLANT ON AQUATIC ORGANISMS

Mollusk Survey, Discussion of Mollusk Survey, and Revised Aquatics Study [Mollusk Survey of the Tennessee River Near [WBN] (Rhea County, Tennessee) (Oct. 28, 2010, Revised Nov. 24, 2010) ("Mollusk Survey") and Aquatic Environmental Conditions in the Vicinity of Watts Bar Nuclear Plant During Two Years of Operation, 1996-1997 (June 1998, Revised June 2010) ("Revised Aquatics Study")]

1. As discussed in Contention 7, TVA's assertion in the FEIS that mussel health is "excellent" because their population is "constant" is contradicted by evidence that mussel populations are declining.... TVA responded to my criticism by hiring a consultant to conduct a new mussel survey utilizing new and expanded methodology. The study site evaluated mussel beds within transects in the same general areas as previous TVA mussel surveys near WBN Plant. Each mussel was identified by species and age. TVA compared the results from the 2010 study with previous mussel studies, including the post-operational mussel surveys in 1996 and 1997. The results from the 1996 and 1997 post-operational surveys are found within the original "Aquatic Study" published in 1998 and the recent "Revised Aquatics Study".

2. TVA no longer asserts that mussel health near WBN1 is excellent. Instead, it states that the studies it conducted "agree that the Chickamauga Reservoir in the WBN vicinity is not the ideal habitat for mussels.".... Nevertheless, TVA's experts state that the survey results demonstrated "that the current mussel community adjacent to WBN is stable and that some species are reproducing." Baxter and Coutant, par. 72. They assert that the mussel community in the WBN vicinity is in "substantially similar condition as it was near the end of the previous operational monitoring period (1996 to 1997), in both species composition and the number of mussels collected." In addition, they state that the 2010 survey "collected juveniles of at least five mussel species, evidencing reproduction of mollusks in the WBN vicinity." Id. Based on these results, TVA contends that "there is no basis to support a finding that the relatively low densities of mussels in the WBN vicinity are the result of operation of WBN Unit 1." (**0017-3-1** [Curran, Diane])

Comment: STATEMENT OF [Dr. Shawn Paul Young] PROFESSIONAL OPINION REGARDING ADEQUACY OF TVA'S RECENT BIOLOGICAL STUDIES TO ADDRESS THE ENVIRONMENTAL IMPACTS OF THE PROPOSED WATTS BAR 2 NUCLEAR POWER PLANT ON AQUATIC ORGANISMS

Mollusk Survey, Discussion of Mollusk Survey, and Revised Aquatics Study [Mollusk Survey of the Tennessee River Near [WBN] (Rhea County, Tennessee) (Oct. 28, 2010, Revised Nov. 24, 2010) ("Mollusk Survey") and Aquatic Environmental Conditions in the Vicinity of Watts Bar

Nuclear Plant During Two Years of Operation, 1996-1997 (June 1998, Revised June 2010) ("Revised Aquatics Study")]

3. I disagree with TVA's assertions. The data collected by TVA show that health of the freshwater mussel community around WBN1 is poor and declining. The data also show a connection between the poor health of the mussel community near WBN1 and the operation of WBN1.

4. There can be no doubt that the health of the mussel community near WBN1 is poor and also declining. The data provided in the Mollusk Survey show that freshwater mussel abundance has declined significantly in the area affected by the SCCW since it began cooling Unit 1 in 1999. TVA failed to address three significant trends reflected in this data. First, the abundance of mussels at the three study sites changed significantly between 1996-97 and 2010. In 1996-97, just before the SCCW went into operation for WBN1 in 1998, 344 mussels were collected from the upper bed located just upriver of WBN. That bed now lies within the SCCW discharge plume (p. 40, Revised Aquatics Study). By 2010 the abundance of mussels at the upper bed had been reduced by approximately half to 175 (p. 4, Mollusk Study). This is a major concern, given that the site is within the mixing zone for the SCCW outfall, which had not been in use for a substantial time prior to or during the 1996-97 surveys.

5. The data also show that mussel abundance in both the middle and lower sites increased since 1996-97 (p. 40, Revised Aquatics Study and p. 4, Mollusk Study. These increases may be due to better sampling techniques employed in 2010, or to better reservoir system management practices implemented at Watts Bar Dam. The Discussion of Mollusk Survey does not explain this development. Quite possibly, the SCCW may be thwarting a rebounding mussel population in the vicinity of WBN. **(0017-3-2** [Curran, Diane])

6. Second, the experimental boulder field to provide increased mussel habitat as a mitigation measure for the use of the SCCW had very few mussels – only five -- indicating this action was a failure. TVA's experts attribute this failure to the force of the water flowing from Watts Bar Dam. Baxter and Coutant, par. 70. But they do not acknowledge that the boulder field is located near the SCCW. The death of most relocated mussels, and the substantial decline of mussel numbers in the upper bed show the SCCW has and will continue to have substantial adverse impacts on the mussels near WBN.

7. Finally, the data indicates that a significant number of mussel species are still unable to reproduce and recruit new members to sustain their local populations. The recent survey found the presence of juveniles for four of the 17 species, indicating some reproduction and recruitment is taking place. However, for the other 13 species -- including two endangered species – no juveniles were present, indicating a lack of reproduction and recruitment capacity, which will lead to eventual local extirpation. In addition, the four reproducing

species that were found near WBN1 are just a fraction of the 64 mussel species known to once inhabit the Tennessee River in the vicinity of present day WBN Plant. Thus, only 6% of the indigenous freshwater mussel species remain viable at this time.....

9. I do not believe TVA has a reasonable basis for placing the blame for mussel decline solely on river impoundment. While it is clear that river impoundment has severely impacted the mussel community, the results of the 2010 surveys show an alarming decline of mussels in the vicinity of the SCCW. This is evidence that current WBN operations have had a large impact on mussel health and that adding another reactor unit will increase and perpetuate these negative impacts. **(0017-3-3** [Curran, Diane])

Response: These comments relate to the health of the mussel population, the decline in mussel abundance, and the effect of SCCW thermal discharges on the populations near the SCCW. In regard to the health of the mussel population, the commenter is correct that the health of the mussel community has been declining and that some species are reproducing while others are not. In Section 2.3.2.1, the NRC staff discusses that "the numbers of native mussels have been declining since the early 1940s when TVA filled the Chickamauga and Watts Bar reservoirs." The commenter is also correct that the number of specimens found in the upper mussel bed (TRM 528-528.9) in 2010 is almost half of the number found in 1997 and in previous years. The NRC staff acknowledges that this is a significant decrease and revised Section 2.3.2.1 to include a discussion of the decrease. However, analysis of the data from 1983 to 1997 shows that, with the exception of the location at TRM 528.9, there have been past years when the quantitative counts of mussels at each of the locations were lower than the count in 2010. Sampling variability can account for part of the decrease; however, the current overall trend for this upper bed is toward fewer individuals. Although juvenile mussels were found in the upper bed, indicating reproduction, there were fewer individuals under age 10 (8) than found in either the middle (31) or lower (20) beds (Third Rock Consultants 2010).

In Sections 4.3.2.3 and 5.1, the NRC staff discusses the monitoring program for the discharge from the SCCW system. This includes the continuous monitoring of water temperature at the stream bottom to ensure that the temperature does not exceed the permitted limit of 33.5°C (92.3°F). TVA also continuously monitors water temperature at the downstream edge of the mixing zone located 610 m (2,000 ft) downriver from the discharge. The monitoring program also includes a biannual temperature survey along a transect in the reservoir at depths from the surface to 2 m (7 ft) below the surface. This survey provides data TVA uses to verify its models. The NRC staff supplemented Section 4.3.2.3 to present hydrothermal data obtained during monitoring studies. Based on the location and depth of the water where the mussels were found, 4 to 6 m (14 to 21 ft), and the buoyancy of the plume, the NRC staff believes it is not likely that the WBN Unit 2 operations (with or without operation of WBN Unit 1) would affect mussel health.

As discussed in Section 2.3.2.1, TVA relocated the mussels that lived near the SCCW discharge (TRM 529.2) in 1997. Further, TVA indicated that the mussels were relocated to the opposite side of the river at TRM 528 to 528.9. The NRC staff revised paragraphs in Section 2.3.2.1 and 4.3.2.6 to indicate that TVA relocated these mussels to the upper mussel bed. TVA relocated a different set of mussels to the boulder field as discussed in SFES Section 2.3.2.1. According to TVA, researchers randomly selected these mussels from the downstream mussel beds and moved them to three experimental plots near the river marker for TRM 528, directly across from the 528.9 boat ramp and mid-channel. This area is downstream of the mixing zone from the SCCW.

Comment: TVA has not taken the necessary steps to evaluate how the effluent from WBN Units 1 and 2 may contribute to the stresses on the fragile health of fish communities, or how these facilities may interfere with mussel reproduction. (**0008-14** [Curran, Diane])

Response: *The NRC staff evaluated the potential effects of thermal (SFES Section 4.3.2.3), chemical (SFES Section 4.3.2.4), and physical (SFES Section 4.3.2.5) discharges from WBN Unit 2 on aquatic biota. The NRC staff also discusses the changes that are anticipated from operation of both units. The NRC staff revised Section 4.3.2.3 and 4.3.2.5 to indicate that TVA received a NPDES permit issued by the State of Tennessee (TDEC 2011) for operation of both units (NPDES Permit TN0020168 on June 30, 2011), as modified (TVA 2011a).*

Comment: TVA [in "Motion for Summary Disposition of Contention 7" (Nov. 21, 2011)] claims that it has conducted studies that resolve the three major deficiencies identified in Contention 7. As discussed in Dr. Young's attached Declaration, this is not correct......With respect to TVA's mischaracterization of the health of the aquatic environment as good, TVA has done nothing to alleviate the concerns raised by Contention 7. Although as discussed above, TVA's data collection is insufficient to present a reasonable picture of the health of the Tennessee River, the data that TVA has collected do not indicate, as TVA claims, that WBN1's impacts on the aquatic ecosystem have been insignificant. Rather, they point to already-significant aquatic impacts by WBN1 that are likely to be significantly exacerbated by the operation of WBN2. (**0015-7** [Curran, Diane])

Comment: SUMMARY OF MY [Dr. Shawn Paul Young] PROFESSIONAL OPINION REGARDING TVA'S ASSERTIONS.

5. With respect to TVA's mischaracterization of the health of the aquatic environment as good, TVA has done nothing to alleviate my concern. Although as discussed above, TVA's data collection is insufficient to present a reasonable picture of the health of the Tennessee River, the data that TVA has collected do not indicate, as TVA claims, that WBN1's impacts on the aquatic ecosystem have been insignificant. Rather, they point to already-significant aquatic impacts by WBN1 that are likely to be significantly exacerbated by the operation of WBN2. (**0017-1-5** [Curran, Diane])

Response: These comments state that TVA's data collection is insufficient to present a reasonable picture of the health of the Tennessee River and that the data points to already-significant aquatic impacts by WBN Unit 1 that would be exacerbated by the operation of WBN Unit 2. The NRC staff reviewed the data provided by TVA and found that quantity of data available is sufficient to draw a conclusion for the purposes of NEPA. In Section 4.3.2.7, the NRC staff concludes that the overall impacts on aquatic biota from operating WBN Unit 2 would be SMALL. The NRC staff notes continued decreases in the population of the upper mussel bed. However, the NRC staff also indicates (in Section 4.3.2.3), that, based on hydrothermal studies and the depth of the mussels located in the upper beds within the passive mixing area, that the WBN operations would not likely affect the health of the mussels in any discernible way. In addition, the NRC staff notes that the State of Tennessee has provided an NPDES permit to TVA for operation of both units that satisfies its concerns relative to the thermal discharges and the effect those discharges may have on the health of the freshwater mussels (TDEC 2011). In addition, the U.S. Fish and Wildlife Service (USFWS 2012) concluded that operations of WBN Unit 2 "may affect, not likely to adversely affect" the pink mucket and would have "no effect" on the other aquatic threatened or endangered species.

Comment: [Southern Alliance for Clean Energy's Statement of Disputed Material Facts]
Description of the Proposed Project
General Information

17. WBN Unit 1 was originally designed to operate only in a closed cycle cooling mode via the Condenser Cooling Water ("CCW") system. After TVA began operation of Unit 1, it determined that a supplemental cooling system would increase the efficiency of the plant. Accordingly, TVA began to use a Supplemental Condenser Cooling Water ("SCCW") system in 1998. Disputed as to the reason TVA began to use the SCCW. The original cooling system was under-designed and would have prevented WB1 from achieving rated power output on hot summer days. Some form of cooling tower enhancement or supplemental cooling was/is necessary for WB1 to achieve rated output on hot summer days (when the highest annual demand is experienced on the TVA system). This is supported by the NRC's Draft SFEIS at page 3-4, which states:

Evaporation of cooling-water system water from the cooling-tower increases the concentration of dissolved solids in the cooling-water system. In most closed-cycle wet cooling systems, a portion of the cooling water is removed and replaced with makeup water from the source (for WBN, the Tennessee River) to limit the concentration of dissolved solids in the cooling system and in the discharge to the receiving water body.

Because the WBN cooling tower cannot remove the desired amount of heat from the circulating water during certain times of the year, TVA added the Supplemental Condenser Cooling Water (SCCW) system to the cooling system for the WBN reactors (TVA 1998). The SCCW draws water from behind Watts Bar Dam and delivers it, by gravity flow, to the cooling-tower basins to supplement cooling of WBN Unit 1. This cooling system would also be used for Unit 2. The temperature of this water is usually lower than the temperature of the water in the cooling-tower basin and, as a result, lowers the temperature of the water being used to cool the steam in the condensers. Slightly less water enters the cooling-tower basins through the SCCW intake than leaves the cooling-tower basins and is discharged to the Tennessee River through the SCCW discharge structure (TVA 2010). Since the SCCW has been operating, elevated total dissolved solids in blowdown water have not been a concern because a large volume of water continually enters and leaves the cooling-tower basins (PNNL 2009). (emphasis added). Had TVA more robust cooling system in the first place, the SCCW would never have been considered necessary by TVA and TVA would not now be proposing to operate WBN2 with the SCCW. (**0016-1-1** [Curran, Diane])

Response: The commenter describes why TVA currently operates the SCCW for WBN Unit 1 and plans to operate the SCCW for WBN Unit 2. The NRC staff describes the current (WBN Unit 1) and proposed operation of the SCCW for WBN Unit 2 in Section 3.2.2 and the current and potential effects of operation in Sections 4.2 and 4.3.2, based on TVA's discussion in its FSEIS (TVA 2008) and in the EA for the SCCW (TVA 1998).

Comment: TVA's finding [in 2007 FSEIS] that WBN Unit 2 will have no significant impacts on aquatic life in the Tennessee River is inadequately supported in the following respects:

3. TVA does not adequately address the cumulative impacts of WBN Unit 2 in conjunction with the impacts of the numerous water impoundments on the Tennessee River, or with other industrial facilities such as the ten fossil fuel-burning plants, the six operating nuclear reactors, and the five additional reactors for which TVA has sought operating licenses. Each of these facilities affects the Tennessee River continuum. That is, each facility not only affects the immediate environment, but those changes are then felt throughout the river as a domino effect.

   The portion of the Tennessee River in the vicinity of WBN is an important part of the river continuum, as are all other segments of the river. Each segment has its own complex ecological balance that is required to support a diverse population of fish and other organisms, providing different habitats needed at different life history stages that must match available food and habitat needs in time and space. Each new industrial facility that is added to the environment will compound the existing disruptions to these interrelated aquatic ecosystems, and further remove the Tennessee River from any semblance of the natural state which would be necessary to restore or even halt the deterioration of the hundreds of declining, threatened, and endangered aquatic species in the Tennessee River Basin. .... The FSEIS is thus inadequate because it does not contain a discussion of these cumulative industrial impacts or the degree to which WBN Unit 2 will contribute to them. (**0008-12** [Curran, Diane])

Comment: [Contention 7: Inadequate Consideration of Aquatic Impacts]
TVA claims that the cumulative impacts of WBN Unit 2 on aquatic ecology will be insignificant (FSEIS Table S-1 at page. S-2, and Table 2-1 at page. 30). TVA's conclusion is not reasonable or adequately supported, and therefore it fails to satisfy 10 C.F.R. § 51.53(b) and NEPA. TVA's discussion of aquatic impacts is deficient in three key respects...... Third, TVA [in 2007 Final Supplemental Environmental Impact Statement] fails completely to analyze the cumulative effects of WBN2 when taken together with the impacts of other industrial facilities and the effects of the many dams on the Tennessee River. (**0008-8** [Curran, Diane])

Comment: TVA [in "Motion for Summary Disposition of Contention 7" (Nov. 21, 2011)] claims that it has conducted studies that resolve the three major deficiencies identified in Contention 7. As discussed in Dr. Young's attached Declaration, this is not correct......Finally, TVA still does not address the cumulative impacts of WBN2 in conjunction with the impacts of the numerous water impoundments on the Tennessee River, or with other industrial facilities such as the ten fossil fuel-burning plants, the six nuclear reactors that are already in operation, and the five additional reactors for which TVA has sought operating licenses. (**0015-10** [Curran, Diane])

Comment: The licensing and operation of WBN2 is just one of many industrial projects that will affect the aquatic health of the Tennessee River. (**0015-12** [Curran, Diane])

Response: The NRC staff based its analysis on the 1997 guidance from the Council on Environmental Quality regarding cumulative impacts and regulations in Part 10 of the Code of Federal Regulations (CFR) § 51.53(b). Section 4.14 of the SFES describes the NRC staff's process for assessing cumulative impacts. In performing its cumulative impact analysis, the NRC staff follows NRC regulations and NEPA guidance. In the SFES, the geographical region for cumulative impacts to aquatic ecology primarily comprises the Watts Bar and Chickamauga reservoirs. The dams built on the Tennessee River and tributaries largely segment the biological communities so that there is no discernible communication of direct effects of operation of WBN Units 1 and 2 beyond one reservoir downstream. The NRC staff included the Watts Bar Reservoir in its analysis because the SCCW is located on the Watts Bar Reservoir. The NRC staff also added text to the SFES in Section 4.14.6 to clarify the geographical region of interest for the aquatic analysis.

The Chickamauga Reservoir extends from TRM 471 to TRM 529.9. The Watts Bar Reservoir extends from TRM 529.9 to TRM 602. The NRC staff included a list of the past, present, and reasonably foreseeable projects and other actions that extend from Fort Loudon Dam to Chickamauga Dam in Table 4-15. In response to this comment, the NRC staff also included the past, present, and reasonably foreseeable projects in the main tributaries to the Tennessee River (Clinch, Little Tennessee, Ocoee, and Hiwassee Rivers) between Fort Loudon Dam and Chickamauga Dam (see Table 4-15).

Comment: [From Southern Alliance for Clean Energy's Statement of Disputed Material Facts] Overview of the Draft SFES Conclusions Regarding TVA's Aquatic Studies

81. Specifically, the Staff concurred with TVA's findings regarding entrainment impacts, concluding in the Draft SFES that hydraulic entrainment would have a very minor impact on the aquatic biota in the vicinity of WBN. The Staff agrees that existing levels of measured entrainment under Unit 1 operation are too low to be readily detected in the aquatic populations in the WBN vicinity, and the additional water withdrawn via the CCW intake will not be noticeable or furthermore destabilizing to the aquatic ecology in the WBN vicinity. Moreover, the Staff concludes that the water withdrawn from the SCCW intake will actually decrease under dual unit operation. In drawing these conclusions, the Staff relies in part on the Revised Aquatics Study and the Peak Spawning Entrainment Study.

Undisputed. It should be noted that the NRC Staff has not conducted any independent studies to support its conclusions. (**0016-4-6** [Curran, Diane])

Comment: [From Southern Alliance for Clean Energy's Statement of Disputed Material Facts] Overview of the Draft SFES Conclusions Regarding TVA's Aquatic Studies

82. The Staff's conclusions regarding impingement impacts are similar. The Staff finds that measured levels of impingement under operation of WBN Unit 1 are low and impingement effects are too minor to be readily detected in aquatic populations in the WBN vicinity. The increased flow rates for the CCW intake under dual unit operation will not alter that conclusion, concludes the Staff, and the decreased flow rates for the SCCW intake will not increase impingement effects. The Staff relied in part on the Impingement Study in drawing these conclusions.

Undisputed. It should be noted that the NRC Staff has not conducted any independent studies to support its conclusions.

83. With respect to thermal impacts from operation of WBN Unit 2, the Staff concludes that this effect also will be undetectable and will not destabilize or noticeably alter the aquatic biota in the WBN vicinity. The Staff based this conclusion in part on the Hydrothermal Study, as well as limits set by the NPDES permit.

Undisputed. It should be noted that the NRC Staff has not conducted any independent studies to support its conclusions. (**0016-4-7** [Curran, Diane])

Comment: [From Southern Alliance for Clean Energy's Statement of Disputed Material Facts] Overview of the Draft SFES Conclusions Regarding TVA's Aquatic Studies

84. The Staff concludes in the Draft SFES that although the impoundments and industrial facilities have a significant cumulative impact on the aquatic biota in the WBN vicinity, "the overall impacts on aquatic biota, including Federally listed threatened and endangered species, from impingement and entrainment at the SCCW and IPS [i.e., CCW] intakes and from thermal . . . discharges as a result of operating Unit 2 on the WBN site are SMALL."

Undisputed. It should be noted that the NRC Staff has not conducted any independent studies to support its conclusions. (**0016-4-8** [Curran, Diane])

Response: The staff understands the word "studies" to mean field or sampling studies. The NRC staff did not conduct independent field studies including sampling of aquatic biota. As documented in the SFES, the NRC staff performed an independent review of applicant data and reports, data obtained from the State of Tennessee, from other Federal agencies, and other published documents and reports.

Comment: TVA [in "Motion for Summary Disposition of Contention 7" (Nov. 21, 2011)] claims that it has conducted studies that resolve the three major deficiencies identified in Contention 7. As discussed in Dr. Young's attached Declaration, this is not correct......In addition, there are still big gaps in the information that TVA has collected. For example, TVA collected entrainment data for the Condenser Cooling Water ("CCW") system only and did not include the Supplemental Condenser Cooling Water ("SCCW") system. In addition, TVA did not collect impingement data for all key locations. (**0015-4** [Curran, Diane])

Comment: SUMMARY OF MY [Dr. Shawn Paul Young] PROFESSIONAL OPINION REGARDING TVA'S ASSERTIONS.

3. In addition, there are still big gaps in the information that TVA has collected. For example, TVA collected entrainment data for the Condenser Cooling Water ("CCW") system only and did not include the Supplemental Condenser Cooling Water ("SCCW") system. See par. III-A.4 below. In addition, TVA did not collect impingement data for all key locations. (**0017-1-2** [Curran, Diane])

Response: The staff reviewed entrainment and impingement data for both intakes (SCCW and IPS) and did not observe the gaps in data that are discussed in the comments. In Sections 2.3.2.1 and 4.3.2.2, the NRC staff discusses the entrainment study performed in 1975 for the Watts Bar Fossil Plant intake in Watts Bar Reservoir, the ichthyoplankton composition and distribution study for the SCCW from the spring of 2000, and the entrainment study for the SCCW between March 7, 2010, through March 25, 2011 (containing new information received after publication of the draft SFES). The NRC staff also discusses impingement data and findings for the SCCW from 1974-1975 (Watts Bar Fossil Plant); additional impingement data from 1999-2000; and finally the 2005-2007 SCCW impingement demonstration, which was part of the 316(b) monitoring program.

The NRC staff also discusses the results of two entrainment studies for the second intake, the IPS. The first study occurred April 8 to June 17, 1996, and March 31 to June 23, 1997. The second entrainment study occurred March 7, 2010, through March 25, 2011. In addition, the staff discusses the results of two impingement studies at the IPS. The first impingement study occurred between March 15, 1996, and February 28, 1997; and March 4, 1997 through September 30, 1997. The second impingement study occurred between March 26, 2010 and March 17, 2011.

Comment: [Contention 7: Inadequate Consideration of Aquatic Impacts]
TVA's discussion of aquatic impacts is deficient in three key respects..... Second, TVA [in 2007 Final Supplemental Environmental Impact Statement] relies on outdated and inadequate data to predict thermal impacts and the impacts of entrainment and impingement of aquatic organisms in the plant's cooling system. (**0008-7** [Curran, Diane])

Response: *This comment refers to the TVA environmental impact statement ("EIS") (TVA 2008), and does not mention the analysis provided by the NRC staff in the draft SFES. Section 4.3.2.2 of the NRC staff's SFES discusses the entrainment, impingement, and thermal studies that have occurred. TVA studies used in the NRC staff's analysis are listed in Table 5-1 by the year in which they were conducted. TVA conducted ichthyoplankton and entrainment studies as recently as 2011. TVA conducts thermal studies biannually. Section 4.3.2.3 contains the discussion of the 2005 and 2006 thermal studies.*

Comment: [From Southern Alliance for Clean Energy's Statement of Disputed Material Facts]
Description of TVA's Aquatics Studies
Analysis of Fish Species Occurrences in Chickamauga Reservoir – A Comparison of Historic and Recent Data (Oct. 2010) ("Fish Species Occurrences Study")

39. SACE claimed in Contention 7 that TVA relies on inadequate and outdated data to form its conclusion that fish populations in the WBN vicinity are in good health, and has not taken steps necessary to evaluate how effluent from WBN may affect fish communities. In direct response, TVA conducted this study to analyze extensive historic and recent fish survey data from the WBN vicinity, and compare the current prevalence of fish species to historic (i.e., pre-operational) values.

   Undisputed except with respect to TVA's characterization of the data as "extensive." As discussed in Dr. Young's Declaration at pars III-E.2-3, there are significant inadequacies in the analyses found in this report.

40. This study uses the extensive fish survey data available for the WBN vicinity, dating back to 1947. Because it also provides recent survey data for the fish populations in the WBN vicinity, this study inherently reflects the impact of the current operation of WBN Unit 1 on those populations.

Undisputed to the extent that TVA states it used fish survey data back to 1947 and provides recent survey data. Disputed with respect to TVA's characterization of the data as "extensive" and TVA's conclusion that this study alone inherently reflects the impact of the current operation of WBN1 on fish populations. See Dr. Young's Declaration throughout. (**0016-3-7** [Curran, Diane])

Response: This comment states that TVA's fish population data is not extensive. Sections 2.3.2, 4.3.2, and 5.5.2 discuss the studies that the NRC staff used to reach its conclusions. The NRC staff reviewed the data provided by TVA and found that quantity of data available was sufficient to draw a conclusion for the purposes of NEPA. The NRC staff did not solely rely on the referenced TVA study (Simmons 2010) to describe the impact of the current operation of WBN Unit 1 on fish populations. As discussed in the SFES, "Analysis of Fish Species Occurrences in Chickamauga Reservoir – A Comparison of Historical and Recent Data," is one of many reports that the NRC staff relied upon to reach its conclusions.

Comment: [From Southern Alliance for Clean Energy's Statement of Disputed Material Facts]

Southern Alliance for Clean Energy ("SACE") respectfully submits the following statement of disputed material facts in response to Tennessee Valley Authority's ("TVA's") Statement of Material Facts on Which No Genuine Issue Exists (Nov. 21, 2011). SACE responds as follows: I. Procedural Background. C. New Information on the Record – TVA's Aquatic Studies and NRC's Draft SFES

10. In direct response to the issues raised by SACE in Contention 7, TVA collected extensive new data on the current health of the aquatic environment and the impact of operation of WBN Unit 1 on that environment, prepared numerous updated and expanded aquatics-related analyses, documented the analyses in published reports and studies, and disclosed these reports and studies to the NRC Staff and SACE.

Undisputed except with respect to TVA's characterization of the data as "extensive." As discussed throughout Dr. Young's Declaration, there are significant gaps and inadequacies in the data.

A complete list of those studies, including the dates that TVA disclosed each to SACE and the NRC Staff, follows:
a. Comparison of Fish Species Occurrence and Trends in Reservoir Fish Assemblage Index Results in Chickamauga Reservoir Before and After [WBN] Unit 1 Operation (June 2010) ("RFAI Study"), which TVA disclosed to SACE and the NRC Staff on July 15, 2010;
b. Analysis of Fish Species Occurrences in Chickamauga Reservoir – A Comparison of Historic and Recent Data (Oct. 2010) ("Fish Species Order (Granting TVA's Unopposed

Motion to Dismiss SACE Contention 1) (June 2, 2010) (unpublished). Occurrences Study"), which TVA disclosed to SACE and the NRC Staff on November 15, 2010;

c. Mollusk Survey of the Tennessee River Near [WBN] (Rhea County, Tennessee) (Nov. 2010) ("Mollusk Survey"), which TVA disclosed to SACE and the NRC Staff on January 18, 2011;

d. Discussion of the Results of the 2010 Mollusk Survey of the Tennessee River Near [WBN] (Rhea County, Tennessee) (Mar. 2011) ("Discussion of Mollusk Survey"), which TVA disclosed to SACE and the NRC Staff on March 15, 2011;

e. Aquatic Environmental Conditions in the Vicinity of [WBN] During Two Years of Operation, 1996-1997 (June 1998, Revised June 2010) ("Revised Aquatics Study"), which TVA disclosed to SACE and the NRC Staff on July 15, 2010;

f. Comparison of 2010 Peak Spawning Seasonal Densities of Ichthyoplankton at [WBN] at Tennessee River Mile 528 with Historical Densities during 1996 and 1997 (Apr. 2011, Revised Nov. 2011) ("Peak Spawning Entrainment Study"), which TVA disclosed to SACE and the NRC Staff on April 15, 2011;

g. Fish Impingement at [WBN] Intake Pumping Station Cooling Water Intake Structure during March 2010 through March 2011 (Mar. 2011, Revised Apr. 2011) ("Impingement Study"), which TVA disclosed to SACE and the NRC Staff on May 16, 2011; and

h. Hydrothermal Effects on the Ichthyoplankton from the [WBN] Supplemental Condenser Cooling Water Outfall in Upper Chickamauga Reservoir (Jan. 2011) ("Hydrothermal Study"), which TVA disclosed to SACE and the NRC Staff on February 15, 2011. Undisputed. (**0016-1-3** [Curran, Diane])

Response: The NRC staff understands this comment to dispute a statement made by TVA in Summary Disposition that the data is "extensive." In addition, the commenter refers generically to gaps and inadequacies in the data.

In Sections 2.3.2, 4.3.2, and 5.5.2, the NRC staff discusses the studies used to reach its conclusions. The NRC staff reviewed the data provided by TVA, including the listed reports. The NRC staff also considered other reports developed by TVA and reports and data developed by the State of Tennessee and Federal agencies such as the U.S. Fish and Wildlife Service or the U.S. Geological Survey. The NRC staff also considered peer-reviewed publications that pertain to the area in the vicinity of the WBN site. The NRC staff found that the quantity of data available was sufficient to draw a conclusion for the purposes of NEPA.

Comment: TVA's finding [in 2007 FSEIS] that WBN Unit 2 will have no significant impacts on aquatic life in the Tennessee River is inadequately supported in the following respects: .....

2. TVA [in 2007 FSEIS] relies on outdated and inadequate data to predict the effects of WBN Unit 2's cooling system on fish, mussels, and other aquatic organisms. In particular, the FSEIS understates the potential impacts of the coolant intake system (i.e., entrainment and impingement) and the thermal impacts of the coolant discharge system on fish and benthic

organisms, by relying on poor or outdated data, distorted interpretations of data, and assumptions and extrapolations in lieu of recent monitoring studies....Given their lack of mobility, fish eggs and most fish larvae cannot escape the intake flow velocity and are sucked into the intake canal and cooling system. Phytoplankton and zooplankton, which constitute important food sources for fish, mussels, and aquatic insects, may also be entrained due to their lack of mobility. Fish and other organisms pass through the plant's cooling system, suffering injury or death though physical contact, rapid pressure or temperature change, and chemical poisoning from biocides and other chemicals introduced into the water. ....Knowledge of the ichthyoplankton population distribution in relation to intakes across time and space is very important to an understanding of entrainment impacts, because ichthyoplankton tend to be patchy (high numbers clumped into a specific portion of the water column). This patchy distribution creates a high level of vulnerability to entrainment mortality if the organisms are located near intakes, because they cannot simply avoid the intakes. But TVA has not collected sufficient data to understand the distribution of icthyoplankton populations or how they are affected by the Watts Bar intakes. That is because TVA has not taken direct measurements of entrainment, even though direct measurements are recommended by the U.S. Environmental Protection Agency. Instead, it has extrapolated entrainment estimates from outdated and inadequate data...... .

TVA's conclusion [in 2007 FSEIS] that entrainment impacts are insignificant is based upon an unsupported assumption: that population densities are uniform across the river channel and from the surface to the bottom of the river. The data do not support this assumption, however, because the numbers are all relative, expressed in percentages. It is therefore impossible to determine what the actual populations of organisms are..... TVA also does not provide any data for fish eggs, which may be found in high abundance during different times of the year and are very vulnerable to entrainment. (**0008-9** [Curran, Diane])

Response: *This comment refers to the TVA EIS (TVA 2008), and not the NRC staff analysis in the draft SFES. The commenter indicates that population densities are not uniform in the river and that direct measurements of entrainment or ichthyoplankton distribution are important in determining impacts from entrainment during plant operations.*

In Section 4.3.2.2 of the SFES, the NRC staff considers the results of entrainment studies conducted at the IPS after the start of operations of WBN Unit 1 (i.e., in 1996-1997 and 2010-2011). The NRC staff compares those results to the species composition found in an ichthyoplankton study in the IPS channel in 1984 and 1985. The NRC staff discusses entrainment studies at the SCCW intake including studies conducted in 1975 (when the facility was used as the intake for the Watts Bar Fossil Plant) and from March 2010 through March 2011 (in the Watts Bar Reservoir). In addition, the NRC staff considers an additional study that included the analysis of ichthyoplankton density (in 2000). The entrainment studies considered fish larvae and fish eggs.

The NRC staff did not base its conclusions on a uniform density of ichthyoplankton in either Watts Bar or Chickamauga reservoirs or in the intake channel. The NRC staff discussed the potential impacts of the coolant intake system (entrainment and impingement) and thermal impacts of the discharge system in Sections 4.3.2.2 and 4.3.2.3.

Comment: STATEMENT OF [Dr. Shawn Paul Young] PROFESSIONAL OPINION REGARDING ADEQUACY OF TVA'S RECENT BIOLOGICAL STUDIES TO ADDRESS THE ENVIRONMENTAL IMPACTS OF THE PROPOSED WATTS BAR 2 NUCLEAR POWER PLANT ON AQUATIC ORGANISMS

Revised Aquatics Study and Peak Entrainment Study [Aquatic Environmental Conditions in the Vicinity of Watts Bar Nuclear Plant During Two Years of Operation, 1996-1997 (June 1998, Revised June 2010) ("Revised Aquatics Study") and Comparison of 2010 Peak Spawning Seasonal Densities of Ichthyoplankton at [WBN] at Tennessee River Mile 528 with Historical Densities During 1996 and 1997 (Apr. 2011, Revised Nov. 2011) ("Peak Spawning Entrainment Study")]

13. As a general matter, TVA also mischaracterizes the relationship between river flow and entrainment. According to TVA, studies show that the hydraulic entrainment from dual unit operation will result in an additional entrained amount of 0.2% of the flow in the Chickamauga Reservoir.... TVA asserts that the resulting total hydraulic entrainment represents approximately 0.5% of the flow in the Chickamauga Reservoir; and that this increased hydraulic entrainment will result in a proportionate increase in entrainment of the ichthyoplankton present in the water column...

14. TVA's calculation is only partly correct, and only accurate at a very specific river flow past WBN Plant. The 0.2% hydraulic entrainment for WB1 is based upon TVA using "a long term average river flow past WBN of 27,000 cfs." See Footnotes 58-60 and Joint Affidavit par. 37. However, the flow past WBN may vary widely depending on seasonal precipitation levels and daily operations of Watts Bar Dam immediately upstream of WBN. Therefore, hydraulic entrainment will vary depending on amount of water in the Tennessee hydrosystem and how much flow is released from Watts Bar Dam. For instance, using CCW water withdrawal rate of 88 cfs (NRC DFES Table 3-1 at page 3-9) and river flow of 3,500 cfs, which is the minimum amount of flow from Watts Bar Dam that permits TVA to discharge thermal and chemical effluent through Outfall 101, the hydraulic entrainment increases to 2.5% (12.5 times higher). Then, with the addition of Unit 2 doubling hydraulic entrainment, the hydraulic entrainment at a flow of 3,500 cfs further increases to approximately 5.0% (25 times higher). Also, with higher hydraulic entrainment, the probability of entraining more ichthyoplankton increases. However, one cannot assume that ichthyoplankton entrainment will increase proportionately. In fact, ichthyoplankton may increase exponentially. The increase depends on the proximity of ichthyoplankton to water intakes. Only data collected by field studies in combination with proper methods for calculation may accurately characterize

ichthyoplankton entrainment under any level of hydraulic entrainment. I note that this is a similar issue in regards to impingement.
(**0017-2-6** [Curran, Diane])

Response: The NRC staff agrees that the 0.2 percent hydraulic entrainment estimate is based on a "long term average river flow" past the site of 778 m^3/s (27,000 cfs) and that the flow past the WBN site will vary depending on multiple factors (e.g., the precipitation and the operation of Watts Bar Dam). The hydraulic entrainment will vary depending on the water flow. Because the water flow varies within a day and over multiple years, the NRC staff used a long-term average to characterize the hydraulic entrainment rate. Likewise, the entrainment of ichthyoplankton also varies depending on the season and the flow of water in the reservoir.

The NRC staff agrees that the best method for characterizing entrainment and impingement is through field studies. The NRC staff reviewed field studies provided by TVA that looked at ichthyoplankton density and estimated entrainment from Watts Bar Reservoir. These studies include an entrainment study performed for the Watts Bar Fossil Plant in 1975, an ichthyoplankton composition and distribution study for the SCCW from the spring of 2000, and an entrainment study for the SCCW from 2010 and 2011. Field studies on entrainment at the IPS in Chickamauga Reservoir include studies after the start of operations of WBN Unit 1 (i.e., in 1996-1997 and 2010-2011). The NRC staff compares those results to the species composition found in a preoperational ichthyoplankton study in the IPS channel in 1984 and 1985 before the start of operations for WBN Unit 1. These studies are discussed in Section 4.3.2.2.

In addition, the commenter also noted similar concerns about data used to characterize impingement. TVA conducted field studies for impingement at both intakes (SCCW and IPS). Studies for the SCCW occurred in 1974-1975 (Watts Bar Fossil Plant), 1999-2000, and as recently as 2005-2007 (the impingement demonstration for the 316(b) monitoring program). The NRC staff also discusses the results of two impingement studies in the SFES for the IPS. The first study occurred between March 1996 and February 1997 and the second between March 2010 and March 2011.

Comment: [Southern Alliance for Clean Energy's Statement of Disputed Material Facts]
Description of the Proposed Project
General Information

22. Studies show that the hydraulic entrainment from dual unit operation will result in an additional entrained amount of 0.2% of the flow in the Chickamauga Reservoir. The resulting total hydraulic entrainment represents approximately 0.5% of the flow in the Chickamauga Reservoir. This increased hydraulic entrainment will result in a proportionate increase in entrainment of the ichthyoplankton present in the water column.

Disputed as to this calculation is only partly correct, and only accurate at a very specific river flow past WBN Plant. As discussed in Dr. Young's Declaration at par. III-A.13-14, the 0.2% hydraulic entrainment for WB1 is based upon TVA using a long term average river flow past WBN of 27,000 cfs. Using 3,500 cfs, which is the minimum amount of flow from Watts Bar Dam that permits TVA to discharge thermal and chemical effluent through Outfall 101, the hydraulic entrainment increases to 2.1% (10 times higher). Then, with the addition of Unit 2 almost doubling hydraulic entrainment, the hydraulic entrainment at a flow of 3,500 cfs further increases to approximately 4.0% (20 times higher). Also, only data collected by field studies in combination with proper methods for calculation may accurately characterize ichthyoplankton entrainment under any level of hydraulic entrainment. (**0016-2-1** [Curran, Diane])

Response: The commenter suggests that the hydraulic entrainment estimate is accurate at only specific river flows. The commenter also states that the assumption that hydraulic entrainment will result in a proportionate increase in entrainment is not always correct and can only be accurately characterized in combination with field data.

The NRC staff agrees that the estimate of hydraulic entrainment as a percentage of the flow of the river will vary depending on the assumed flow rate for the river. The NRC staff used the mean annual flow from Watts Bar Dam as the river flow rate, rather than 3500 cfs as suggested by the commenter. As stated in Section 2.2.1.1 of the SFES, "TVA has recorded average daily flows of less than 280 m^3/s (10,000 cfs) only 4.8 percent of the time and less than 140 m^3/s (5,000 cfs) only 0.9 percent of the time at the site."

The NRC staff used the mean annual flow of 778 m^3/s (27,500 cfs) in Sections 4.2.2.1 and 4.3.2.1 to characterize the percentage of the flow removed at the intakes and the percentage lost to consumptive use over an entire year.

The NRC staff acknowledges that ichthyoplankton does not have a uniform distribution in the reservoir and that the distribution does not remain constant throughout the year. For this reason, in Section 4.3.2.2, the NRC staff reviewed and reported on TVA entrainment studies that measure ichthyoplankton density along a transect across the river and entrainment rates across an entire year. These studies serve to characterize the quantity of ichthyoplankton entrained under various levels of hydraulic entrainment and at various times of the year. During the entrainment studies at the IPS, the river flow ranged from 171 m^3/s to 1,320 m^3/s (6,037 to 46,650 cfs) (TVA 2012).

Comment: [From Southern Alliance for Clean Energy's Statement of Disputed Material Facts] Description of TVA's Aquatics Studies
Aquatic Environmental Conditions in the Vicinity of Watts Bar Nuclear Plant During Two Years of Operation, 1996-1997 (June 1998, Revised June 2010) ("Revised Aquatics Study")

57. TVA revised this study in direct response to concerns raised by SACE in Contention 7, and by Dr. Young in support of Contention 7, that TVA's methods for estimating entrainment were flawed. Dr. Young claimed that TVA erroneously assumed that distribution of ichthyoplankton across the reservoir is uniform, and did not take into account variations in seasonal abundance of ichthyoplankton. Dr. Young also alleged that TVA should estimate entrainment using actual intake water demand and river flow values. Undisputed as to stated information. Disputed as to the Aquatics Study was also revised after Dr. Young identified major clerical and mathematical errors that had gone unnoticed for over a decade.

59. After conducting the revised entrainment estimates, TVA found that its overall conclusions regarding entrainment were unchanged. Estimated entrainment rates remained very low. For samples collected in 1996, percent entrainment in the revised analysis was estimated to be 0.29% for fish eggs and 0.57% for fish larvae. For samples collected in 1997, percent entrainment in the revised analysis was estimated to be 0.02% for fish eggs and 0.22% for fish larvae.

Undisputed that TVA has describe the results of the study. Disputed is the accuracy and validity of these results. See Young Declaration, ¶ III-A.2.

60. TVA's experts concluded that these rates are "low" and therefore there is no impact to the ichthyoplankton populations of Chickamauga Reservoir as a result of operation of WBN Unit 1.

Undisputed as to the description of the conclusion by TVA's experts. Disputed as to the reasonableness of the conclusion. The data were not sufficient to support the conclusion as this study was only for a 3-month period during only 2 years, one of which Unit 1 was not even operational or only at partial-capacity for a majority of time. The Revised Aquatics Study has the same shortcomings and still arrives at the same conclusions that are disputed in Contention 7. See Young Declaration, ¶ III-A.2 and III-A.12. (**0016-4-1** [Curran, Diane])

Response: These comments refer to errors in the entrainment analysis for the 1996-1997 Aquatic Environmental Conditions report and express a concern that the period of time when WBN Unit 1 was not operating at full power was not considered.

The NRC staff agrees with that there were errors in the original Aquatic Environmental Conditions study. The NRC staff also noticed discrepancies and errors, specifically in the entrainment calculations for the IPS, in the TVA report, Aquatic Environmental Conditions in the Vicinity of Watts Bar Nuclear Plant During Two Years of Operation (TVA 1998). The NRC staff formally requested clarification, which TVA provided in a letter dated July 2, 2010 as a revision to the report (TVA 2010). The NRC staff has revised Section 4.3.2.1 to include the recalculated entrainment percentages (quotient of total estimated number entrained and total estimated number transported, rather than an average of the weekly percent entrained values) from TVA (2012).

The commenter is also correct that WBN Unit 1 did not operate at full power the entire period of the operational sampling from April 8 through June 17, 1996. Although TVA conducted operational testing in March of 1996, the plant did not reach 100 percent power until May 9, 1996, and did not begin commercial operation until May 27, 1996. WBN Unit 1 operated at 84 percent capacity for over 15 months, which includes the April/June sampling periods for 1996 and 1997 operational entrainment studies. The NRC staff was aware of this information during the development of the SFES.

Comment: [From Southern Alliance for Clean Energy's Statement of Disputed Material Facts]
Description of TVA's Aquatics Studies
Aquatic Environmental Conditions in the Vicinity of Watts Bar Nuclear Plant During Two Years of Operation, 1996-1997 (June 1998, Revised June 2010) ("Revised Aquatics Study")

56. The original study concluded that ichthyoplankton were present in relatively low densities in the vicinity of the WBN intake, and that those that were present had passed through the turbines of the Watts Bar Dam. The study also found that most spawning that occurs in Chickamauga Reservoir occurs downstream of the WBN intake. In other words, relatively few ichthyoplankton were available to be entrained at the WBN intake. The original study concluded that the percent of ichthyoplankton entrained was very low, and that WBN entrainment has no impact on the fish populations in the WBN vicinity.

   Undisputed with respect to TVA's description of the study. Disputed in Contention 7. Disputed as to accuracy of results and conclusions. See Young Declaration, ¶ III-A.2 and III-A.12.

Response: *The commenter doubts the accuracy of the results and disagrees with the conclusions of the ichthyoplankton study performed in 1996 and 1997 (Baxter et al. 2010). The NRC staff discusses this study in Section 4.3.2.2 of the SFES. The NRC staff also discusses the results of a second entrainment study conducted by the applicant between March 2010 and March 2011 (TVA 2012). As reported in Section 4.3.2.2, entrainment rates for fish eggs during comparable periods of time were 0.02 percent for both 1996 and 1997 and 0.12 percent in 2010. The entrainment rates for larvae in 1996, 1997, and 2010 were 0.88, 0.22, and 0.4 percent, respectively. Although the rates are higher for fish eggs in 2010 than in previous years, the NRC staff considers the overall entrainment rate for fish eggs of 0.12 percent as low.*

Comment: [From Southern Alliance for Clean Energy's Statement of Disputed Material Facts]
Description of TVA's Aquatics Studies
Aquatic Environmental Conditions in the Vicinity of Watts Bar Nuclear Plant During Two Years of Operation, 1996-1997 (June 1998, Revised June 2010) ("Revised Aquatics Study")

58. In response to Dr. Young's concerns, TVA revised the entrainment analysis to account for seasonality of ichthyoplankton occurrence and reservoir releases from Watts Bar Dam. TVA also used actual intake water demand and reservoir flow values.

Undisputed that TVA revised its entrainment analysis to account for seasonality of ichthyoplankton occurrence and reservoir releases and that TVA used actual intake water demand and reservoir flow values. Disputed as to whether TVA did, in fact, account for seasonality of ichthyoplankton occurrence prior to the Peak Entrainment Study in 2010. See Young Declaration, ¶ III-A.2.

Response: The commenter is concerned that the entrainment studies conducted prior to 2010 do not take into account that ichthyoplankton occurs seasonally in reservoirs. In Section 2.3.2.1, the NRC staff discusses the commercially, recreationally, and biologically important fish species and provides the spawning season and time between spawning and hatching for most of the species. In response to this comment, the NRC staff updated the discussion to include spawning seasons for all commercially, recreationally, and biologically important fish species. In all cases, most of the spawning occurs between early spring and into the summer. Most species spawn in April through June; thus, the NRC staff found the April through June operational sampling period for the 1996 and 1997 entrainment study (Baxter et al. 2010) to be appropriate.

Comment: [From Southern Alliance for Clean Energy's Statement of Disputed Material Facts] Description of TVA's Aquatics Studies
Comparison of 2010 Peak Spawning Seasonal Densities of Ichthyoplankton at [WBN] at Tennessee River Mile 528 with Historical Densities During 1996 and 1997 (Apr. 2011, Revised Nov. 2011) ("Peak Spawning Entrainment Study")

62. TVA conducted this study to respond to SACE and Dr. Young's concerns that TVA's methods for estimating entrainment were flawed, and that TVA should have taken direct measurements of entrainment. Undisputed. TVA collected raw data on actual entrainment at WBN during Unit 1 operation from March 2010 through March 2011, to ensure that all of SACE and Dr. Young's concerns regarding entrainment estimates were addressed, and in direct response to requests from SACE and Dr. Young for recent actual entrainment monitoring at WBN during operation of WBN Unit 1.

Undisputed with respect to the assertion that TVA collected raw data on actual entrainment at WBN1 in 2010-11. Disputed as to whether the data collected were sufficient to resolve Dr. Young's concerns. See Young Declaration, ¶ III-A.4.

63. This study reports entrainment resulting from operation of WBN Unit 1, as measured during the peak spawning period of April through June, 2010. TVA used this timeframe to address SACE and Dr. Young's concern that TVA account for the spawning patterns of fish species

in the Chickamauga Reservoir and the high abundance of ichthyoplankton during certain times of year.

Disputed with respect to the assertion that the study reports entrainment from operation of WBN1 as measured through the peak spawning period in 2010. This study only reports entrainment at the CCW, and does not report entrainment by the SCCW. Thus, the cumulative entrainment due to operation of WBN Unit1 is not known. Disputed with respect to whether the data collected were sufficient to resolve Dr. Young's concerns. See Young Declaration, ¶ III-A.5.

64. This study concluded that measured entrainment rates at the WBN in 2010 were below one half of one percent of the ichthyoplankton population in the WBN vicinity, and consistent with those calculated for the same period during the first two years of operation of Unit 1, 1996 to 1997, when consistent calculation methods were applied. Specifically, the study found that the percent of entrained eggs in 2010 (0.12%) was within the range for 1996 (0.2%) and 1997 (0.2%). Likewise, the study found that the percent of entrained larvae in 2010 (0.40%) was within the range for 1996 (0.88%) and 1997 (0.22%).

Undisputed that TVA correctly describes the study's results. Disputed with respect to the accuracy of the results. See Young Declaration, ¶¶ III-A.2, III-A.5, and III-A.10-11.

65. TVA's experts concluded that these entrainment rates are "very low," and are not adversely affecting the fish population in the WBN vicinity. Undisputed that this is the conclusion by TVA's experts.

Disputed as to the accuracy and reasonableness of the conclusion. See Young Declaration, ¶ III-A.1-12.

66. The increased water intake demand for the CCW caused by dual unit operation will result in an estimated increase in hydraulic entrainment of approximately 0.2%. This study found that ichthyoplankton entrainment will increase proportionately with hydraulic entrainment. This increase will result in entrainment percentages that are still less than 1% of the ichthyoplankton population. This study concluded that, as a result, dual unit operation will not result in a material change in entrainment impacts.

Disputed as to the accuracy and reasonableness of this conclusion, and the rationale/methodology to arrive at this conclusion. See Young Declaration, ¶ III-A.13-14. (**0016-4-2** [Curran, Diane])

Response: These comments question the rational, methodology, sufficiency, accuracy, and conclusions of the Peak Spawning Entrainment Study.

After publication of the draft SFES, the NRC staff received a copy of a yearlong entrainment study for the SCCW and the IPS from TVA (TVA 2012). As a result of the NRC staff's review of this study, the NRC staff replaced the discussion in Section 4.3.2.2 with the results of the March 2010 through March 2011 study (TVA 2012). Further, in Section 4.14.6, the NRC staff added text to discuss the cumulative effects of operating the intakes for both units simultaneously.

The NRC staff acknowledges that TVA incorrectly calculated the density of fish eggs and larvae entrained during 1996 as reported by TVA (1998) and Baxter et al. (2010). TVA (TVA 2012) provided corrected information related to the density of fish eggs and larvae, as discussed by the NRC staff in Section 4.3.2.2 of this SFES.

Comment: STATEMENT OF [Dr. Shawn Paul Young] PROFESSIONAL OPINION REGARDING ADEQUACY OF TVA'S RECENT BIOLOGICAL STUDIES TO ADDRESS THE ENVIRONMENTAL IMPACTS OF THE PROPOSED WATTS BAR 2 NUCLEAR POWER PLANT ON AQUATIC ORGANISMS
Revised Aquatics Study and Peak Entrainment Study [Aquatic Environmental Conditions in the Vicinity of Watts Bar Nuclear Plant During Two Years of Operation, 1996-1997 (June 1998, Revised June 2010) ("Revised Aquatics Study") and Comparison of 2010 Peak Spawning Seasonal Densities of Ichthyoplankton at [WBN] at Tennessee River Mile 528 with Historical Densities During 1996 and 1997 (Apr. 2011, Revised Nov. 2011) ("Peak Spawning Entrainment Study")]

1. TVA asserts that it has revised its method for estimating entrainment impacts and has also collected raw data on actual entrainment associated with WBN1 for one year. TVA Motion at 14-15. TVA asserts that these studies show the rate of entrainment is very low. Id. In my professional opinion, however, TVA's studies do not provide a reasonable degree of support for the conclusion that the rate of entrainment is low. In fact, they indicate a rate of entrainment that is unacceptable.

2. The Revised Aquatics Study is a revision of the "Aquatics Study" for which TVA collected ichthyoplankton data in order to estimate entrainment at WBN Unit 1 only during April – June 1996 and 1997, not the entire year, a major shortcoming. The timing of the original Aquatic Study corresponded to the commencement of operation of WBN Unit 1. The study results were published in 1998. TVA concluded that WBN Unit 1 ichthyoplankton entrainment was low and had insignificant impacts on the fish community. In 2009, I identified major errors in this document that had major implications. TVA revised this study, and released a revision in 2010 that did not include an additional level of detail for data presentation and analysis to assess whether the errors were properly rectified. Further, TVA's conclusions remained unchanged. Based upon the original erroneous document, in

1998, TVA convinced the Tennessee Department of Environment and Conservation ("TDEC") to allow termination of the entrainment monitoring program mandated in the original NPDES permit. Therefore, since 1997, TVA had not collected any post-operational entrainment study at Unit 1.

3. After SACE's contention 7 was admitted, TVA conducted one year of entrainment monitoring during 2010 to compare the results against 1996 and 1997 entrainment data. The Peak Entrainment Study was a survey of the ichthyoplankton drift past the Supplemental Condenser Cooling Water ("SCCW") discharge (Outfall 113) and the Unit 1 water intake pumping structure for the CCW system. The Peak Entrainment study was conducted in conjunction with the "Hydrothermal Study" in order to also determine ichthyoplankton abundance at the SCCW intake, and in the SCCW discharge under two different thermal mixing zone scenarios.

4. In the Peak Entrainment Study, TVA collected ichthyoplankton along a transect from riverbank to riverbank below the SCCW discharge plume and above the intake pumping structure (IPS) for the CCW. As such, the study provides only a minimal account of the conditions in the Tennessee River. In order to make a reasonable analysis of the impacts of WBN1 on the river and the likely impacts of WBN2, TVA should have been collecting entrainment data regularly since WBN1 went online in 1996. For any reasonable biologist, two measurements taken thirteen years apart would not provide a sufficient basis for an analysis of entrainment impacts. TVA should have collected data for at least three years after WBN1 began operating in order to determine any annual variability of ichthyoplankton abundance. And TVA should have updated those measurements after it decided to pursue an operating license for WBN2, with at least two more years of measurements. (**0017-2-1** [Curran, Diane])

Response: The commenter provides a discussion of entrainment studies for the two intakes at the WBN site. The commenter suggests that additional entrainment data beyond the 2010 study (preferably 3 years of post operational studies on WBN Unit 1) is needed to determine annual variability of ichthyoplankton abundance for WBN Unit 2.

In evaluating entrainment for the IPS, the NRC staff reviewed two entrainment studies. The first occurred in 1996-1997 (approximately 16 years old) after the start of operations of Unit 1. The second occurred in April through June 2010. The NRC staff discusses both reports in Section 4.3.2.2 of the SFES. In June 2012, the NRC staff received an additional report from TVA that provided the results of a yearlong entrainment study for the SCCW and the IPS (TVA 2012). The NRC staff deleted the text in Section 4.3.2.2 that discussed the report referred to by SACE's comment (TVA 2011b) and inserted a discussion of its review of the March 2010 through March 2011 study (TVA 2012). The NRC staff believes that sufficient information is available to perform an assessment of the impacts of WBN Unit 2 operation and to arrive at an impact level conclusion.

Comment: STATEMENT OF [Dr. Shawn Paul Young] PROFESSIONAL OPINION REGARDING ADEQUACY OF TVA'S RECENT BIOLOGICAL STUDIES TO ADDRESS THE ENVIRONMENTAL IMPACTS OF THE PROPOSED WATTS BAR 2 NUCLEAR POWER PLANT ON AQUATIC ORGANISMS
Revised Aquatics Study and Peak Entrainment Study [Aquatic Environmental Conditions in the Vicinity of Watts Bar Nuclear Plant During Two Years of Operation, 1996-1997 (June 1998, Revised June 2010) ("Revised Aquatics Study") and Comparison of 2010 Peak Spawning Seasonal Densities of Ichthyoplankton at [WBN] at Tennessee River Mile 528 with Historical Densities During 1996 and 1997 (Apr. 2011, Revised Nov. 2011) ("Peak Spawning Entrainment Study")]

5. TVA's data collection for the Peak Entrainment Study was incomplete because TVA reported entrainment measurements only for the CCW intake. Even though TVA collected ichthyoplankton samples at the SCCW intake and in Watts Bar forebay, TVA did not present the data or calculate entrainment rates for the SCCW within the Peak Entrainment Study. Instead, TVA only presented data on ichthyoplankton abundance near the SCCW intake within the Hydrothermal Study, and again did not present any entrainment rates. Thus, TVA failed to adequately estimate total entrainment at the WBN1 water intake structures. The omission is significant because Tables 2 and 3 of the "Hydrothermal Study" list the results of ichthyoplankton abundance at and near the SCCW intake in Watts Bar Reservoir forebay. The results listed in the hydrothermal study show that 300% more fish larvae were captured at the SCCW intake on May 11-12, 2010 (Table 3) than were captured in the forebay nearby (Table 2). This indicates that a very high level of entrainment may be occurring at the SCCW intake. TVA, however, failed to recognize this significant material fact. (**0017-2-2** [Curran, Diane])

Response: The commenter expresses a concern that TVA has not conducted adequate entrainment studies on the SCCW intake on Watts Bar Reservoir and that the hydrothermal study resulted in high levels of entrainment on a single day that might be typical. In addition, the commenter states that TVA did not estimate total entrainment at both WBN1 intake structures.

The NRC staff discussed two entrainment studies at the SCCW intake in Section 4.3.2.2. Since the time that NRC received this comment, TVA released a second year-long entrainment study that included measurements at the SCCW (TVA 2012). The NRC staff deleted the text in Section 4.3.2.2 that discussed the report referred to by SACE's comment (TVA 2011c) and inserted a discussion of the results of the March 2010 through March 2011 study (TVA 2012).

The NRC staff acknowledges the high level of fish larvae captured near the SCCW intake on May 11-12, 2010, and notes that 1,181 of the 1,185 larvae captured in the night samples were clupeids (threadfin and gizzard shad are the numerically dominant clupeids in Watts Bar Reservoir). This shows that the distribution of ichthyoplankton is not uniform in the reservoir. It also shows that threadfin and gizzard shad are known to spawn in mid-May to mid-June and as a group as discussed in Section 2.3.2. The yearlong entrainment study gave numbers of 1.98 percent for larvae and 2.23 percent for fish eggs as reported in Section 4.3.2.2. The NRC staff determined that sufficient entrainment data is available to perform an assessment of the impacts of operation of WBN Unit 2 and to arrive at an impact level conclusion.

The NRC staff added text in Section 4.14.6 to discuss the cumulative effects of operating the intakes for both units simultaneously.

Comment: STATEMENT OF [Dr. Shawn Paul Young] PROFESSIONAL OPINION REGARDING ADEQUACY OF TVA'S RECENT BIOLOGICAL STUDIES TO ADDRESS THE ENVIRONMENTAL IMPACTS OF THE PROPOSED WATTS BAR 2 NUCLEAR POWER PLANT ON AQUATIC ORGANISMS
Hydrothermal Study [Hydrothermal Effects of the Ichthyoplankton from the Watts Bar Plant Supplemental Condenser Cooling Water Outfall in Upper Chickamauga Reservoir (Jan. 2011) ("Hydrothermal Study")]

[TVA] should also evaluate changes in nighttime operations to reduce the rate of entrainment of aquatic organisms. (**0017-2-13** [Curran, Diane])

Response: *The NRC staff discussed entrainment at the SCCW in Section 4.3.2.2 of the SFES. The NRC staff did not discuss mitigation for the SCCW intake because the average monthly flow through the SCCW intake for the operation of WBN Units 1 and 2 will be slightly less than the flow through the SCCW while operating Unit 1 only (within the uncertainty in the estimate for flow while operating either one or two units).*

Comment: TVA's finding [in 2007 FSEIS] that WBN Unit 2 will have no significant impacts on aquatic life in the Tennessee River is inadequately supported in the following respects:
.....TVA's impingement data are likewise inadequate to support the FSEIS' finding of no significant impact. For instance, TVA failed to follow-up on a survey conducted at the SCCW intake that found an increased level of impingement in comparison to earlier surveys. ...... TVA also failed to update the thirty-five-year-old data on which it relied for its conclusions about impingement impacts at the WBN Unit 1 intake. Additionally, TVA inappropriately treats its impingement data for the Lake Chickamauga and Watts Bar Reservoir intakes as if they were the same. The vicinities of the two intakes, however, have very different habitat characteristics and are therefore likely to support very different populations of aquatic organisms. (**0008-10** [Curran, Diane])

Comment: [From Southern Alliance for Clean Energy's Statement of Disputed Material Facts] Description of TVA's Aquatics Studies
Fish Impingement at [WBN] Intake Pumping Station Cooling Water Intake Structure During March 2010 through March 2011 (Mar. 2011, Revised Apr. 2011) ("Impingement Study")

68. This study analyzes raw impingement data collected at the CCW intake during operation of WBN Unit 1 from March 2010 through March 2011. Undisputed. TVA used this data, in combination with the existing recent SCCW impingement data, to estimate the annual impingement mortality of fish in the vicinity of WBN as the result of operation of WBN Unit 1, and to predict the impact from operation of Unit 2.

Disputed as to the fact that TVA did not update the SCCW impingement in conjunction with the CCW impingement in this study. TVA conducted this study in response to allegations by SACE and Dr. Young that TVA's analysis of the effects of WBN operation on the aquatic community was deficient because TVA had not conducted recent studies of actual impingement at the CCW intake. Undisputed with respect to the assertion that TVA conducted the study. Disputed as to whether the study was sufficient to resolve Dr. Young's concerns. See Young Declaration, ¶ III-B.1-5. (**0016-4-3** [Curran, Diane])

Comment: STATEMENT OF [Dr. Shawn Paul Young] PROFESSIONAL OPINION REGARDING ADEQUACY OF TVA'S RECENT BIOLOGICAL STUDIES TO ADDRESS THE ENVIRONMENTAL IMPACTS OF THE PROPOSED WATTS BAR 2 NUCLEAR POWER PLANT ON AQUATIC ORGANISMS

Impingement Study [Fish Impingement at [WBN] Intake Pumping Station Cooling Water Intake Structure During March 2010 through March 2011 (Mar. 2011) ("Impingement Study")]

5. TVA also failed to take impingement measurements for all key locations. The Impingement Study sampled fish impingement at the IPS for the CCW only, and did not include the SCCW. A study was conducted in 2000 to evaluate impingement at the SCCW intake above Watts Bar Dam; however, this study did not monitor an entire year. This study still showed that impingements may also occur at the SCCW intake (p. 6, Watts Bar Nuclear Plant Supplemental Condenser Cooling System Fish Monitoring Program, January 2001); yet, TVA still did not conduct impingement monitoring at the SCCW during 2010 in conjunction with the CCW study to determine the cumulative impingement by current operations of WBN Unit 1. (**0017-2-8** [Curran, Diane])

Response: The commenter is concerned that there is a lack of recent data on impingement at the SCCW and that TVA considered impingement on the two reservoirs as if they were from the same intake.

The commenter is correct that TVA did not conduct impingement monitoring at the SCCW during 2010 in conjunction with the CCW study. However, the NRC staff discusses data and findings in Section 4.3.2.2 from three different impingement studies at the SCCW intake. The first occurred between August 8, 1974, and May 29, 1975; the second between August 31, 1999, and September 29, 1999, and again between March 7, 2000, and April 26, 2000; and the third and most recent from August 16, 2005, through August 9, 2006, and again August 16,

2006, through August 7, 2007. The 2005-2007 study was part of the 316(b) monitoring program for the SCCW system and was a full two-year impingement study completed in the same year that TVA submitted its initial request for an operating license.

The commenter is correct that the two intakes are located in areas with "very different habitat characteristics and are therefore likely to support very different populations of aquatic organisms." The NRC staff discusses the differences in the aquatic habitats of the Watts Bar Reservoir forebay and the Chickamauga Reservoir inflow in Section 2.3.2. In Section 4.3.2.2, the NRC staff discusses the results of the impingement studies at the SCCW and the IPS for the CCW. The NRC staff differentiates its analyses of impingement occurring at the SCCW intake from its analysis of impingement occurring at the IPS. The NRC staff believes that sufficient impingement data are available to perform an assessment of the impacts of operation of WBN Unit 2 and to arrive at an impact level conclusion for the purposes of NEPA. The NRC staff added text in Section 4.14.6 to discuss the cumulative effects of operating the intakes for both units simultaneously.

Comment: Southern Alliance for Clean Energy's Statement of Disputed Material Facts
Description of the Proposed Project
General Information

23. Studies show that CCW flow rates resulting from dual unit operation will average 134 cubic feet per second ("cfs") at summer pool levels and 113 cfs at winter pool levels, an increase from those rates observed under operation of Unit 1 alone: 73 cfs and 68 cfs, respectively. (The maximum intake velocities will not change under dual unit operation because of the additional IPS openings available to accommodate increased flow.) The increased flow rates in the CCW intake channel resulting from dual unit operation will result in a proportionate increase in the rates of fish impingement.

   Disputed. It is important to note that TVA identifies the makeup flow through the IPS as 174 fps, double the withdrawal from the Tennessee River that would occur with only WBN1 online, and an increase in warm blowdown discharge to the Tennessee River from 135 cfs to 170 cfs, a 26 percent increase. These are substantial increases, independent of the role of the SCCW. See Table 3-1 of the DFES, at page 3-9:

   The rates of fish impingement may exponentially increase. Similar to the issue of hydraulic versus ichthyoplankton entrainment, only field monitoring will accurately determine impingement rates. See Young Declaration, ¶¶ III-A.13-14. **(0016-2-2** [Curran, Diane])

Response: This comment concerns the withdrawal rate of water at the IPS from the river, the discharge of warm water back to the river, and a potential increase in fish impingement resulting in the need for monitoring of impingement rates.

In Section 3.2.2, the NRC staff discusses the cooling system including the makeup flow rates and discharge rates. Table 3-1 provides the makeup flow rate through the IPS for summer and winter conditions. The NRC staff revised these numbers to quantify a normal withdrawal rather than the maximum withdrawal for accident conditions. The NRC staff agrees that the IPS makeup flow will increase with the addition of WBN Unit 2, although the flow does not double. The NRC staff also agrees that the maximum volume of water discharged through the CCW as blowdown would increase from 3.82 m³/s (135 cfs) to 4.81 m³/s (170 cfs), as shown in Table 3-4.

In Section 4.3.2.2, the NRC staff discusses and compares the results of two impingement studies at the IPS. The first occurred from March 1996 through September 1997 and the second between March 2010 and March 2011. The NRC staff is aware that a number of factors can increase the impingement rates. However, the field studies the NRC staff reviewed show very low rates of impingement for fish other than shad. The high rates of impingement for shad appear to be the result of cold water temperatures rather than from the operation of the IPS, as discussed in Section 4.3.2.2. While fish impingement rates could increase with the addition of the second unit, the NRC staff found no evidence to support the hypothesis that fish impingement would increase exponentially.

Comment: STATEMENT OF [Dr. Shawn Paul Young] PROFESSIONAL OPINION REGARDING ADEQUACY OF TVA'S RECENT BIOLOGICAL STUDIES TO ADDRESS THE ENVIRONMENTAL IMPACTS OF THE PROPOSED WATTS BAR 2 NUCLEAR POWER PLANT ON AQUATIC ORGANISMS

Revised Aquatics Study and Peak Entrainment Study [Aquatic Environmental Conditions in the Vicinity of Watts Bar Nuclear Plant During Two Years of Operation, 1996-1997 (June 1998, Revised June 2010) ("Revised Aquatics Study") and Comparison of 2010 Peak Spawning Seasonal Densities of Ichthyoplankton at [WBN] at Tennessee River Mile 528 with Historical Densities During 1996 and 1997 (Apr. 2011, Revised Nov. 2011) ("Peak Spawning Entrainment Study")]

6. In any event, the results that TVA reported for the CCW intake show that WBN1 has had significant impacts on the aquatic environment and that operation of WBN2 is also likely to impose significant additional impacts. First, the Peak Entrainment Study shows that ichthyoplankton abundance in the vicinity of WBN has declined significantly since operation of WBN1 commenced. The abundance of ichthyoplankton was substantially lower in 2010 than in post-operational surveys during years 1996 and 1997 as calculated and listed by TVA in the Revised Aquatics Study. As stated in the Peak Entrainment Study at page 3 with respect to fish larvae:

Average densities (525, 924, 282), peak seasonal densities (1,387; 1,699; 828) and dates of peak densities (06/03, 05/15, 05/16) for larvae during April through June 1996, 1997, and 2010, respectively, are presented in Table 5. All of these values for samples collected during 2010 were slightly lower than the range of the two previous years (1996 and 1997) of monitoring. (emphasis added). TVA and the NRC Staff failed to properly acknowledge the significant decline as a very important material fact in their respective analyses and conclusions. (**0017-2-3** [Curran, Diane])

Comment: STATEMENT OF [Dr. Shawn Paul Young] PROFESSIONAL OPINION REGARDING ADEQUACY OF TVA'S RECENT BIOLOGICAL STUDIES TO ADDRESS THE ENVIRONMENTAL IMPACTS OF THE PROPOSED WATTS BAR 2 NUCLEAR POWER PLANT ON AQUATIC ORGANISMS

Revised Aquatics Study and Peak Entrainment Study [Aquatic Environmental Conditions in the Vicinity of Watts Bar Nuclear Plant During Two Years of Operation, 1996-1997 (June 1998, Revised June 2010) ("Revised Aquatics Study") and Comparison of 2010 Peak Spawning Seasonal Densities of Ichthyoplankton at [WBN] at Tennessee River Mile 528 with Historical Densities During 1996 and 1997 (Apr. 2011, Revised Nov. 2011) ("Peak Spawning Entrainment Study")]

7. The Peak Entrainment Study also reported a decline in the number of fish eggs between 1996 and 2010: average densities were reported as 262, 150, and 75 and peak seasonal densities were reported as 1,095, 1,004, and 811 for April through June 1996, 1997, and 2010, respectively. The significance of this decline is not discussed by either TVA in its Motion or the NRC Staff in the DES.

8. Based on the data reported in the Peak Entrainment Study, (Table 7, p. 19), larger than anticipated entrainment events occurred at WBN1. Daily entrainment rates of fish larvae were as high as 8.65% (June 21, 2010) during peak ichthyoplankton abundance. In my professional opinion, such a high rate of entrainment may have adverse impacts on the fish community. This measurement is very significant, given that hydraulic entrainment will double at the IPS for the CCW with the addition of WBN2, likely doubling ichthyoplankton entrainment. Larval fish entrainment events may double from 8.5% to 17%, a rate of entrainment that would certainly have a significant impact on the health of the fish population.

9. The Peak Entrainment Study also reported in Table 7 that daily entrainment rates of fish eggs were as high as 4.08% (May 16, 2010) during peak ichthyoplankton abundance. In my professional opinion, an egg entrainment rate of 4% is high enough to have a potentially adverse impact on the fish community. This measurement is very significant, given that hydraulic entrainment will double at the IPS for the CCW with the addition of WBN2, likely doubling fish egg entrainment events from 4.0% to 8.0%. At 8%, the impacts would indeed be significant. (**0017-2-4** [Curran, Diane])

Response: The commenter is concerned about an apparent decline in ichthyoplankton abundance between 1996 and 2010. The commenter is also concerned that daily entrainment rates reached as high as 8.65 percent and that the addition of WBN Unit 2 could double the entrainment rate.

The commenter is correct that the data between 1996 and 2010 show a decline in the density of fish eggs and larvae when averaged across the intake and the reservoir samples during entrainment studies in Chickamauga Reservoir. The density of ichthyoplankton, when averaged across the intake and reservoir transect samples, however, is within the range of the preoperational data obtained in years 1976 to 1985 as reported by Baxter et al. (2010). The NRC staff revised Section 2.3.2.1 to include a discussion of the lower densities in 2010 and the variation in densities between 1976 and 1985. The NRC staff also revised Section 4.3.2.2 with the new entrainment estimates from 2010 as presented in TVA (2012).

The additional incremental flow through the IPS would be less than it is for WBN Unit 1. Larval entrainment rates of up to 10.34 percent were observed on July 25, 2010. As discussed in Section 4.3.2.2, this date corresponded to a roughly 14-week period that saw higher densities of centrarchid larvae in the IPS channel than in the reservoir samples. Centrarchids are known to use intake channel shorelines as spawning and nursery habitat. The NRC staff concludes in the SFES that the overall estimate of entrainment for fish larvae of 0.22 to 0.88 percent from the two field studies after the start of WBN Unit 1 operations to be so low as to have no noticeable effect and to not destabilize the aquatic populations near the WBN site.

Comment: STATEMENT OF [Dr. Shawn Paul Young] PROFESSIONAL OPINION REGARDING ADEQUACY OF TVA'S RECENT BIOLOGICAL STUDIES TO ADDRESS THE ENVIRONMENTAL IMPACTS OF THE PROPOSED WATTS BAR 2 NUCLEAR POWER PLANT ON AQUATIC ORGANISMS

Revised Aquatics Study and Peak Entrainment Study [Aquatic Environmental Conditions in the Vicinity of Watts Bar Nuclear Plant During Two Years of Operation, 1996-1997 (June 1998, Revised June 2010) ("Revised Aquatics Study") and Comparison of 2010 Peak Spawning Seasonal Densities of Ichthyoplankton at [WBN] at Tennessee River Mile 528 with Historical Densities During 1996 and 1997 (Apr. 2011, Revised Nov. 2011) ("Peak Spawning Entrainment Study")]

10. I am also concerned about potential errors in the Peak Entrainment Study. At page i, TVA stated that another revision should be released sometime this month, December 2011. This indicates to me that there may be more errors in the study.

11. Further, I identified errors in methodology TVA used to complete calculations in the "Hydrothermal Study" which may have consequences for the Peak Entrainment Study. Both studies should have used the same formula to calculate the number of ichthyoplankton

within 1,000 m3 of source water from the number of organisms actually captured in the volume of water actually sampled to catch those organisms. Within the Hydrothermal Study, the number of ichthyoplankton density per 1,000 m3 of water was estimated to determine how many fish eggs and fish larvae were exposed to high water temperatures in the SCCW thermal plume during the day and during the night. To arrive at an estimate of the daily abundance per 1,000 m3 of water, the day and night estimates should have been averaged, not added together. See pars. III-C. 6-9, below in this declaration. Thus, results for daily ichthyoplankton abundance at the SCCW intake are incorrect; and since the two studies incorporate similar methods to estimate ichthyoplankton densities, similar errors in calculations may have been made in the Peak Entrainment Study also. However, the entrainment study lists results in a different manner that does not allow one to determine this.

12. In conclusion, the Revised Aquatic Study and the Peak Entrainment Study do not support TVA's conclusion that the environmental impacts from entrainment at the current IPS for the CCW intake with one reactor are insignificant, nor do they support a conclusion that the additional impacts of WBN2 would be insignificant. To the contrary, the data reported shows that the impacts from entrainment from the IPS for the CCW from one reactor unit alone may be large and warrants further investigation. Further, the Hydrothermal Study suggests that entrainment at the current SCCW intake may be also be significant with large impacts to the fish community. (**0017-2-5** [Curran, Diane])

Response: *The commenter expresses concern that there are errors in the Hydrothermal Study (TVA 2011c) specifically related to adding, rather than averaging the ichthyoplankton densities for day and night samples to obtain a daily average density. The commenter also expresses concern about potential errors in the Peak Entrainment Study (TVA 2011b). The commenter includes that the results of the two studies show that entrainment from WBN Unit 1 may be large at both intakes.*

The commenter is correct that an estimate of the ichthyoplankton density per 1,000 m³ of water for the day and night estimates should not be combined unless the resulting density is given for 2,000 m³ of water. TVA's follow-on report submitted in June 2012 (TVA 2012) corrected this error. That report also extended the results from the Peak Entrainment Study for both intakes. The NRC staff updated Section 4.3.2.2 of the SFES with the results of the March 2010 through March 2011 entrainment study submitted in June 2012. The NRC staff concluded that entrainment from operation of the IPS and SCCW for WBN Unit 2 would not have a noticeable effect and would not destabilize the population of aquatic biota near the WBN site.

Comment: STATEMENT OF [Dr. Shawn Paul Young] PROFESSIONAL OPINION REGARDING ADEQUACY OF TVA'S RECENT BIOLOGICAL STUDIES TO ADDRESS THE ENVIRONMENTAL IMPACTS OF THE PROPOSED WATTS BAR 2 NUCLEAR POWER PLANT ON AQUATIC ORGANISMS

Hydrothermal Study [Hydrothermal Effects of the Ichthyoplankton from the Watts Bar Plant Supplemental Condenser Cooling Water Outfall in Upper Chickamauga Reservoir (Jan. 2011) ("Hydrothermal Study")]

4. In addition, contrary to statements in the Motion for Summary Disposition and the DFES, the Hydrothermal Study did not list nor discuss ichthyoplankton exposure rates i.e., the amount of time fish eggs and larvae remain in the thermal plume. The omission of this information is significant because the early life stages of fish, especially eggs and larvae are vulnerable to abrupt temperature change such as those found at Outfall 113 and 101, and exposure to such water temperature changes caused by WBN heat waste discharge may cause high mortality rates. Abrupt temperature changes are detrimental to fish eggs and larvae. Also, abrupt temperature change affects species differently. This is an important omission because a rapid increase of 5-10°F can kill fish eggs and fish larvae, and from the data presented, most of the ichthyoplankton likely experienced this as they drifted through the SCCW mixing zone. Further, not only are ichthyoplankton exposed to the SCCW thermal plume, but these same fish eggs and larvae then drift through the CCW diffuser thermal plume below. A second abrupt temperature increase further elevates risk of mortality from the heat discharged from WBN.

5. The Hydrothermal Study is also deficient because TVA failed to report and discuss the fact that an alarming number of ichthyoplankton were likely entrained by the SCCW and subsequently killed by heat within the SCCW system before being discharged back into the river. This is an extremely important consideration in this matter. Further, the portion of ichthyoplankton in the Watts Bar Reservoir forebay not directly entrained and killed by the SCCW would likely pass through the dam and then still would be subjected and potentially killed by the waste heat in the SCCW and CCW (Outfalls 113 and 101) thermal plumes. The use of the SCCW creates a "double whammy" for fish eggs and larvae, likely causing an alarming level of mortality. TVA does not adequately describe this situation or adequately analyze presented data that shows significant mortality may be occurring via both pathways. (**0017-2-10** [Curran, Diane])

Comment: STATEMENT OF [Dr. Shawn Paul Young] PROFESSIONAL OPINION REGARDING ADEQUACY OF TVA'S RECENT BIOLOGICAL STUDIES TO ADDRESS THE ENVIRONMENTAL IMPACTS OF THE PROPOSED WATTS BAR 2 NUCLEAR POWER PLANT ON AQUATIC ORGANISMS. Hydrothermal Study [Hydrothermal Effects of the Ichthyoplankton from the Watts Bar Plant Supplemental Condenser Cooling Water Outfall in Upper Chickamauga Reservoir (Jan. 2011) ("Hydrothermal Study")]

1. In the Hydrothermal Study, TVA reports the results of monitoring the water temperatures in the thermal plume of the SCCW (Outfall 113) during May and August 2010. TVA recorded water temperatures during the two mixing zone scenarios that occur daily, the active mixing zone when Watts Bar Dam releases water and the passive mixing zone when Watts Bar

Dam does not release water. TVA also completed ichthyoplankton sampling at and near the SCCW above Watts Bar Dam, and downriver of Watts Bar Dam below the actual thermal plume during both day and night. TVA asserts that the Hydrothermal Study shows that thermal discharges from WBN1 and WBN2 will not have a significant impact on aquatic organisms....

2. TVA should have conducted the study over several years to characterize thermal plume water temperatures and ichthyoplankton abundance that may vary across years due to variable climatic conditions, and due to variable operations of Watts Bar Dam caused by variable hydrological conditions in the Tennessee River Basin.

3. The Hydrothermal Study also failed to address important parameters. For instance, it did not include any data or analysis for Outfall 101 (discharge at the CCW diffuser), which releases heated effluent when the dam discharge exceeds 3,500 cfs. Outfall 101 should have been included, especially in light of the fact that ichthyoplankton may drift through Outfall 113 mixing zone and then into the Outfall 101 mixing zone. This omission is significant. (**0017-2-9** [Curran, Diane])

Response: The commenter is concerned that the draft SFES did not discuss the amount of time that ichthyoplankton spend in the thermal plume. The commenter suggests that ichthyoplankton are either entrained by the SCCW and killed by the heat within the SCCW system, or that they may pass through the dam and then be killed by the waste heat in the SCCW and the CCW thermal plumes. In addition, the commenter recommends that TVA should continue the hydrothermal study over multiple years and should include Outfall 101.

In Section 4.3.2.3, the NRC staff considers the effect of Outfalls 101 and 113. Both outfalls and their active mixing zones (as well as the passive mixing zone for Outfall 113) are illustrated in Figure 4-1. In Section 4.3.2.2, the NRC staff clarifies the assumption that 100 percent mortality of ichthyoplankton entrained by the SCCW system. The cumulative analysis in Section 4.14.6 considers both intakes and addresses the potential for a loss of eggs and larvae due to transport through the dam.

The commenter is correct that the NRC staff's analysis did not include a discussion of the amount of time that the ichthyoplankton are in the thermal plume. As discussed in Section 4.2, TVA has obtained an NPDES permit for operation of WBN Units 1 and 2 (TDEC 2011). The NPDES permit states that the instream temperature rise, as measured at the edge of the mixing zone, cannot exceed 3°C (5.4°F) relative to an upstream control point. It also defines temperature limits as shown in Table 4-1 of the SFES. These limits are identical to those set in the previous NPDES permit, which was solely for operation of WBN Unit 1. TDEC dictates the design and frequency of studies relating to the effects of thermal discharge. The NPDES permit also specifies that TVA perform two instream temperature surveys each year. The NRC staff inserted additional information related to these studies in Section 4.3.2.3. The NRC relies on the TDEC's NPDES program to monitor thermal discharges and to protect the aquatic biota from the effects of the thermal discharge.

Comment: [From Southern Alliance for Clean Energy's Statement of Disputed Material Facts]
II. Description of the Proposed Project. C. WBN Cooling System Output

25. The thermal discharge from WBN operation is bound by thermal limits established by TVA's NPDES permit. The NPDES system establishes legally enforceable, aquatic health-based limits on hydrothermal discharges, in accordance with state and federal statutes. The Tennessee Water Pollution Control Division ("TDEC") issued a new NPDES permit for the operation of WBN Units 1 and 2 on June 30, 2011, most recently revised on August 31, 2011.

Undisputed, except for the facts that the existence of a legal limit does not ensure there will be no significant impacts to aquatic organisms and is not a guarantee that the operation will stay within the limit. **(0016-2-4** [Curran, Diane])

Response: The commenter is concerned that the limit set by the NPDES permit does not ensure that there would be no significant impacts to aquatic organisms. As discussed in Section 4.2, the State of Tennessee has issued a NPDES permit to TVA for operation of both units that satisfies its concerns relative to the thermal discharges and the effect those discharges may have on the health of the freshwater mussels (TDEC 2011). In addition, the U.S. Fish and Wildlife Service (USFWS 2012) concluded that operations of WBN Unit 2 "may affect, not likely to adversely affect" the pink mucket and would have "no effect" on the other aquatic threatened or endangered species.

The State of Tennessee has provided an NPDES permit to TVA for operation of both units that satisfies its concerns relative to the thermal discharges and the effect those discharges may have on the health of the freshwater mussels (TDEC 2011). The NPDES permit specifies the allowable instream temperature rise and defines temperature limits. TDEC dictates the design and frequency of studies relating to the effects of thermal discharge. The NRC relies on the TDEC's NPDES program to monitor thermal discharges and to protect the aquatic biota from the effects of the thermal discharge. The NPDES permit addresses noncompliance with the conditions of the permit, the process for reporting noncompliance and liability for noncompliance.

Comment: [From Southern Alliance for Clean Energy's Statement of Disputed Material Facts]
I. Description of TVA's Aquatics Studies. Comparison of 2010 Peak Spawning Seasonal Densities of Ichthyoplankton at [WBN] at Tennessee River Mile 528 with Historical Densities During 1996 and 1997 (Apr. 2011, Revised Nov. 2011) ("Peak Spawning Entrainment Study")

69. This study found that total impingement values in 1996 to 1997 (161) were less than those measured in 2010 to 2011 (13,573). This study also found, however, that mortality resulting from a cold shock event dominated impingement mortality at WBN in 2010 to 2011. Shad in the Southeastern United States, including the Chickamauga Reservoir, are susceptible to cold shock. When temperatures fall below 50°F, they become lethargic and more susceptible to impingement. The study found that the most significant impingement events observed at WBN in 2010 to 2011 were the result of cold shock.

   Undisputed as to the accuracy of TVA's description of the study's conclusions. Disputed as to the implication that cold shock, not the operation of WBN1, is the most significant cause of impingement mortality. See Young Declaration, ¶ III-B.1-4.

70. Excluding the cold shock event, this study found that fewer fish and number of species were impinged in 2010 to 2011, than in 1996 to 1997. The EPA endorses an impingement modeling approach that excludes the effects of extreme environmental conditions. The EPA also acknowledges the effects of cold shocks on shad.

   Disputed as to the cause of mortality. The mortality was caused by impingement against a man-made structure due to intake flow velocities not just the physiological consequences of cold temperatures. See Young Declaration, ¶ III-B.1-4.

71. This study concludes that low numbers of impinged fish in both 1996-97 and 2010-11 indicate that impingement resulting from operation of WBN Unit 1 will not materially affect fish populations in the WBN vicinity.

   Disputed as to the reasonableness of the study duration being adequate to determine this conclusion, and as to the reasonableness of the conclusion. See Young Declaration, ¶ III-B.1-5.

72. Dual unit operation will result in increased withdrawal of water through the CCW intake channel. Impingement will likewise increase at a rate that is proportional to the increase in flow rate. This study concluded that the impingement increase from dual unit operation would still be very small when compared to the effects of cold shock and winter kills on shad. As a result, TVA's experts concluded that operation of Unit 2 will not result in material increases in impingement at WBN.

   Disputed as to this methodology that was also used similarly by TVA to arrive at conclusions of entrainment from the combined operation of Unit 1 and 2. See Young Declaration, ¶ III-A.13-14. (**0016-4-4** [Curran, Diane])

Comment: STATEMENT OF [Dr. Shawn Paul Young] PROFESSIONAL OPINION REGARDING ADEQUACY OF TVA'S RECENT BIOLOGICAL STUDIES TO ADDRESS THE ENVIRONMENTAL IMPACTS OF THE PROPOSED WATTS BAR 2 NUCLEAR POWER PLANT ON AQUATIC ORGANISMS. Impingement Study [Fish Impingement at [WBN] Intake Pumping Station Cooling Water Intake Structure During March 2010 through March 2011 (Mar. 2011) ("Impingement Study")]

1. TVA claims that impingement data it collected between March 2010 and March 2011 at the CCW intake show that impingement rates under normal conditions were unchanged from those that TVA historically measured at the CCW intake, but that unusually cold weather in the winter of 2011 produced high impingement rates. TVA also cites the DES for the proposition that impingement impacts during operation of both WBN1 and WBN2 would be "too low to noticeably alter the aquatic community"....

2. I disagree with TVA that the Impingement Study provides sufficient data on which to reach a conclusion about impingement impacts of either WBN1 or dual operation of WBN1 and WBN2.

3. Although WBN1 has been operating since 1996, the last time TVA took an impingement measurement for the CCW was in 1997. Although TVA has planned for some time to finish building and operate WBN2, it made no effort to measure impingement rates until 2010, after Contention 7 was admitted for a hearing. For any reasonable fish biologist, two measurements taken more than ten years apart would not suffice to provide the basis for any analysis of the impingement impacts of WBN1.

4. The circumstances of the 2010 measurements illustrate my point. In comparison to the 161 fish impinged in March 1996 through 1997, 13,573 were impinged in 2010... TVA attributes this exponential increase to cold weather in 2010. But it is also possible that the through-screen velocity of water flowing into the CCW intake is partially responsible for the high impingement rate. At page 1, the Impingement Study lists the through-screen velocity as 0.67 fps. The EPA recommends that through-screen velocity be kept below 0.5 fps, however, in order to reduce entrainment and impingement. Without more data over a period of several years, the contribution of the cold and plant operating conditions to the rate of impingement can only be guessed at. In short, it is not possible for TVA to make up for years of neglect in only one year. (**0017-2-7** [Curran, Diane])

Response: This comment expresses concern about the sufficiency of impingement data, the higher levels of impingement in the most recent 2010-2011 study (TVA 2011d), and the evidence that cold shock, not operation of Unit 1, was the cause of impingement. The commenter's concerns are based, in part, on the larger number of individuals impinged in 2010-2011 as compared to 1996-1997 study.

In Section 4.3.2.2, the NRC staff discusses the impingement studies at the IPS and compares the timing of impingement and the corresponding water temperatures. The majority of the fish were impinged (99.6 percent) between January and the first week of March. Because the impingements occurred during the winter months when the water temperature was the coldest and because all but eight of the fish impinged were shad, the NRC staff thinks that the shad were likely affected by the cold and rendered in a moribund or dead state at the time they were impinged. If a higher through-screen velocity caused the additional impingements in 2010-2011, then the impingements should likely have occurred at higher rates throughout the study, and not only in the winter months.

In Section 4.3.2.2, the NRC staff also considered the velocity of water past the site in comparison to the velocity of water through the intake screens. The velocity of water in the reservoir as it flows past the site averages 0.7 m/s (2.3 ft/s) under normal winter conditions and 0.3 m/s (1.0 ft/s) in the summer months (as a result of lower flows and higher reservoir elevations in the summer). Fish living near the site that are large enough to be impinged are accustomed to an environment with flow rates greater than 0.21 m/s (0.67 ft/s) and are not likely to be impinged while they are healthy.

Comment: STATEMENT OF [Dr. Shawn Paul Young] PROFESSIONAL OPINION REGARDING ADEQUACY OF TVA'S RECENT BIOLOGICAL STUDIES TO ADDRESS THE ENVIRONMENTAL IMPACTS OF THE PROPOSED WATTS BAR 2 NUCLEAR POWER PLANT ON AQUATIC ORGANISMS. Hydrothermal Study [Hydrothermal Effects of the Ichthyoplankton from the Watts Bar Plant Supplemental Condenser Cooling Water Outfall in Upper Chickamauga Reservoir (Jan. 2011) ("Hydrothermal Study")]

10. TVA also failed to note the significance of the great discrepancy between the daytime and night-time population measurements, or to analyze how they may be affected by daily variations in thermal plume temperature. In light of the size of the discrepancy, TVA should have undertaken more studies of the differences between daytime and nighttime fish populations.....

11. The Hydrothermal Study showed that thermal discharge observed for current operation of Unit 1 is already near the limits set in the NPDES permit. TVA's temperature data shows that it is staying within its permit limit of a 5°F daily average change from upriver temperature at the downstream edge of the mixing zone; however, the results from the May and August 2010 tests show that it is operating on the edge of those limits with only Unit 1 operating. ... the maximum difference between ambient and surface temperature reached 5°F during the May night test, 5.34°F during the May day test, and 5.36°F during the August day test. Also, at the point of discharge, the Hydrothermal Study shows that SCCW discharge water is 10°F hotter than the water above the SCCW thermal plume and above Watts Bar Dam. Organisms drifting downriver nearest the point of discharge will likely suffer

from this abrupt temperature change, especially fish eggs and larvae. These impacts were not considered by TVA. (**0017-2-12** [Curran, Diane])

Response: The commenter is concerned that the variation in daytime and night-time population measurements are not considered for ichthyoplankton and other fish populations. The commenter also expresses concern regarding the ability of TVA to operate two units and remain within the limits set by the NPDES permit.

In Section 2.3.2.1, the NRC staff acknowledges that threadfin and gizzard shad synchronize their spawning behavior and that threadfin shad generally congregate before spawning a few hours after sunrise. However, the NRC staff believes that the information available is sufficient to meet the intent of NEPA to perform the impact assessment for operation of WBN Unit 2 without additional studies related to differences in daytime and nighttime fish populations.

TDEC regulates the effects of cooling-water intake structures and effluents through NPDES permits, and TVA currently holds a valid NPDES permit to operate WBN Units 1 and 2 (TDEC 2011). The NRC relies on the TDEC's NPDES program to monitor thermal discharges and to protect the aquatic biota from the effects of thermal discharge. The NPDES permit addresses noncompliance with the conditions of the permit, the process for reporting noncompliance and liability for noncompliance.

Comment: TVA's finding [in 2007 FSEIS] that WBN Unit 2 will have no significant impacts on aquatic life in the Tennessee River is inadequately supported in the following respects: .....TVA further states that thermal impacts will be insignificant, even though TVA's conclusions are contradicted by its own acknowledgement of the need to relocate mussels in the vicinity of the SCCW discharge to avoid mortality from elevated temperatures. .... And TVA provides no evidence, such as scientific studies or field observations, to justify its conclusion. ... For instance, TVA is missing a number of basic data sets with respect to thermal impacts, including data on overall drift communities, and data on spatial and temporal distribution of ichthyoplankton in relation to thermal mixing zones. .... Other factors neglected by TVA (which must be understood in order to properly assess thermal impacts on aquatic life), include characteristics of the thermal plume; variations in the size and temperature profile of the mixing zone; the temperatures in the core of the thermal plume (rather than at the edge) and whether they have an effect on aquatic organisms; and the effects of high temperatures on fish eggs and larvae, which are highly vulnerable to elevated and rapidly changing temperature. ....Finally, TVA fails to show that it has accounted for the impacts of overflow from the holding ponds, where excess cooling water may be stored at very high temperatures. (**0008-11** [Curran, Diane])

Comment: TVA [in "Motion for Summary Disposition of Contention 7" (Nov. 21, 2011)] claims that it has conducted studies that resolve the three major deficiencies identified in Contention 7. As discussed in Dr. Young's attached Declaration, this is not correct.....TVA's Hydrothermal

Study does not address important parameters such as Outfall 101 or the amount of time that fish larvae remain in the thermal plume. (**0015-5** [Curran, Diane])

Comment: [From Southern Alliance for Clean Energy's Statement of Disputed Material Facts] Description of TVA's Aquatics Studies

Hydrothermal Effects of the Ichthyoplankton from the Watts Bar Nuclear Plant Supplemental Condenser Cooling Water Outfall in Upper Chickamauga Reservoir (Jan. 2011) ("Hydrothermal Study")

75. In direct response to these claims, TVA designed this study to document the flow patterns and characteristics of the thermal plume from WBN, and track the thermal plume in conjunction with ichthyoplankton sampling. This allowed TVA to understand the temporal and spatial distribution of ichthyoplankton and exposure rates to thermal discharges.

    Disputed as to the fact that TVA failed to study the thermal discharge from Outfall 101 in conjunction with Outfall 113 to encompass the cumulative thermal discharge from WBN, and failed to address exposure rates and the effects of abrupt temperature changes on ichthyoplankton in this study. See Young Declaration, ¶ III-C.3-4.

76. TVA conducted this study in May and August, 2010, because those time frames represented extreme conditions: peak abundance of fish eggs and larvae, near maximum ambient water temperatures, and no release from the upstream Watts Bar Dam.

    Undisputed as to timeframe of study. Disputed as to whether this would be representative over time as this study only represents a few points in time, not adequately addressing environmental variability. See Young Declaration, ¶ III-C.2.

77. This study found that, even under these extreme conditions, water temperatures did not approach the limits established by TVA's NPDES permit for operation of WBN Units 1 and 2.

    Disputed as study results directly stated to the contrary. See Young Declaration, ¶ III-C.11. Because discharge temperatures did not exceed those set in TVA's NPDES permit, this study concluded that there was no risk of thermal damage to ichthyoplankton from operation of WBN. Disputed as to accuracy and reasonableness of these conclusions. See Young Declaration, ¶ III-C.1-11.

78. Even if operation of WBN Units 1 and 2 causes effluent temperatures to rise above those measured even under extreme conditions for Unit 1, TVA is bound by its NPDES discharge limits. Accordingly, dual unit operation does not pose any greater risk of thermal damage to the aquatic community in the WBN vicinity than does operation of Unit 1 alone.

Disputed as to accuracy and reasonableness of these conclusions. See Young Declaration, ¶ III-C.1-11. (**0016-4-5** [Curran, Diane])

Comment: SUMMARY OF MY [Dr. Shawn Paul Young] PROFESSIONAL OPINION REGARDING TVA'S ASSERTIONS.

[3.] ....And TVA's Hydrothermal Study does not address important paramaters such as Outfall 101 or the amount of time that fish larvae remain in the thermal plume. (**0017-1-3** [Curran, Diane])

Response: The commenter lists information such as temperature exposure rates for ichthyoplankton and overflow from the holding ponds that should be included in the determination of thermal impacts. In addition, the commenter states that the cumulative impacts from the diffusers and the SCCW discharge are not considered.

The NRC relies on the TDEC's NPDES program to monitor thermal discharges and to protect the aquatic biota from the effects of the thermal discharge. The TDEC regulates the effects of cooling-water intake structures and effluents through NPDES permits, and TVA currently holds a valid NPDES permit to operate WBN Units 1 and 2 (TDEC 2011). The Clean Water Act gives States the responsibility to place controls on thermal discharges that are stringent enough to ensure protection and propagation of a balanced indigenous population of shellfish, fish, and wildlife. The NRC relies on the NPDES permitting process to protect aquatic resources.

Section 3.2.2.4, contains a description of the operation of the YHP discharge, the SCCW, and the diffuser discharge structures. In Section 4.2.2.2, the NRC staff discusses impacts from thermal and chemical discharges on surface-water-quality. As discussed in Section 4.2.2.2, TVA continuously monitors the Outfall 101 temperature (diffuser discharge). If the temperature reaches 35°C (95°F), a signal in the control room alerts operators of the condition and they divert discharge to the YHP. Discharges from the YHP are through Outfall 101 or 102 and conform to the NPDES permit requirements. Thermal discharges from the SCCW are continuously monitored at the discharge stream bottom to ensure that the temperature does not exceed the permitted limit of 33.5°C (92.3°F). TVA also continuously monitors water temperature at the downstream edge of the mixing zone (located 610 m [2,000 ft] downriver from the discharge). The monitoring program includes a biannual temperature survey along a transect in the reservoir at depths from the surface to 2 m (7 ft) below the surface. In Sections 4.3.2.2 and 4.2.3.5, the NRC staff discusses the thermal and chemical impacts on aquatic organisms from thermal discharges. Section 4.14.4.2 discusses the cumulative impacts of thermal discharges from Outfall 101 and Outfall 113 on river temperature.

Comment: [From Southern Alliance for Clean Energy's Statement of Disputed Material Facts]
Description of the Proposed Project
WBN Cooling System Output

30. Because the thermal discharge limits established by TVA's NPDES permit for dual unit operation are unchanged from those for Unit 1 operation, thermal impacts on the aquatic environment resulting from WBN operation will not be materially different under dual unit operation than they are for operation of Unit 1 alone.

Disputed. There will be substantial increases in discharge from the CCW and SCCW. See Table 3-1 of the DFES, at page 3-9:... Also, as discussed in Dr. Young's Declaration at pars. III-C.1-11, the already-stressed Tennessee River aquatic environment will be further stressed by additional CCW and SCCW thermal discharge from cumulative cooling tower blowdown discharge to the Tennessee River as a result of WB2 operation. (**0016-2-3** [Curran, Diane])

Response: This comment states that the discharge from the CCW and SCCW would be substantial and would add further stress to an already stressed river. TDEC regulates the effects of cooling-water intake structures and effluents through NPDES permits, and TVA currently holds a valid NPDES permit to operate WBN Units 1 and 2 (TDEC 2011). The NRC staff's assessment relies on the TDEC's NPDES program to monitor thermal discharges and to protect the aquatic biota from the effects of the thermal discharge. Sections 3.2.2 and 4.2.2.2 contain discussions of the thermal discharge in terms of the amount of heat released under single- or dual-unit operation. In Sections 4.14.4.2, the NRC staff considers the cumulative impacts—including thermal discharges. In Section 4.3.2.3, the NRC staff concludes that any additional impact to aquatic biota from thermal discharges resulting from operation of WBN Unit 2 would be undetectable. The NRC staff bases this determination on the incremental rise in thermal discharge anticipated from Outfalls 101, 102, and 113 from operation of both WBN Units 1 and 2.

Comment: TVA [in "Motion for Summary Disposition of Contention 7" (Nov. 21, 2011)] claims that it has conducted studies that resolve the three major deficiencies identified in Contention 7. As discussed in Dr. Young's attached Declaration, this is not correct......TVA's description of its method of analyzing aquatic impacts indicates a troubling lack of care or competence. For example, by adding widely divergent diurnal and nocturnal entrainment measurement, TVA violates guidance of the U.S. Environmental Protection Agency ("EPA") and grossly overstates the size and diversity of the fish population. Some of the studies relied on by TVA had to be revised after they were released, indicating that TVA has significant problems ensuring the quality of its measurements and analyses. It is reasonable to expect that the results from TVA's biological studies will be accurate in order to support TVA's conclusions. In too many instances, however, TVA makes significant mistakes. (**0015-6** [Curran, Diane])

Comment: SUMMARY OF MY [Dr. Shawn Paul Young] PROFESSIONAL OPINION REGARDING TVA'S ASSERTIONS.

4. Finally, TVA's description of its method of analyzing aquatic impacts indicates a troubling lack of care or competence. For example, by adding widely divergent diurnal and nocturnal entrainment measurement, TVA violates guidance of the U.S. Environmental Protection Agency ("EPA") and grossly overstates the size and diversity of the fish population. See pars. III-C.6 and III-C.10 below. Some of the studies relied on by TVA had to be revised after they were released, indicating that TVA has significant problems ensuring the quality of its measurements and analyses. See pars. III-A.2 and III-A.11, and pars. III-C.6-9 below. It is reasonable to expect that the results from TVA's biological studies will be accurate in order to support TVA's conclusions. In too many instances, however, TVA makes significant mistakes. (**0017-1-4** [Curran, Diane])

Response: The commenter states that some TVA studies contained errors. The NRC staff also noticed discrepancies and errors in its review (e.g., differences in the numbers of fish impinged in two different reports, discrepancies in tables, and errors in entrainment calculations). These errors, inconsistencies and other topics for clarification were the subject of Requests for Additional Information (RAIs) sent from the NRC to TVA.

Comment: STATEMENT OF [Dr. Shawn Paul Young] PROFESSIONAL OPINION REGARDING ADEQUACY OF TVA'S RECENT BIOLOGICAL STUDIES TO ADDRESS THE ENVIRONMENTAL IMPACTS OF THE PROPOSED WATTS BAR 2 NUCLEAR POWER PLANT ON AQUATIC ORGANISMS

Hydrothermal Study [Hydrothermal Effects of the Ichthyoplankton from the Watts Bar Plant Supplemental Condenser Cooling Water Outfall in Upper Chickamauga Reservoir (Jan. 2011) ("Hydrothermal Study")]

6. The conclusions of the Hydrothermal Study are also based on incorrect methodology that leads to distorted results. In reporting the results of ichthyoplankton sampling, TVA added the daytime and nighttime measurements rather than averaging them, thus giving a distortedly high population reading. For instance Table 4 on page 25 of the Hydrothermal Survey shows that on May 11-12, 2010, during daytime sampling, TVA estimated 75 organisms per 1,000 m^3 of water at the SCCW outfall. During the nighttime sampling, TVA estimated 8,232 organisms for the same volume of water. TVA then reported the number of organisms per volume of 1,000 m^3 of water for the sampling period as 8,307. In actuality, however, the number of organisms ranged between 75 and 8,232, with an average of approximately 4,153 fish larvae per 1000 m^3 of water during a 24-hour diel cycle.

7. There is no controversy about what method TVA should have employed – it is listed in the "Materials and Methods" section of TVA's April 2011 "Peak Entrainment Study." For TVA not to notice another significant error in its own reporting raises fundamental questions regarding TVA's methodology for all of its studies.

8. TVA's methodological error has several implications in the analyses of impacts on the fish community. This error results in the overstatement of the size of the fish population in the river, which in turn will lead to an understatement of the percentage of fish that are affected by entrainment. This has major implications for the validity of the "Entrainment Study" because it results in an incorrect estimate of the percentage of organisms that were entrained at the CCW. If the same error found in the Hydrothermal Study was made during calculations of ichthyoplankton abundance for the Peak Entrainment Study, the results listed in the Peak Entrainment Study are not accurate, and TVA conclusions are not based on accurate material facts. In addition, the original Aquatics Study also had major errors, and one cannot be sure those errors have been remedied in the Revised Aquatics Study. Both documents used to compare post-operation entrainment and the associated impacts have had major errors casting doubt on the validity of TVA's analyses and conclusions.

9. Another significant error can be found in Tables 5 through 10. Table 10 lists the total ichthyoplankton abundance found at the five different sampling stations across the survey transect. However, the reported total number of ichthyoplankton captured is less than the reported number of ichthyoplankton that were captured at just one of the individual sampling stations. This error raises serious questions about the actual results of the study, not to mention TVA's competence and quality assurance procedures for conduct of biological monitoring and anlaysis. **(0017-2-11** [Curran, Diane])

Response: The commenter is correct in that the appropriate analysis for the hydrothermal study would have been a weighted average of the day and night densities, rather than a sum of the densities. Also, the data in Table 9 were not added correctly for week 2, night sampling. The NRC staff used the results of a yearlong entrainment study submitted by TVA (2012) rather than the results of the hydrothermal study. The NRC staff deleted the discussion in Section 4.3.2.2 of the draft SFES that discussed the hydrothermal study in relation to entrainment at the SCCW.

E.2.9   Comments Concerning Socioeconomics

Comment: Page/Line 2-69 I 13: In Section 2.4.2.7, Tax Revenues, consider adding information about TVA's Mitigation payments (see SEIS pg. 68, Section 3.8.7, second paragraph) under which TVA makes additional payments to local governments impacted by TVA activities during the period of constructing WBN. TVA notes that these mitigation payments are mentioned on pages 4-42 and 4-43 of the DFES. Adding a statement about mitigation payments in Section 2.4.2.7 would clarify that areas around WBN receive funds in addition to the annual in-lieu of tax payments. **(0010-5** [Brickhouse, Brenda])

Response: This comment requests consideration of adding information to Section 2.4.2.7, Tax Revenues, regarding TVA Mitigation payments (referred to as tax-equivalent "impact" payments in this SFES), which are payments made by TVA to local governments impacted by the construction of WBN Unit 2. This SFES text in Section 2.4.2.7, Tax Revenues, was revised to clarify that local governments and counties also receive TVA "impact" payments in addition to the annual in-lieu of tax payments.

E.2.10 Comments Concerning Historic and Cultural Resources

Comment: Thank you for your letter regarding the Notice of Availability of the draft supplement to the Final Environmental Statement for Watts Bar Nuclear Plant Unit 2 for public comment. Through your notification the Chickasaw Nation is made aware that the US Nuclear Regulatory Commission is reviewing an application submitted by the Tennessee Valley Authority for an operating license for Watts Bar Nuclear Plant (WBN), Unit 2 located in Rhea County, Tennessee. We are in agreement with the NRC to review the permit.

After reviewing your information, we are in agreement with the assessment and have no objections to the proposed undertaking. We concur with your finding of no adverse effect to historic properties and we accept the special conditions set forth in this report. We do not presently know of any specific historic properties or properties of significant religious or sacred value. In the event your agency becomes aware of the need to enforce other statutes we request to be notified under NEPA, NAGPRA, AIRFA and ARPA. (**0013-1** [Keel, Jefferson])

Response: This comment states that the Chickasaw Nation concurs with the finding of no adverse effect to historic properties, and has no objections to the project. The Historic and Cultural Resources Consultation Section 2.5.3 of this SFES was modified as a result of this comment.

Comment: We [the Choctaw Nation of Oklahoma] have reviewed the following proposed project(s) as to its effect regarding religious and/or cultural significance to historic properties that may be affected by an undertaking of the projects area of potential effect.

> **RE: Draft supplement to the Final Environmental Statement for Watts Bar Nuclear Plant Unit 2.**

Comments: After further review of the above mentioned project(s), and based on the information provided it has come to our attention that the project is out of the Choctaw Nation of Oklahoma areas of interest. A list of states and counties has been provided. (**0014-1** [Thompson, Ian])

Response: This comment states that the Choctaw Nation of Oklahoma determined that the project is outside their area of interest. The Historic and Cultural Resources Consultation Section 2.5.3 of this SFES was modified as a result of this comment.

E.2.11 Comments Concerning Health – Radiological

Comment: [Two plants in the same place makes twice as much risk] for radiation in the water, for tritium in the water (**0003-4-4** [Kurtz, Sandy])

Response: NRC regulations require that licensees limit the amount of radioactive material, including tritium, released to the environment (the air and water) during normal operations. All licensees that operate a nuclear power plant must keep releases of radioactive material to unrestricted areas during normal operations as low as is reasonably achievable (ALARA) and must comply with the dose limits for members of the public given in 10 CFR Part 20 and with the ALARA design criteria cited in 10 CFR Part 50.34a and Appendix I to 10 CFR Part 50. In addition, NRC regulations require licensees to have various effluent and environmental monitoring programs, so that the impacts from plant operations are minimized and radioactive releases are accurately recorded and reported. The NRC requires licensees to report plant radioactive effluent releases and results of radiological environmental monitoring around their plants to ensure that potential impacts are detected and reviewed. Accurate records are required on radioactive releases to the air and water. Annual radioactive effluent release reports document the amount of radioactive liquid and airborne effluents discharged from plants and the associated doses. Also, annual radiological environmental monitoring reports document the environmental radioactivity levels around the plant. These reports, which are available to the public, include data from thermoluminescent dosimeters (which measure radiation dose levels); airborne radioiodine and particulate samplers; samples of surface, groundwater, drinking water, and downstream shoreline sediment from existing or potential recreational facilities; and samples of ingestion sources (e.g., milk, fish, invertebrates, and broad-leaf vegetation). The NRC conducts periodic onsite inspections of the effluent and environmental monitoring programs to ensure compliance with NRC requirements. The NRC documents the results of its compliance inspections of each licensee's effluent and environmental monitoring programs in inspection reports, which are available to the public. The NRC staff evaluated doses to members of the public from potential radioactive effluent releases from WBN Units 1 and 2, including any tritium releases. The calculated doses are within NRC and EPA radiation protection standards. The radiological evaluation is discussed in Section 4.6 and Appendix I of this SFES. In addition, the NRC staff addressed the TVA Groundwater Protection Program in Section 2.6, "Radiological Environment" of this SFES. No changes have been made to this SFES in response to this comment.

Comment: [Two plants in the same place makes twice as much risk for] the health of our people themselves. (**0003-4-6** [Kurtz, Sandy])

Comment: And then the idea that there is a swimming hole within a thousand feet of that plant is just sick. (**0003-5-3** [Harris, Ann])

Response: *Recreational activities, including swimming, are taken into consideration in determining the dose to the maximally exposed individual residing near the plant. Section 4.6 of this SFES discusses the impact to the public from the liquid effluent, gaseous effluent, and direct radiation pathways. The NRC staff's independent dose assessment determined that doses to the public from liquid effluents would be below the ALARA dose design criteria specified in Appendix I to 10 CFR Part 50. No changes have been made to this SFES in response to these comments.*

Comment: I didn't see the word Chernobyl one time. I understand from the discussion earlier in the informal period with Mr. Susco that it's included in the computer model. But including 6,000 thyroid cancers into the computer model is not quite the same thing as stating that when that reactor blew up the effect was 6,000 at least documented cases of thyroid cancer. And if we understood us to say, oh, it was a minor thing. The effects aren't that great. But you know, they're still monitoring a lot of the agricultural products in that region. There are still great areas that are uninhabitable for many years. It's the *cesium that's going to stay positive for 300 to 500 years. That's what we're talking about here. (**0003-3-3** [Safer, Don])

Response: *The environmental impact of potential design basis accidents was assessed by the NRC staff in the WBN Unit 2 plant-specific FES and was determined to be small. Since the licensee is required to maintain the plant within acceptable design and performance criteria, including during any extended life operation, these impacts are not expected to change. In addition, the NRC's ongoing safety and inspection program focuses on early identification and prevention of safety problems so that potential operating issues do not lead to accidents. The NRC continuously monitors the performance of licensees and operators, including frequent onsite inspections and the use of resident inspectors. The NRC Reactor Oversight Process integrates NRC inspection, assessment, and enforcement programs. The operating reactor assessment program evaluates the overall safety performance of operating commercial nuclear reactors and communicates those results to licensee management, members of the public, and other government agencies. The assessment program collects information from inspections and performance indicators to enable the agency to arrive at objective conclusions about a licensee's safety performance. Based on this assessment information, the NRC determines the appropriate level of agency response, including supplemental inspection and pertinent regulatory actions ranging from management meetings up to and including orders for plant shutdown. The NRC conducts follow-up actions, as applicable, to ensure that the corrective actions designed to address performance weaknesses were effective. The NRC developed requirements to ensure adequate protection or no undue risk to public health and safety through design, construction operation, maintenance, modification, and quality assurance measures.*

The release of cesium from the Chernobyl accident was significantly larger than would occur in the improbable event of a severe accident at WBN Unit 2. This is because the design and operation of the Chernobyl reactor was fundamentally different from the types of nuclear power reactors licensed and permitted in the United States. The Chernobyl accident is not directly comparable to a loss-of-coolant accident at WNB Unit 2 due to two key differences. First, Chernobyl did not have a steel and concrete containment building enveloping its reactor like that used at WBN Unit 2, and second, Chernobyl's fuel was moderated by graphite, a combustible material, whereas WBN Unit 2 uses uranium pellets encased in steel rods arranged in fuel bundles contained within a robust reactor vessel. However, the NRC staff considers health effects, primarily cancers, as one of the consequences of a severe accident. The NRC staff uses probability- weighted consequence (risk) to evaluate reactor severe accidents. Risk is the product consequence times the probability that the consequence would occur. While the consequences of severe accidents are large, the probability of occurrence is extremely small. In Section 6.2, the NRC staff evaluated the consequences of severe accidents for WBN Unit 2 and concluded that the probability-weighted consequences of severe accidents are small. No changes have been made to this SFES in response to this comment.

Comment: [F]our years ago we discovered we had low level waste going into our landfill. That landfill is right on the Stones River, which provides the drinking water for most of Rutherford County. That includes the city of Murfreesboro. My first interest was in water and it's about water that I want to speak today. It's a source of life. If our drinking water is polluted and we're taking radiation or chemicals into our bodies, I think that can very well account for the epidemic of cancer that we have in this country today. **(0003-6-1** [Ferris, Kathleen])

Response: The scope of this environmental assessment is specific to the potential environmental impacts associated with the operation of the WBN Unit 2. The evaluation did not investigate the potential impacts of the landfill referenced in the comment because Murfreesboro is over 50 mi (80 km) from the WBN site and is outside the scope and area evaluated for this analysis. All analysis on doses and cumulative dose associated with the operation of WBN Unit 2 are performed for areas within 50 mi (80 km) of the site. Radioactive effluent releases from WBN Units 1 and 2, as discussed above in the response to comment 0003-4-4, are required by NRC regulations to be controlled to limit releases to the environment (the air and water). As part of the NRC requirements for operating a nuclear power facility, licensees must keep releases of radioactive material to unrestricted areas during normal operation as low as is reasonably achievable (as described in the NRC's regulations in 10 CFR Part 50.34a) and comply with radiation dose limits for the public as given in the regulations in 10 CFR Part 20. In addition, NRC regulations require licensees to have various effluent and environmental monitoring programs so that the impacts from plant operations are minimized and radioactive releases are accurately recorded and reported. To ensure that U.S. nuclear power plants are operated safely, the NRC licenses the nuclear power plants to operate, licenses the plant operators, and establishes license conditions for the safe operation of each plant. The NRC provides continuous oversight of plants through its Reactor Oversight Process to verify that the plants are being operated in accordance with NRC regulations. The NRC has authority to take whatever action is necessary to protect public health and safety and the environment and may require immediate licensee actions, up to and including a plant shutdown. The amount of radioactive material released from normal operations at nuclear power plants is well measured, well monitored, and is very small. Cancers in members of the public attributed to the low doses of radiation as a result of exposure to nuclear power facilities have not been observed and would not be expected. No changes have been made to this SFES in response to this comment.

E.2.12   Comments Concerning Accidents – Severe

Comment: Two plants in the same place makes twice as much risk for accidents (**0003-4-2** [Kurtz, Sandy])

Comment: [Two plants in the same place makes twice as much risk] for human error (**0003-4-3** [Kurtz, Sandy])

Comment: And whether it's Chernobyl or Fukushima or the fire at Browns Ferry or Three Mile Island, these nuclear reactors are just as prone to accidents, mistakes, failures, human error, whatever, terrorism, or whatever it might be as any other human enterprise and any other human activity. They will have problems. They will have worse-case scenarios. It may not happen very often, but that doesn't mean it can't happen tomorrow here. And if you have two reactors, it just doubles the chance. **0003-9-3** [Safer, Don])

Response: The comments are concerned about an increased risk for severe accidents. Section 6.2.1 of this SFES states that human error is considered in establishing the core damage frequencies used to determine severe accident risks. For multi-reactor sites where there are purposely few shared support systems to minimize the likelihood of a severe accident at one unit adversely affecting other onsite units, the severe accident risk for the site is the sum of the risks for the individual units. The last paragraph of SFES Section 6.2 makes this point. No changes have been made to this SFES in response to these comments.

Comment: I just have to make note that the definition of risk in this document is it's the product of frequency and the consequences of an accident. Work on that one for a while. I don't know. As I said, it's in 6.2.4, where the staff concludes that the environmental consequences of a severe accidents are small, 6.2.4. I don't know how you get that. (**0003-7-9** [Safer, Don])

Response: The comment points out an inconsistency in the terminology related to severe accidents in this SFES. Risk is defined as the product of the frequency of an accident and the consequences of the accident. The NRC staff performed an independent review. The conclusion in SFES Section 6.2.4 has been revised to state that the probability-weighted environmental consequences of severe accidents are small.

Comment: SAMA [Severe Accident Mitigation Alternatives] is incomplete. It is based on a PRA [probabilistic risk assessment] that does not comply with applicable RG [Regulatory Guide] or ASME/ANS [American Society of Mechanical Engineers/American Nuclear Society] standard. (**0006-1** [Anonymous])

Response: The comment concerns the adequacy of the probabilistic risk assessment (PRA) used in the NRC staff's severe accident mitigation design alternatives (SAMDA) analysis. The comment also states that the PRA does not comply with a Regulatory Guide (RG) or ASME/ANS standard. The NRC staff has reviewed the PRAs submitted by TVA. The NRC staff review of the PRAs is discussed in Section 6.3 and Appendix H of this SFES. The NRC staff concludes that the PRAs provide acceptable bases for evaluating the benefits associated with various SAMDAs. RGs describe to the public and applicants or licensees, methods for complying with NRC regulations that the staff finds acceptable. RGs are not a substitute for regulations and compliance with them is not required. Compliance with ASME/ANS standards is also not required. No changes have been made to this SFES in response to this comment.

Comment: Page H-3, Appendix H is the Severe Accident Mitigation Design Alternatives. I'll quote, "TVA did not include the contribution from external events in the Watts Bar Nuclear Plant risk estimates." External events. Now whether those is a tornado like the one that almost hit Browns Ferry, a F-5 tornado. The very same day, April 27th, tornadoes came through here and they got pretty close to Watts Bar. They went right across Bellefonte, where that reactor was going to be built as soon as they finish Watts Bar. And it's just kind of -- to me it's kind of eerie, but in the same day all of TVA's nuclear power plants were affected by tornadoes. (**0003-7-5** [Safer, Don])

Comment: So those kinds of things [tornadoes] .... that's an external event. They just don't even factor that into this. (**0003-7-6** [Safer, Don])

Response: This commenter expressed concerns regarding tornadoes being considered as initiating events for severe accidents at WBN. Nuclear power plant structures important to safe operation and shutdown of the nuclear reactor are designed to withstand tornado strikes and tornado missiles. Section 6.3.2.4 of this SFES describes a change to the WBN auxiliary building to prevent damage by tornado missiles. Further, this SFES notes that the severe accident analysis in Section 6.2 only considers internal events. However, in SFES Section 6.3 and Appendix H, the NRC staff notes that the effects of external events are included in the SAMDA analysis. Section 6.3.1 states that the benefits of potential SAMDAs are estimated by multiplying the estimated benefits for internal events by a factor of 2.28 to account for external events. Further, the quote in the comment is incomplete. The sentence on page H-3 in Appendix H is "TVA did not include the contribution from external events in the WBN risk estimates; however, it did account for the potential risk-reduction benefits associated with external events by multiplying the estimated benefits for internal events by a factor of 2. This factor was subsequently increased to 2.28 in response to an NRC staff RAI (TVA 2011a)." No changes have been made to this SFES in response to these comments.

Comment: Severe accident mitigation alternatives in 6.3, there's a quote in here. I'll read it directly because I didn't copy it all down. They eliminate the severe accident scenarios that were "excessively costly to implement such that the estimated cost would exceed the dollar value associated with completely eliminating all severe accident risks at WBN 2". I take that to mean that some of the risks it was just too costly to mitigate those risks, so we just threw them out because, if it costs too much, we couldn't possibly deal with it. So we'll just deal with the risks. That's what that says to me. (**0003-7-10** [Safer, Don])

Response: This comment expresses concern about the identification of SAMDA that should be evaluated in detail. As described in Section 6.3.2.3 of this SFES, the purpose of SAMDA analysis is to evaluate postulated design alternatives to determine if the risk-reduction benefit of the design alternative expressed in dollars is greater than the cost of implementing the design alternative. An initial screening is performed to eliminate from consideration those alternatives that are clearly too costly. This initial screening assumes that each alternative would eliminate all risk. It is not necessary to consider further any alternative with an implementation cost that exceeds the benefit of eliminating all risk because no alternative can eliminate all risk and therefore the design alternative under consideration would not be cost beneficial in a more realistic evaluation of risk reduction. No changes have been made to this SFES in response to this comment.

Comment: I think we all have seen what has gone on at Fukushima and it's terrible. People are having to leave their homes. School children's tennis shoes are contaminated with radiation. Whole areas are evacuated. Farms that have been in families for generations are now abandoned and probably never to be returned to. So when you talk about the environmental impact of this type of reactor, of any nuclear reactor, you have to realize that the effects are not small. And that was what in this document what they came up with was that the environmental impacts are small. (**0003-3-1** [Safer, Don])

Comment: Now Fukushima you've got enormous releases of radiation. You've got just today in the newspaper the announcement that baby food is being recalled, 90 containers of baby food because the Japanese are catching it. But the rice has been contaminated. It's a nightmare. The economic effects, the human effects, and the ecological effects are going to reverberate around the planet till kingdom come basically. And that's what's at stake here. (**0003-3-4** [Safer, Don])

Comment: The concerns raised by the Fukushima Task Force and our contention [Motion to Admit new Contention Regarding the Safety and Environmental Implications of the Nuclear

Regulatory Commission Task Force Report on the Fukushima Dai-ichi Accident (August 11, 2011)] remain unaddressed by the DEIS, which does not even mention the Fukushima accident. (**0008-2** [Curran, Diane])

Comment: The NRC's failure to address the environmental implications of the Fukushima Task Force is also extremely grave, given that the Task Force has called for a complete upgrade of the NRC's program for mandatory safety regulations and has targeted WBN2 for specific recommendations. (**0008-4** [Curran, Diane])

Response: On March 11, 2011, and for an extended period thereafter, several nuclear power plants in Japan experienced the loss of important equipment necessary to maintain reactor cooling after the combined effects of severe natural phenomena (i.e., an earthquake followed by a tsunami). In response to these events, the Commission established a task force to review the current regulatory framework in place in the United States and to make recommendations for improvements. The task force reported the results of its review (NRC 2011b) and presented its recommendations to the Commission on July 12 and July 19, 2011, respectively. As part of the short-term review, the task force concluded that while improvements are expected to be made as a result of the lessons learned, the continued operation of nuclear power plants and licensing activities for new plants did not pose an imminent risk to public health and safety. A number of areas were recommended to the Commission for long-term consideration. Collectively, these recommendations are intended to clarify and strengthen the regulatory framework for protection against severe natural phenomena, mitigation of the effects of such events, coping with emergencies, and improving the effectiveness of NRC programs. To the extent that any revisions are made to NRC safety or environmental regulatory requirements, they would be made applicable to nuclear power reactors regardless of whether the utility possesses a renewed license or an operating license. Therefore, no additional analyses have been performed in the SFES as a result of the Fukushima events. No changes have been made to this SFES in response to these comments.

E.2.13 Comments Concerning the Uranium Fuel Cycle

Comment: The more I know [about high level waste], the more it troubles me. So don't be saying that, oh, all you need to do is know more about it and you won't be so concerned. The more I know, the more concerned I get. (**0003-3-5** [Safer, Don])

Comment: [T]he highly irradiated used fuel that's often called spent fuel is being stored in fuel pools as long as possible. And that's just the biggest danger that it can be. They need to be moved to a hardened onsite storage. And I know they have to be in the pools for about five years till they cool down. But beyond that point the packing of these pools with more and more rods that way beyond what they were designed for is a real huge risk that the community is taking on and needs to be aware of. And the community needs to support the idea of hardened onsite storage. (**0003-7-2** [Safer, Don])

Comment: But the reality when it comes to these nuclear materials that are being manufactured at these sites is that they're going to have to be dealt with. But if the federal government refuses to do it, it's going to be the community's problem just to safeguard that plant from now until kingdom come again, till eternity. And I think we've all been sort of shaken in our confidence of the federal government being able to continue its obligations into the future. And these materials, these radioactive materials, need to be kept out of the environment for half a million years. Now if you think the federal government is going to be here half a million years from now, I'd like to see your information. (**0003-9-2** [Safer, Don])

Comment: And again the whole clean-up thing is a whole nother issue that whether that money is really going to be there. The same is true for the high level waste that's in those fuel pools. The federal government is trying to figure out how to take care of its responsibilities on that and they're struggling with that. They have struggled with it for over 50 years. There's not a single deep repository for radioactive waste that I know of in the world. So talking about the amount of high level radioactive waste we've been generating in the United States, it's a lot of material and we simply don't have any place to put it. The Blue Ribbon Commission has been working on it for over a year. They're due to release their report which is highly controversial in my opinion. But you can't take it on face value that these materials are going to be able to be handled the way they need to have been. (**0003-9-4** [Safer, Don])

Response: *The NRC is committed to ensuring that both spent nuclear fuel and high-level radioactive wastes are managed to prevent health impacts to the public. Spent nuclear fuel is currently stored safely at reactor sites in the spent fuel pools and/or independent spent fuel storage installations (ISFSIs). This practice is expected to continue until the U.S. Department of Energy is ready to take possession of the spent nuclear fuel. At this time, it is uncertain when this will happen. Interim storage needs for spent nuclear fuel vary among plants, with older units having less available pool storage capacity than newer ones. However, given the uncertainty as to when a geologic repository will open and lack of other options, it is likely that some sort of expanded spent fuel storage capacity beyond the original design capacity will be needed at all nuclear power plants.*

As discussed in Section 4.10.1, the impacts of storing spent nuclear fuel when the reactor is operating are addressed on an ongoing basis as part of reactor operations or under a separate license for an ISFSI. Current and potential environmental impacts from spent fuel storage onsite during the licensed operating period at the current reactor sites have been studied extensively by the NRC and are well understood.

The offsite radiological impacts resulting from spent fuel and high-level waste disposal and the onsite storage of spent fuel, that will occur after the reactors have been permanently shut down, are addressed in the Commission's Waste Confidence Decision Rule (WCD), 10 CFR 51.23. In 2010, the Commission revised the WCD (i.e., WCD Update) to reflect information gained based on experience in the storage of spent nuclear fuel and the increased uncertainty in the siting and construction of a permanent geologic repository for the disposal of spent nuclear fuel.

On June 8, 2012, the United States Court of Appeals for the District of Columbia (D.C.) Circuit (New York v. NRC 2012), in response to a legal challenge to the WCD, vacated the NRC's WCD Update (75 FR 81032 and 75 FR 81037). The court decision was based on grounds relating to aspects of the NEPA. The court decision held that the WCD Update is a major Federal action necessitating either an EIS or a finding of no significant environmental impact (FONSI), and the Commission's evaluation of the risks associated with the storage of spent nuclear fuel for at least 60 years beyond the licensed life for reactor operation is deficient.

The Commission directed (SRM-COMSECY-12-0016 [NRC 2012]) the NRC staff to proceed with a rulemaking that includes the development of an EIS to support an updated WCD within 24 months (by September 2014). The Commission indicated that the EIS used to support the revised rule should build on the information already documented in various NRC studies and reports on the impacts associated with the storage of spent nuclear fuel that were developed as part of the 2010 WCD Update. It should primarily focus additional analyses on the deficiencies identified in the D.C. Circuit's decision. The NRC considers the WCD to be a generic issue that is best addressed through rulemaking and that the NRC rulemaking process provides an appropriate forum for public review and comment on both the draft EIS and the proposed WCD.

The updated rule and supporting EIS is expected to provide the necessary NEPA analyses of waste confidence related human health and environmental issues. As directed by the Commission, the NRC will not issue a new or renewed license for a nuclear power plant or for an Independent Spent Fuel Storage Installation, prior to the resolution of waste confidence related issues. This will ensure that there would be no irretrievable or irreversible resource commitments or potential harm to the environment before waste confidence impacts have been addressed. Section 4.10.1 of this SFES has been revised to include this information.

If the results of the WCD EIS identify information that requires a supplement to this EIS, the NRC staff will perform any appropriate additional NEPA review for those issues before the NRC makes a final licensing decision.

Comment: [T]here are environmental concerns with additional information requested in the FSEIS. Specifically, as outlined in EPA's comment letter dated May 14, 2007, referenced subject, TVA's Draft Supplement Environmental Impact Statement Watts Bar Nuclear Plant Unit 2. [excerpt from EPA's May 14, 2007 letter:] "protecting the environment involves the continuing need for appropriate storage and ultimate disposition of radioactive wastes generated on-site." (**0005-4** [Mueller, Heinz])

Comment: [T]here are environmental concerns with additional information requested in the FSEIS. Specifically, as outlined in EPA's comment letter dated May 14, 2007, referenced subject, TVA's Draft Supplement Environmental Impact Statement Watts Bar Nuclear Plant Unit 2. [excerpt from EPA's May 14, 2007 letter:] "please address the following concerns in the FSEIS: Solid Radioactive Wastes (page 81): The shipping arrangements for Unit 2 after 2008 appear uncertain with Barnwell's closing. Please provide more information on the availability and disposal costs options for Clive, Utah facility, Sequoyah Nuclear Plant or other disposition options under consideration." (**0005-6** [Mueller, Heinz])

Response: Low-level waste (LLW) generated at the WBN site is discussed in Section 4.10.1 of this SFES. LLW includes items that have become contaminated with radioactive material or have become radioactive through exposure to neutron radiation. This waste is typically contaminated protective shoe covers and clothing, wiping rags, mops, filters, reactor water treatment residues, equipment and tools, and laboratory glassware. The LLWs with higher radioactivity are typically found in the water treatment residues, piping that contained reactor coolant and small gauges containing radioactive material. The NRC's waste classification system for LLW is based on the waste's potential hazards, and has specified disposal and waste form requirements for each of the general classes of waste: Class A, Class B and Class C waste. Although the classification of waste can be complex, Class A waste generally contains lower concentrations of long half-lived radioactive material than Class B and C wastes. TVA ships its Class A LLW to a vendor in Oak Ridge, Tennessee, for processing and compaction. Once the waste is processed, it is shipped to a licensed facility in Clive, Utah, for disposal. The NRC anticipates that licensees that do not currently have a disposal pathway would temporarily store Class B and C LLW onsite until offsite disposal locations are available. Several operating nuclear power plants have successfully increased onsite storage capacity in the past in accordance with existing NRC regulations. In addition, the NRC issued information for extended onsite interim storage of LLW in two Regulatory Issue Summaries 2008-12 and 2008-32. TVA has storage capacity for WBN Unit 2's Class B and C LLW at its Sequoyah Nuclear Plant, which is located close to WBN.

As a result of the comment, the NRC staff conducted a search of potential LLW disposal sites and found information on a new disposal facility, the Texas Low-Level Radioactive Waste Disposal Compact Facility, located in Andrews County, Texas which opened on November 10, 2011. The facility is licensed by the State of Texas to dispose of Class A, B, and C LLW. This LLW disposal facility is expected to be available to WBN Unit 2 for the disposal of its LLW if TVA applies for and receives approval from the Texas Low-Level Radioactive Waste Disposal Compact Commission. With the potential availability of this disposal facility, the current LLW handling and storage facilities are expected to be adequate to handle LLW waste generated at WBN Unit 2 without the need to expand storage capacity or ship the waste to Sequoyah Nuclear Plant for storage. The NRC concludes that the storage of LLW at the WBN site will be within current regulatory requirements. There should be no significant issues or environmental impacts associated with interim storage of LLW generated at the WBN site. Section 4.10.1 of this SFES has been revised to include this information.

Comment: [T]here are environmental concerns with additional information requested in the FSEIS. Specifically, as outlined in EPA's comment letter dated May 14, 2007, referenced subject, TVA's Draft Supplement Environmental Impact Statement Watts Bar Nuclear Plant Unit 2. [excerpt from EPA's May 14, 2007 letter:] "please address the following concerns in the FSEIS: Spent Fuel Storage (page 83): Clarify whether the referenced dry cask facility is being processed as a definite project with funding to construct it. Is Unit 2 operation contingent on this facility being constructed? Clarify where the current Unit 1 spent fuel is being stored. Does the capacity for this new facility consider the contingency of Yucca Mountain being indefinitely postponed? Is the data in Table 3-24 in addition to the data given for Unit 1, or the cumulative dimensions, capacity, etc.? **(0005-7** [Mueller, Heinz])

Response: ISFSIs that use dry casks for storage of spent fuel must comply with NRC's safety requirements in 10 CFR Part 72, "Licensing Requirements for the Independent Storage of Spent Nuclear Fuel, High-Level Radioactive Waste, and Reactor-Related Greater Than Class C Waste." As such, ISFSIs safety issues are addressed by these regulatory requirements. Subpart H of 10 CFR Part 72, "Physical Protection," addresses the physical protection requirements for the physical security and safeguards of the ISFSI. In addition, Subpart E of 10 CFR Part 72, "Siting Evaluation Factors," requires a licensee to investigate and assess site characteristics that may affect the safety or environmental impact of the ISFSI. Specifically, the ISFSI must be examined with respect to the frequency and the severity of external natural and man-induced events that could affect the safe operation of the ISFSI. Thus, if an ISFSI were to be sited under the electric transmission lines, the licensee's evaluation would have to include the potential impact to the storage casks from the power lines to ensure that it would not affect the safe operation of the ISFSI. According to the TVA ER (TVA 2008), WBN Unit 1 is currently storing fuel in the spent fuel pool, with plans to construct and operate and ISFSI. The operation of WBN Unit 2 would accelerate the need for TVA to license and ISFSI at the Watts Bar site. An ISFSI would need to be licensed sometime around 2015. The licensing and operation of an ISFSI was not addressed in this SFES. If TVA submits an application to build an ISFSI, then TVA must demonstrate that it has the financial qualifications to build, operate, and decommission an ISFSI, as required by 10 CFR Part 72.22(e) and Part 72.30. No changes were made to the SFES as a result of this comment.

E.2.14 Comments Concerning Decommissioning

Comment: I think that one of the things that the community needs to realize is that decommissioning of these reactors, both Unit 1 and Unit 2, is going to be a huge task on down the road. Now the decommissioning fund, TVA and the other utilities put money into it all the time, but they invest that money just like anybody that has a little money tries to invest it in the stock market or wherever to do the best they can. Well, when the stock market took a big hit, that decommissioning fund took the same hit. And there's just simply not as much money as you need to decommission these reactors. **(0003-9-1** [Safer, Don])

Response: According to Title 10 of the Code of Federal Regulations (10 CFR) Part 50.75(b)(2), the amount of funding necessary for decommissioning must be updated annually. 10 CFR Part 50.75(b)(3) requires that the adjusted amount must be covered as described in the regulations. Therefore, the applicant, and subsequently the licensee, will be required to certify annually that the decommissioning funds will be available when the reactors are decommissioned. No changes have been made to this SFES in response to this comment.

E.2.15 Comments Concerning Related Federal Projects

Comment: Page/Line 2-88/7: The following statement is no longer accurate: "The [Watts Bar] fossil plant currently is not operating, but could be reactivated in the future." TVA demolished the Watts Bar Fossil plant in December 2011. Accordingly, TVA recommends this section be updated with information available in TVA's environmental assessment of Watts Bar Fossil Plant Deconstruction http://www.tva.gov/environment/reports/wbf_deconstruction/ (**0010-6** [Brickhouse, Brenda])

Response: Sections 1.1, 2.1, 2.5.2, 2.7.2, 2.9, 3.2.4, 4.5, 4.14.7, and 7.2 of this SFES have been updated to reflect the recent demolition of the Watts Bar Fossil Plant.

E.2.16 Comments Concerning Cumulative Impacts

Comment: The Watts Bar – this book says that no other new nuclear facilities within 50 miles are being considered. That's just false. Oak Ridge is within 50 miles. TVA is on record as being far beyond considering small modular reactors. They're in communication with the NRC, other branches of the NRC, daily. There's meetings; there's been several meetings this month on SMRs and TVA. They plan to build -- first they planned to build six up at the Clinch River site; now I think it's down to two. These are 125 to 150 megawatt reactors that are modular built and they're sunk into a hole 150 feet deep in the ground, the same karst geology. So I wish that would be corrected or you check that. I know there was some discussion about whether that's accurate or not. I don't see -- if you speak English, considered means considered. And they're certainly being considered. (**0003-7-1** [Safer, Don])

Response: Section 4.14 of this SFES considers cumulative impacts of other past, present, and reasonably foreseeable future actions. Although, at this time, no application is pending before the NRC that relates to small modular reactors at Oak Ridge, the SFES lists the small modular reactors at Oak Ridge in Chapter 4 as a potential future project.

Comment: I get this everywhere I go across the United States for the last 35 plus years. Will that thing blow up?

And my answer has always been no, there is no nuclear special grades material in our commercial reactors in the United States that is used to make nuclear weapons grade material.

That is all made at Savannah River. Don't even concern yourself.

Well, now I have to eat crow because the Department of Energy got into an agreement with NRC and TVA to shut down Savannah River and make their special nuclear grade material at Watts Bar....But I happen to know back in 1978 we didn't even consider having a special core in Watts Bar Unit 1 or 2. It wouldn't in the Environmental Impact Statement to even look at that question of that special nuclear grade material in the rods to go into the core for the Department of Energy then to reprocess.

So what I would expect or what NRC taught me to do back in 1978, come up with some good words. And so even though I haven't had a chance to read this, one of the good things was you are here to find out if you missed an issue or has something changed. Well, something has changed. And it could be for the good. I'm all for it; I'm not against it.

What I'd like to see are words added to the Final Environmental Impact Statement that address the issue of the relationship with the Department of Energy and the core that's being used that everything's okay. That's what I'd like to see. Good words to address it. And I don't know, maybe it's addressed in here. But if it's not addressed at all, then I think that leaves the door open for somebody later on to raise an issue and I don't like issues that fall under the area of assumptions or unverified assumptions. I would like to know that NRC addressed it and everything is okay. (**0004-1-3** [Riden, David])

Response: The commenter requests that the NRC address tritium production that is currently taking place at WBN Unit 1 under a contract with the U.S. Department of Energy (DOE). As noted in Section 4.14.8 of this SFES, there are no plans for WBN Unit 2 to produce tritium for DOE. A request on the part of TVA to produce tritium in WBN Unit 2 would require a change to the operating license for WBN Unit 2 and result in an environmental assessment prior to the license change. The NRC staff has evaluated tritium production at WBN Unit 1. On September 15, 1997, NRC issued an amendment (NRC 1997) which authorized the use of tritium test assemblies at WBN Unit 1. In 2001, TVA submitted a request to amend the operating license of WBN Unit 1 to allow tritium production at the facility (NRC 2001). The NRC staff reviewed the application and on August 26, 2002, published an environmental assessment and finding of no significant impact in the Federal Register (67 FR 54826). On September 23, 2002, the NRC issued an amendment (NRC 2002) to the WBN Unit 1 operating license addressing the changes that needed to be made to safely produce tritium at WBN. No changes have been made to this SFES in response to this comment.

E.2.17 Comments Concerning Alternatives – Energy

Comment: Now if you look at the chart, here is nuclear, here is coal, oil, gas. Solar thermal uses a lot of water, so does biofuel. However, solar photovoltaic and wind, look, you can't even see a line for how much water is required to produce energy in those ways. **(0003-6-3** [Ferris, Kathleen])

Comment: And one way we can make sure that other industry and agriculture and people have enough clean water is to use clean means of producing energy. And we've got them. I couldn't get up here if I didn't know the technology already exists for wind and solar energy production. And I get this argument with people all the time, both at these meetings and elsewhere. Say, oh, they can't produce enough. Well, if we put the billions of dollars into solar and wind energy that we are putting into nuclear energy, we could do it. We went to the moon. We're sending off explorers into space. The technology is there. What's lacking is the will. **(0003-6-7** [Ferris, Kathleen])

Response: *These comments concern the water use and water-quality impacts associated with nuclear power and other energy-generation alternatives. Both water-use and water-quality impacts from the operation of WBN Unit 2 are discussed in Section 4.2.2 of this SFES. Comment 0003-6-3 correctly notes that nuclear, coal-, oil-, biofuel- and natural-gas-fired generation facilities generally require water for cooling, while solar photovoltaic and wind-powered generation do not. Section 7.2.4.2 of this SFES discusses water use and water quality impacts associated with the natural-gas-fired generation alternative, and Section 7.2.5.3 discusses water-use and water-quality impacts associated with a combination of natural-gas-fired generation, biomass-fired generation, and wind power generation. These impacts are summarized in Tables 7-1 and 7-2 of this SFES. Section 7.2.3 discusses the viability and environmental impacts of other power-generating alternatives including coal, biofuels, and solar and wind power. Insofar as the comments call for developmental funding of wind- and solar-power deployment, the comments are not relevant to the NRC staff's environmental review. The NRC is not involved in funding energy development. No changes have been made to this SFES in response to these comments.*

E.2.18 General Comments in Opposition to the Licensing Action

Comment: I realize that there's nothing that I can say here today that's going to turn this thing around. I feel like David and Goliath. And I don't have the magic stone that he had and I wish I did. And I wouldn't throw it to hurt anybody, but I would sure throw it to stop this reactor because I think the effects of it -- unfortunately if something goes wrong, something go monumentally wrong. **(0003-3-2** [Safer, Don])

Comment: And of course we [Bellefonte Efficiency and Sustainability Team...a chapter of the Blue Ridge Environmental Defense League] don't care much for the idea of building yet another nuclear plant in the Chattanooga region. (**0003-4-1** [Kurtz, Sandy])

Comment: And I just think that we should not continue to add more risk to what's happening already. We already had six nuclear reactors in this area and we don't really need another one. (**0003-4-7** [Kurtz, Sandy])

Response: These comments express opposition to the operation of WBN Unit 2. No changes have been made to this SFES in response to these comments.

E.2.19 Comments Concerning Issues Outside Scope – NRC Oversight

Comment: And the other thing is you're relying on these local state agencies through your agreement state letter to do a lot of your work for you. That's just so that you're further removed. The state is not testing for a lot of this stuff. So there's no records and no benchmark for you to start with. And you're using old data to do that. Well, somebody's got to go back in there and do some real work instead of just dropping it into this. (**0003-2-6** [Harris, Ann])

Comment: [H]ow far TVA can go that you won't let them go any further? You're saying in here I see about the tritium. TVA managed to fill a leak so they're containing the tritium. ..... Now the State of Tennessee, TDEC, they're not even testing for anything like that. They don't look at it. They say it's not -- they don't have the money to test for it. Well, whenever I question an NRC person, they look at me and say, "That's TDEC and we don't have authority over them." Well, you gave them a letter of authority under the agreement state letter in ..... But the thing is that there has to be a limit of how far they can go. (**0003-5-1** [Harris, Ann])

Response: These comments question the adequacy of the NRC oversight and reliance on local or state agencies. Nuclear power plants are required to have radiological environmental monitoring programs and effluent monitoring programs that measure the amount of radioactive material that is released from each site. The licensees are required to submit annual reports that describe the amount of radioactive material released in the previous year and a dose assessment to demonstrate that the licensees are complying with the NRC dose rate regulations cited in 10 CFR Part 20. The environmental and effluent programs are also subject to inspections as part of the NRC reactor oversight process. The NRC has not relinquished jurisdiction over nuclear power plant operations to the State of Tennessee through the Agreement State Program. For more information on State monitoring programs, contact the Tennessee Department of Environment and Conservation.

E.2.20   Comments Concerning Issues Outside Scope – Safety

Comment: It is my understanding that TVA's nuclear power plants cannot withstand a 7 or great earthquake. Please rethink continuing to build/refurbish such a power plant until such time as one can be built to withstand a severe earthquake. Watts Bar is not so very far away from the New Madrid fault line. (**0001-1** [Budnick, Donna])

Response: Seismic safety and emergency planning issues are outside the scope of the environmental review, but they are addressed in NRC's parallel safety review which is documented in a Final Safety Evaluation Report (FSER). Nuclear power plants are built to withstand environmental hazards, including earthquakes. Even those plants that are located outside of areas with extensive seismic activity are designed for safety in the event of such a natural disaster. The NRC requires all of its licensees to take seismic activity into account when designing and maintaining nuclear power plants. When new seismic hazard information becomes available, the NRC evaluates the new data and models and determines whether any changes are needed at plants. On March 9, 2012, the Near Term Task Force transmitted Orders and 5054F demands concerning seismic activities from its review of insights from the Fukushima Dai-Ichi accident.

E.2.21   Comments Concerning Issues Outside Scope – Security and Terrorism

Comment: The other thing is that I want to know what kind of security is around the intake pumping station. And I'm talking about serious security and about the outfalls. (**0003-5-2** [Harris, Ann])

Comment: If you read this document, you'll see – I don't think you'll see the word Fukushima or even Chernobyl in there. As I said earlier, they say it's factored into the computer models, but that sure is sanitizing the realities. Page 6-15, "It is noted that the risks from deliberate aircraft impacts were explicitly excluded since this was being considered in other forms along with other sources of sabotage." I don't know where the other form is. I asked an individual, a couple of individuals, with the NRC here. They were going to get back to me on that. But again deliberate aircraft impacts were explicitly excluded from this document. (**0003-7-11** [Safer, Don])

Comment: So those kinds of things, plus terrorist attacks, that's an external event. They just don't even factor that into this. And I'm sorry that we live in a world where that has to be factored in, but we all know that it does. And that these things are -- they're the biggest target for a terrorist that you can imagine and the effects would be -- they'd put 9/11 into a footnote of history almost. (**0003-7-7** [Safer, Don])

Comment: So these things are huge target for external events and that really needs to be factored into the design. And believe me, in 1970, they weren't factoring in the possibility of a terrorist attack on U.S. soil in the design of the containment structure which, as I've said, is already thinner than other nuclear power plants of that era. **(0003-7-8** [Safer, Don])

Comment: There was a Newsweek article either last week or the week before of an FBI undercover agent who was working with extremist groups in the South. And some of the people he encountered were planning an attack on Browns Ferry. So this isn't some hypothetical. I mean there are people out there who would love to attack these reactors and probably many of them don't have any idea of what the real consequences would be. **(0003-8-1** [Ferris, Kathleen])

Response: Comments related to security and terrorism are safety-related issues and are outside the scope of the NRC staff's environmental review. The NRC is devoting substantial time and attention to terrorism-related matters, including coordination with the U.S. Department of Homeland Security. As part of its mission to protect public health and safety and the common defense and security pursuant to the Atomic Energy Act, NRC staff is conducting vulnerability assessments for the domestic utilization of radioactive material. Since the events of September 11, 2001, the NRC has identified the need for license holders to implement compensatory measures and has issued several orders to license holders imposing enhanced security requirements. Finally, the NRC has taken actions to ensure that applicants and license holders maintain vigilance and a high degree of security awareness. Accordingly, the NRC will continue to consider measures to prevent and mitigate the consequences of acts of terrorism in fulfilling its safety mission. Additional information about the NRC staff's actions regarding physical security since September 11, 2001, can be found on the NRC's public web site http://www.nrc.gov.

E.2.22 General Editorial Comments

Comment: Page/Line xviii /3: Insert space between "used" and "information" **(0010-1** [Brickhouse, Brenda])

Comment: Page/Line 2-80/27: In the statement "...NPDES temperature limits for WBN outfalls to the Tennessee River are at or below 95ºC, which...." the unit of measure should be changed to Fahrenheit. **(0010-8** [Brickhouse, Brenda])

Response: These changes were made in this SFES.

Comment: Page/Line 1-5 I 8: Text indicates there is a 45-day comment period on the DFES. On page xix, line 8, text indicates a 75-day comment period. **(0010-3** [Brickhouse, Brenda])

Response: The text in Section 1.2 was updated to reflect the addition of the 30-day extension to the original 45-day comment period.

Comment: The electronic version of many of the figures in chapters 2, 3, and 4 are difficult to interpret. In particular, figures 2-4 and 2-5 (pgs. 2-15 and 2-16) are inadequate to allow the reviewer to identify the important water related features. We suggest that the final version include legible versions of these figures; this will allow the public to better assess the adequacy of the information provided in the text. (**0009-1** [Hogue, Gregory])

Response: The final electronic SFES document production process was modified to ensure that figures are much more legible than in the electronic draft SFES, and several figures were replaced with updated versions. Figure 2-4 was modified to show only the landcover information important to the review; Figure 2-5 (wetlands and streams) was modified to more clearly identify the water-related features.

Comment: Page/Line 2-101/10-12 and 18-21: As the DFES states, in 2008, TVA submitted its Final Supplemental Environmental Impact Statement (FSEIS) dated 2007, as an environmental report (ER) in the application. In the list of references for Chapter 2, "TVA 2007a" and "TVA 2008a" are both citations for the same FSEIS. In the DFES, TVA's document is referred to as TVA's FSEIS and TVA's ER. TVA suggests that a single reference for TVA's 2007 FSEIS would improve clarity for the reader. (**0010-2** [Brickhouse, Brenda])

Response: This SFES text was modified as suggested: to reference TVA 2008a instead of TVA 2007a. The reference for TVA 2007a was dropped from the list of references in Chapter 2. The document is now consistently referred to as the TVA ER throughout the SFES.

E.3 References

10 CFR Part 20. Code of Federal Regulations, Title 10, *Energy*, Part 20, "Standards for Protection Against Radiation."

10 CFR Part 50. Code of Federal Regulations, Title 10, *Energy*, Part 50, "Domestic Licensing of Production and Utilization Facilities."

10 CFR Part 51. Code of Federal Regulations, Title 10, *Energy*, Part 51, "Environmental Protection Regulations for Domestic Licensing and Related Regulatory Functions."

10 CFR Part 72. Code of Federal Regulations, Title 10, Energy, Part 72, "Licensing Requirements for the Independent Storage of Spent Nuclear Fuel, High-Level Radioactive Waste, and Reactor-Related Greater Than Class C Waste."

67 FR 54826. August 26, 2002. "Tennessee Valley Authority, Watts Bar Nuclear Plant, Unit 1; Environmental Assessment and Finding of No Significant Impact." *Federal Register.* U.S. Nuclear Regulatory Commission.

75 FR 81032. December 23, 2010. "10 CFR Part 51 Consideration of Environmental Impacts of Temporary Storage of Spent Fuel after Cessation of Reactor Operation; Waste Confidence Decision Update; Final Rules." *Federal Register.* U.S. Nuclear Regulatory Commission.

75 FR 81037. December 23, 2010. "Waste Confidence Decision Update." *Federal Register.* U.S. Nuclear Regulatory Commission.

76 FR 3392. January 19, 2011. "Endangered and Threatened Wildlife and Plants; Endangered Status for the Sheepnose and Spectaclecase Mussels; Proposed Rule." *Federal Register.* Department of the Interior, Fish and Wildlife Service.

76 FR 48722. August 9, 2011. "Endangered and Threatened Wildlife and Plants; Endangered Status for the Cumberland Darter, Rush Darter, Yellowcheek Darter, Chucky Madtom, and Laurel Dace." *Federal Register.* Department of the Interior, Fish and Wildlife Service.

76 FR 70130. November 10, 2011. "Environmental Impact Statements; Notice of Availability." *Federal Register.* U.S. Environmental Protection Agency.

76 FR 70169. November 10, 2011. "Draft Supplement 2 to Final Environmental Statement Related to the Operation of Watts Bar Nuclear Plant, Unit 2; Tennessee Valley Authority." *Federal Register.* U.S. Nuclear Regulatory Commission.

76 FR 71559. November 18, 2011. "Environmental Impact Statements; Notice of Availability." *Federal Register.* U.S. Environmental Protection Agency.

76 FR 80409. December 23, 2011. "Draft Supplement 2 to Final Environmental Statement Related to the Operation of Watts Bar Nuclear Plant, Unit 2; Tennessee Valley Authority." *Federal Register.* U.S. Nuclear Regulatory Commission.

Baxter, D.S., J.P. Buchanan, G.D. Hickman, J. Jenkinson, J.D. Milligan, and C.J. O'Bara. 2010. *Aquatic Environmental Conditions in the Vicinity of Watts Bar Nuclear Plant During Two Years of Operation, 1996-1997.* June 1998, Revised June 7, 2010. Tennessee Valley Authority, Norris, Tennessee. Accession No. ML11325A411.

New York v. NRC, 681 F.3d 471 (D.C. Cir. 2012). Accession No. ML12191A407.

[NRC] U.S. Nuclear Regulatory Commission. 1997. Letter from R.E. Martin to Tennessee Valley Authority dated September 15, 1997, regarding "Issuance of Amendment on Tritium Producing Burnable Absorber Rod Lead Test Assemblies (TAC No. M98615)." Accession No. ML020780128.

[NRC] U.S. Nuclear Regulatory Commission. 2001. *Watts Bar Nuclear Plant (WBN) – Unit 1 – Revision of Boron Concentration Limits and Reactor Core Limitations for Tritium Production*

Cores (TPCs) – Technical Specification (TS) Change No. TVA-WBN-TS-00-015. Accession No. ML012390115.

[NRC] U.S. Nuclear Regulatory Commission. 2002. Letter from L.M. Padovan to Tennessee Valley Authority dated September 23, 2002, regarding "Watts Bar Nuclear Plant, Unit 1– Issuence of Amendment to Irradiate up to 2304 Tritium-Producing Burnable Absorber Rods in the Reactor Core (TAC No. MB1884)." Accession No. ML022540925.

[NRC] U.S. Nuclear Regulatory Commission. 2010. Letter from P.D. Milano to Nuclear Generation Development and Construction dated July 2, 2010, regarding "Watts Bar Nuclear Plant, Unit 2 – Program for Construction Refurbishment (TAC No. ME1708)." Accession No. ML101720050.

[NRC] U.S. Nuclear Regulatory Commission. 2011a. *Draft Final Environmental Statement Related to the Operation of Watts Bar Nuclear Plant, Unit 2.* Supplement 2. Docket Number 50-391. NUREG-0498 Supplement 2, Washington, D.C. Accession Nos. ML11298A094, ML11298A095.

[NRC] U.S. Nuclear Regulatory Commission. 2011b. *Recommendations for Enhancing Reactor Safety in the 21st Century: The Near-Term Task Force Review of Insights from the Fukushima Dai-Ichi Accident.* SECY-11-0093, Washington, D.C. Accession No. ML111861807.

[NRC] U.S. Nuclear Regulatory Commission. 2012. "Approach for Addressing Policy Issues Resulting from Court Decision to Vacate Waste Confidence Decision and Rule." COMSECY-12-0016, Washington, D.C. July 9, 2012. Accession No. ML12180A424.

Simmons, J.W. 2010. *Analysis of Fish Species Occurrences in Chickamauga Reservoir – A Comparison of Historical and Recent Data.* Tennessee Valley Authority, Biological and Water Resources, Chattanooga Tennessee. Accession No. ML11325A411.

Simmons, J.W.. 2011. *Biological Monitoring of the Tennessee River Near Watts Bar Nuclear Plant Discharge, Autumn 2010.* Tennessee Valley Authority, Aquatic Monitoring and Management, Chattanooga, Tennessee. Accession No. ML12073A367.

Simmons, J.W. and D.S. Baxter. 2009. *Biological Monitoring of the Tennessee River Near Watts Bar Nuclear Plant Discharge, 2008.* Tennessee Valley Authority, Aquatic Monitoring and Management, Chattanooga, Tennessee. Accession No. ML093510802.

Simmons, J.W., D.S. Baxter, and G.P. Shaffer. 2010. *Biological Monitoring of the Tennessee River Near Watts Bar Nuclear Plant Discharge, Autumn 2009.* Tennessee Valley Authority, Aquatic Monitoring and Management, Chattanooga, Tennessee. Accession No. ML12073A365.

[TDEC] Tennessee Department of Environment and Conservation. 2011. *State of Tennessee NPDES Permit No. TN0020168*. Accession No. ML11215A099.

Third Rock Consultants, LLC. 2010. *Mollusk Survey of the Tennessee River near Watts Bar Nuclear Plant (Rhea County, Tennessee)*. Prepared for Tennessee Valley Authority, October 28, 2010, revised November 24, 2010. Lexington, Kentucky. Accession No. ML11325A411.

[TVA] Tennessee Valley Authority. 1998. *Watts Bar Nuclear Plant Supplemental Condenser Cooling Water Project, Environmental Assessment*. Knoxville, Tennessee. Accession No. ML12170A104.

[TVA] Tennessee Valley Authority. 2004. *Reservoir Operations Study Final Programmatic Environmental Impact Statement*. Tennessee Valley Authority In cooperation with U.S. Army Corps of Engineers and U.S. Fish and Wildlife Service. Accession Nos. ML041100586, ML041100585, ML041100590, ML041100588.

[TVA] Tennessee Valley Authority. 2008. *Final Supplemental Environmental Impact Statement; Completion and Operation of Watts Bar Nuclear Plant Unit 2, Rhea County, Tennessee* submitted to NRC as the TVA Environmental Report for an Operating License, Knoxville, Tennessee. Accession No. ML080510469.

[TVA] Tennessee Valley Authority. 2009. *Watts Bar Nuclear Plant (WBN) - Unit 2 - Final Safety Analysis Report (FSAR)*, Amendment 94. Spring City, Tennessee. Accession No. ML092460774.

[TVA] Tennessee Valley Authority. 2010. Letter from Masoud Bajestani (Watts Bar Unit 2, Vice President) to U.S. Nuclear Regulatory Commission, dated July 2, 2010, "Watts Bar, Unit 2 – Submittal of Additional Information Requested during May 12, 2010, Request for Additional Information (RAI) Clarification Teleconference Regarding Environmental Review." Accession No. ML101930470.

[TVA] Tennessee Valley Authority. 2011a. Letter from J.W. Shea (Tennessee Valley Authority) to U.S. Nuclear Regulatory Commission dated December 27, 2011, "Notification of National Pollutant Discharge Elimination System Permit Modification." Chattanooga, Tennessee. Accession No. ML12009A072.

[TVA] Tennessee Valley Authority. 2011b. *Comparison of 2010 Peak Spawning Seasonal Densities of Ichthyoplankton at Watts Bar Nuclear Plant at TRM 528 with Historical Densities during 1996 and 1997*. Tennessee Valley Authority, Environmental Science and Resources. April 2011, revised November 2011. Accession No. ML11329A001.

[TVA] Tennessee Valley Authority. 2011c. *Hydrothermal Effects on the Ichthyoplankton from the Watts Bar Nuclear Plant Supplemental Condenser Cooling Water Outfall in Upper Chickamauga Reservoir' Report.* Accession No. ML110400384.

[TVA] Tennessee Valley Authority. 2011d. *Fish Impingement at Watts Bar Nuclear Plant Intake Pumping Station Cooling Water Intake Structure during March 2010 through March 2011.* Tennessee Valley Authority. April 2011. Accession No. ML11124A025.

[TVA] Tennessee Valley Authority. 2011e. Letter from David Stinson (Watts Bar Unit 2) to U.S. Nuclear Regulatory Commission dated May 19, 2011, "Watts Bar Nuclear Plant (WBN) Unit 2 – Intake Pumping Station Water Velocity – Response to Request for Additional Information." Accession No. ML11143A083.

TVA] Tennessee Valley Authority. 2012. *Estimates of Entrainment of Fish Eggs and Larvae at Watts Bar Nuclear Plant at Tennessee River Mile 528 from March 2010 through March 2011. May 2012.* Accession No. ML12166A068.

[USFWS] U.S. Fish and Wildlife Service. 2012. Letter from M. E. Jennings, USFWS, to U.S. Nuclear Regulatory Commission. December 20, 2011. "FWS# 12-CPA-00081. Watts Bar Nuclear Plant, Unit 2 – Biological Assessment for Section 7 Consultation Related to the Operating License Application, Rhea County, Tennessee." Accession No. ML12004A167.

[USGS] U.S. Geological Survey. 2006. *The Cranes: Status Survey and Conservation Action Plan for the Sandhill Crane (*Grus canadensis*).* Northern Prairie Wildlife Research Center. Online Report available at http://www.npwrc.usgs.gov/resource/birds/cranes/gruscana.htm.

E.4 Copies of Transcripts and Correspondence Containing Public Comments

Official Transcript of Proceedings

NUCLEAR REGULATORY COMMISSION

Title: Watts Bar Nuclear Plant EIS
                     Public Meeting: Afternoon Session

Docket Number: (n/a)

Location: Sweetwater, Tennessee

Date: Thursday, December 8, 2011

Work Order No.:   NRC-1313   Pages 1-53

NEAL R. GROSS AND CO., INC.
Court Reporters and Transcribers
1323 Rhode Island Avenue, N.W.
Washington, D.C. 20005
(202) 234-4433

1

1   UNITED STATES OF AMERICA

2   NUCLEAR REGULATORY COMMISSION

3   + + + + +

4   PUBLIC MEETING TO DISCUSS

5   DRAFT SUPPLEMENT TO THE

6   FINAL ENVIRONMENTAL STATEMENT

7   FOR WATTS BAR NUCLEAR PLANT, UNIT 2

8   + + + + +

9   Thursday, December 8, 2011

10 Magnuson Hotel

11 1421 Murrays Chapel Road

12 Sweetwater, Tennessee

13 + + + + +

14 Afternoon Session

15 2:00 p.m. to 4:00 p.m.

16

17

18

19

20

21

22

23

24

25

1   you to sign in. And the individuals who signed in and

2   told us before the meeting started that they wanted to

3   ask questions put a little check mark beside their

4   name. I've got the list.

5   So Ms. Ferris, we're going to start with

6   you.   And then Mr. Safer, you'll be the second

7   question.   The reason for that is simply that Ms.

8   Ferris' name is first on there and you're the second

9   name on there. Okay.

10   And what we'll do, if you don't mind,

11   we'll going to start this with a three-minute time

12   limit for questions and answers. That's to give

13   anyone else here a chance. If we go through the first

14   set of questions and there's nobody else, we'll throw

15   it open for a second round, a third round, however

16   many we need. And then we'll go into -- once we're

17   through with the questions and answers, we'll go into

18   the comment period. Okay? Very good.

19   Justin here is going to help me. He will

20   raise his hand whenever we get to three minutes. So

21   again, just give everybody an opportunity to have

22   their time to talk. Okay?

23   Ms. Ferris. Would you like to stand up?

24   Would you like me to hold that for you?

25   MS. FERRIS: No. My question has to do

NEAL R. GROSS
COURT REPORTERS AND TRANSCRIBERS
1323 RHODE ISLAND AVE., N.W.
(202) 234-4433   WASHINGTON, D.C. 20005-3701   (202) 234-4433

May 2013   E-107   NUREG-0498, Supp 2

21

1 with the geology, the underground structures that this

2 plant has been built upon. And my question is whether

3 this is karst, k-a-r-s-t. Don't ask me what that

4 stands for. But it's limestone. And I'm wondering

5 whether this is being built and has been built on

6 limestone topography?

7 MR. MILANO: Sorry about that. There were

8 too many switches. Although today we're not here to

9 discuss the safety analysis that was done for the

10 plant, the aspects of geology were discussed and can

11 be seen in Section 2 of the staff's Safety Evaluation

12 Report of which is in it's also an NRC Regulation.

13 No, that's the Final Environmental

14 Statement, Gene. The Safety Evaluation Report is also

15 another NUREG document. And it's NUREG-0847, zero

16 eight four seven. And it's -- you can observe it on

17 the NRC's website. And both the original that was

18 done to support the operation of Watts Bar Units 1 and

19 2 when TVA at the time was proposing licensing both

20 units at the same time.

21 And it has been supplemented. It was

22 supplemented through Supplement 20 to support Watts

23 Bar Unit 1 and right now we're at Supplement 25 --

24 Supplements 21 through 25 have been specifically for

25 Watts Bar Unit 2 operation. And I'm sorry. I'm not a

NEAL R. GROSS
COURT REPORTERS AND TRANSCRIBERS
1323 RHODE ISLAND AVE., N.W.
(202) 234-4433   WASHINGTON, D.C. 20005-3701   (202) 234-4433

NUREG-0498, Supp 2   E-108   May 2013

22

1 geologist or seismologist. So all I can do is refer

2 you to those documents and it's discussed in there.

3 And also there's some -- there is information in TVA's

4 Final Safety Analysis Report and also in Section 2 on

5 site characteristics that describes that.

6 　　　　　MR. CARPENTER: Mr. Safer.

7 　　　　　MR. SAFER: I have a couple of questions.

8 I don't think it will take three minutes. The first

9 one is since the public comment period is over

10 December 27th, right in the middle of the holidays, I

11 think that's extremely inconvenient. Those of us that

12 think the NRC is not that cooperative to the public

13 comments feel like it's by design. But we would ask

14 for an extension of 45 days so that people have an

15 opportunity to comment on this outside of the holiday

16 period. And I don't know that that can be granted

17 today, but I think that's a formal request, as formal

18 as I can get right here.

19 　　　　　So I don't know if there's a response to

20 that. And I have another question.

21 　　　　　MR. SUSCO: Very reasonable request, and

22 we've in many other proceedings we've entertained

23 extensions. Andrea, can you speak to -- we've one of

24 our lawyers here -- what the process is as far as

25 requesting an extension?

NEAL R. GROSS
COURT REPORTERS AND TRANSCRIBERS
1323 RHODE ISLAND AVE., N.W.
(202) 234-4433　　　WASHINGTON, D.C. 20005-3701　　　(202) 234-4433

May 2013　　　　　　　　E-109　　　　　　　　NUREG-0498, Supp 2

23

1   MS. JONES: Sir, I am a lawyer for the
2 NRC. I'm actually not very clear on what the
3 extension process is. But I do know that the timeline
4 -- I don't mean this to patronize you either. But I
5 do know that the timeline is set according to
6 Regulation. But as far as extensions are concerned, I
7 think that would be a matter we'd have to take back to
8 the office and ask them. That would be a decision for
9 the office to make.

10   MR. SAFER: And how would we get the
11 answer? When would we expect an answer?

12   MS. JONES: I'm not sure, because I'm not
13 sure what the process for getting an extension would
14 actually be. We don't necessarily have a process for
15 that in our Regulations, not that I can recall.

16   MR. SUSCO: We will grab your information
17 and we can discuss that afterwards about the exact
18 process for making that occur.

19   MR. SAFER: Okay, well, that would be
20 something that of course other members of the public
21 would be interested in.

22   Then another technical question. In 2.6,
23 the radiological environment, it references a report,
24 Annual Radiological Environmental Operating Report,
25 RAMP, and also the Annual Radioactive Affluent Release

24

Report. I believe those are from TVA, but -- and I know you all are not TVA. But I'm just wondering how to get a hold of this document.

MR. MILANO: Both of those documents are available through -- they were submitted on the docket by TVA. And if you're familiar with our Agency-wide Document Access and Management System, ADAMS, you can find them in ADAMS. They're probably -- since both of those documents are generally to support operation of the facility, probably use the Watts Bar Unit 1 docket number which is 50-390. So in ADAMS they use a bunch of zeros so you go like zero five zero zero zero three nine zero.

THE REPORTER: Zero five --

MR. MILANO: Zero five zero zero zero three nine zero.

MR. SUSCO: If you look in the references to that particular chapter -- actually any chapter -- almost everything that we reference will give that specific ADAMS number. So if you go on our website and you go in ADAMS, type in that number, it'll pull up that report for you. So look in the references for Chapter 2 for the ADAMS number.

MR. CARPENTER: Thank you. Other questions?

25

1   MS. HARRIS: We've got a new guy here. We

2   have a lot of people from headquarters that should

3   come here more often to see what's really going on.

4   The other thing is -- my name is Ann Harris. And I'm

5   with We the People. And I want to second Mr. Safer's

6   request that an extension be given to this because

7   that right now I'm looking at 13 different comments

8   that has been requested that I make on. And all of

9   them are due within like 6 days of each other.

10   And then when we can't get the documents,

11   because this document is -- it's a nightmare. I'm

12   seeing a lot of information that has not been updated

13   from 1972. I'm a local resident so I know.

14   So I don't know where you got your

15   information. You may have gotten it from different

16   agencies. You said federal, state, and local. Well,

17   some of these with information in here that whenever I

18   went, I got different information. So I'm having a

19   hard time dealing with your numbers and the

20   information that you're giving as opposed to what I'm

21   getting from the same agencies.

22   Because they said, "Oh, they've already

23   been here." And I asked. I said, "Give me the

24   information you gave them." And they said, "Oh, we

25   can't do that." So I said, "Okay." So they're making

1 me do FOIA requests over your documents that you

2 requested to put into this. So this is a nightmare.

3 And I don't know how that somebody that

4 can read and write out of the third grade could

5 discover what you've done here. It's so convoluted.

6 It's really, really a written nightmare. Now I've

7 been through about half way through it. And it's

8 taken me two and a half weeks. And I spent at least

9 two to four hours at night trying to go through it.

10 But some of the information -- let's go

11 over one issue. You talk about the tritium in the

12 water. And I know nobody don't want to hear about it.

13 And you're here sick and tired of hearing me talk

14 about it. I'm sick and tired of having to deal with

15 it. But the other thing is, you've not dealt with the

16 tritium. You call it a spill.

17 Three years of over the limit and then you

18 didn't even do anything to TVA about it to begin with.

19 That is still sitting out there. Don't tell me the

20 tritium is gone because I know better.

21 And the other thing is you're relying on

22 these local state agencies through your agreement

23 state letter to do a lot of your work for you. That's

24 just so that you're further removed. The state is not

25 testing for a lot of this stuff. So there's no

NEAL R. GROSS
COURT REPORTERS AND TRANSCRIBERS
1323 RHODE ISLAND AVE., N.W.
(202) 234-4433   WASHINGTON, D.C. 20005-3701   (202) 234-4433

May 2013   E-113   NUREG-0498, Supp 2

27

1 records and no benchmark for you to start with. And

2 you're using old data to do that. Well, somebody's

3 got to go back in there and do some real work instead

4 of just dropping it into this.

5   So I'm making a formal request that we

6 have an extension for 45 days. And this is strictly

7 up to the staff, ma'am. I don't know how long you've

8 been in General Counsel's office. But this is just --

9 it's not something that's a big deal. These guys

10 here, they can do it today or Bob Petty has the

11 authority to say I will see that the extension goes

12 through.

13   Thank you.

14   MR. CARPENTER: Thank you. Appreciate the

15 comments.

16   Any questions? No other questions for the

17 staff?

18   And now we'll go on to the comment section

19 since we've basically already have been in comment's

20 section. So do we have comments?

21   MS. FERRIS: Can we comment more than

22 once?

23   MR. CARPENTER: You may comment as many

24 times as you wish. But we'll go one time each. And

25 then everybody else has a fair term before we go to

NEAL R. GROSS
COURT REPORTERS AND TRANSCRIBERS
1323 RHODE ISLAND AVE., N.W.
(202) 234-4433   WASHINGTON, D.C. 20005-3701   (202) 234-4433

NUREG-0498, Supp 2   E-114   May 2013

28

1   the second.

2   MR. SAFER: Is there a time limit on these

3   comments?

4   MR. CARPENTER: Want to say again, three

5   minutes.

6   MS. FERRIS: Well, my first comment is

7   that last Christmas, I had 20 people coming to dinner.

8   And I was spending time trying to get the NRC to give

9   us a public hearing on the German waste that's being

10   imported into Tennessee to be burned at Oak Ridge.

11   And I had -- I was working on my computer by night and

12   cooking by day.

13   It is not a dot away. If you put these

14   comments and these deadlines right at Christmas time,

15   you're going to get a lot fewer of them. And you

16   probably are aware of that. Donnie took it from

17   there. The time my company arrived, he worked on the

18   proposal yet.

19   Of course we were told we had no standing

20   after spending untold hours trying to get an appeal to

21   the NRC to at least give us a public hearing on the

22   fact that radioactive waste is being imported into the

23   state.

24   So I think this matter of a deadline is

25   extremely important.

NEAL R. GROSS
COURT REPORTERS AND TRANSCRIBERS
1323 RHODE ISLAND AVE., N.W.
(202) 234-4433   WASHINGTON, D.C. 20005-3701   (202) 234-4433

May 2013   E-115   NUREG-0498, Supp 2

29

1   MR. CARPENTER: Any comments?

2   MR. SAFER: Hello, everybody. I think we

3   were here in, what was it, 2007, you said, or '09 --

4   nine. I read some of my comments in the book. If you

5   didn't listen the first time, you probably don't

6   listen this time. But I'll say it again because maybe

7   it makes a few people squirm in their seats.

8   I speak today for all the future

9   generations that have no voice in this proceeding but

10   will be terribly affected by the decisions that are

11   being made that are allowing this plant to be built.

12   I think we all have seen what has gone on

13   at Fukushima and it's terrible. People are having to

14   leave their homes. School children's tennis shoes are

15   contaminated with radiation. Whole areas are

16   evacuated. Farms that have been in families for

17   generations are now abandoned and probably never to be

18   returned to.

19   So when you talk about the environmental

20   impact of this type of reactor, of any nuclear

21   reactor, you have to realize that the effects are not

22   small. And that was what in this document what they

23   came up with was that the environmental impacts are

24   small. Now tell that to the people in Fukushima.

25   Tell that to the people in Chernobyl. That's really

30

1   the crux of it.

2   I realize that there's nothing that I can

3   say here today that's going to turn this thing around.

4   I feel like David and Goliath. And I don't have the

5   magic stone that he had and I wish I did. And I

6   wouldn't throw it to hurt anybody, but I would sure

7   throw it to stop this reactor because I think the

8   effects of it -- unfortunately if something goes

9   wrong, something go monumentally wrong.

10   And I do find it troubling that in this

11 document there is -- I didn't see the word Chernobyl

12 one time. I understand from the discussion earlier in

13 the informal period with Mr. Susco that it's included

14 in the computer model. But including 6,000 thyroid

15 cancers into the computer model is not quite the same

16 thing as stating that when that reactor blew up the

17 effect was 6,000 at least documented cases of thyroid

18 cancer. And if we understood us to say, oh, it was a

19 minor thing. The effects aren't that great.

20   But you know, they're still monitoring a

21 lot of the agricultural products in that region.

22 There are still great areas that are uninhabitable for

23 many years. It's the *cesium that's going to stay

24 positive for 300 to 500 years. That's what we're

25 talking about here.

31

1   Now Fukushima you've got enormous releases

2   of radiation. You've got just today in the newspaper

3   the announcement that baby food is being recalled, 90

4   containers of baby food because the Japanese are

5   catching it. But the rice has been contaminated.

6   It's a nightmare. The economic effects, the human

7   effects, and the ecological effects are going to

8   reverberate around the planet till kingdom come

9   basically. And that's what's at stake here.

10   And I'm just going to keep saying it even

11   though it seems to fall on deaf ears because

12   everything that was brought up leading up to that

13   document there was always an answer to it and don't

14   worry, pat us on the back and say --well, when I went

15   to the Atlanta hearing on the Blue Ribbon Commission

16   for high level waste, they kept on saying that the

17   more -- it's just you're not educated.

18   Listen, I've been studying this stuff for

19   about 15, 20, 30 years. I mean I was involved in this

20   process back the first time around with TVA and the

21   nuclear plants. The more I know, the more it troubles

22   me. So don't be saying that, oh, all you need to do

23   is know more about it and you won't be so concerned.

24   The more I know, the more concerned I get.

25   MR. CARPENTER: Any other comments?

NEAL R. GROSS
COURT REPORTERS AND TRANSCRIBERS
1323 RHODE ISLAND AVE., N.W.
(202) 234-4433   WASHINGTON, D.C. 20005-3701   (202) 234-4433

NUREG-0498, Supp 2   E-118   May 2013

32

1   MR. KURTZ: I feel like I'm in the middle
2   here and I ought to stand somewhere where everybody
3   can see me. Where shall I go? I'll go up here with
4   Donnie.
5   I'm Sandy Kurtz. I'm with Bellefonte
6   Efficiency and Sustainability Team and we are a
7   concerned citizens group, a chapter of the Blue Ridge
8   Environmental Defense League. And of course we don't
9   care much for the idea of building yet another nuclear
10   plant in the Chattanooga region. I live in
11   Chattanooga.
12   And it just -- I guess I can summarize my
13   comments with these words, more is not better. And as
14   I told the press, it makes no sense to say there will
15   be no significant environmental impacts when you
16   double the number of nuclear plants at the same site.
17   The only reason I can think that they could possibly
18   say that is because the environment has already been
19   ruined with the first plant. Two plants in the same
20   place makes twice as much risk for accidents, for
21   human error, for radiation in the water, for tritium
22   in the water, and for ongoing aquatic danger to the
23   aquatic species, not to mention the health of our
24   people themselves.
25   And I just think that we should not

NEAL R. GROSS
COURT REPORTERS AND TRANSCRIBERS
1323 RHODE ISLAND AVE., N.W.
(202) 234-4433   WASHINGTON, D.C. 20005-3701   (202) 234-4433

May 2013   E-119   NUREG-0498, Supp 2

33

1   continue to add more risk to what's happening already.

2   We already had six nuclear reactors in this area and

3   we don't really need another one.

4   The Environmental Impact Statement, I just

5   can't believe that there would be no more impacts when

6   you're actually doubling the possibilities.

7   Thank you.

8   MR. CARPENTER: Other comments?

9   MS. HARRIS: Questions?

10   MR. CARPENTER: You have a question? Yes.

11   MS. HARRIS: At what point -- cause we

12 don't seem to have a benchmark of what -- how far TVA

13 can go that you won't let them go any further? You're

14 saying in here I see about the tritium. TVA managed

15 to fill a leak so they're containing the tritium.

16   Now whatever that magic thing is, Mr.

17 Stinson (phonetic), I hope, will sell it to the other

18 100 nuclear plants in this country which leak every

19 day during their operations. So it's worth bazillions

20 of dollars to stop the tritium leaking into the river.

21   Now the State of Tennessee, TDEC, they're

22 not even testing for anything like that. They don't

23 look at it. They say it's not -- they don't have the

24 money to test for it. Well, whenever I question an

25 NRC person, they look at me and say, "That's TDEC and

34

we don't have authority over them." Well, you gave
them a letter of authority under the agreement state
letter in -- I'm sorry, Roger. Yeah, it went into
your hair on your head, into your ears. But the thing
is that there has to be a limit of how far they can
go.

The other thing is that I want to know
what kind of security is around the intake pumping
station. And I'm talking about serious security and
about the outfalls. Now everybody says, oh, nobody
don't know where they're at. They don't know what's
going on with them. Get real. These things are not
secrets.

I mean when you live here on this river
like I have all my life, you know all the secrets on
this river. They ain't secrets. And then the idea
that there is a swimming hole within a thousand feet
of that plant is just sick.

I have -- the media that has come in here,
they wanted to talk and use this plant as a poster
child, some of it for good and some of it for bad.
But some of it has been good for the NRC. But I don't
see the NRC -- is it you don't have the regulations in
place? Is that the point?

Do I need to start pounding on Senator

1 Boxer's door more often? I mean I'm up there on a

2 regular basis with information and talking about the

3 problems and the things that are good and bad about

4 here because I want a safe plant cause I still have

5 relatives that lives in the evacuation zone. I just

6 buried my mother with colon cancer.

7   These things are significant to people

8 like myself who live here. I have children; I have

9 grandchildren; I have great-grandchildren.   My

10 grandson just came back from Baghdad. He's fighting

11 for us to have clean air and clean water and go by the

12 rules and have rules to go by.

13   But I don't see the cooperation. I just

14 got told that whenever I make a statement about a

15 problem at this plant that if I won't give up my

16 sources, NRC just flips it over in the garbage can.

17 Now that pissed me off. I can tell you it did.

18   Because whenever I tell you something, I

19 don't have a problem; my credibility is not on the

20 line here. Whenever I tell you that there's a problem

21 in the area and you don't deal with it, then it

22 aggravates me to no end. And I'm not seeing when you

23 just fluff them off. It's like you're swatting at

24 gnats. This is our life and our community and our

25 future and our whole future of these communities and

36

1 these mountains.

2 I'm deeply resentful that there is a fan

3 crane 2,500 acre island down here that is so

4 contaminated that the geese are even -- they're not

5 even coming in there anymore. The cranes don't want

6 to go there. You can't entice them; you can't put

7 enough food on them to entice them in is what I'm

8 seeing.

9 Now come on, guys, let's get real about

10 this. Can we please tell me what rules you go by? I

11 mean some of the things that you say I can't even

12 find. So somewhere along the line you've got to put

13 some reality in here instead of all this fluff and pie

14 in the sky.

15 MR. CARPENTER: Any other questions or

16 comments?

17 MS. FARRIS: My name is Kathleen Farris.

18 I'm from Rutherford County, Murfreesboro, Tennessee.

19 And if you're wondering why I'm here, it's because

20 four years ago we discovered we had low level waste

21 going into our landfill. That landfill is right on

22 the Stones River, which provides the drinking water

23 for most of Rutherford County. That includes the city

24 of Murfreesboro.

25 My first interest was in water and it's

NEAL R. GROSS
COURT REPORTERS AND TRANSCRIBERS
1323 RHODE ISLAND AVE., N.W.
(202) 234-4433   WASHINGTON, D.C. 20005-3701   (202) 234-4433

May 2013   E-123   NUREG-0498, Supp 2

1 about water that I want to speak today. It's a source

2 of life. If our drinking water is polluted and we're

3 taking radiation or chemicals into our bodies, I think

4 that can very well account for the epidemic of cancer

5 that we have in this country today.

6 And furthermore, it's not just a question

7 of pollution. It's a question of consumption. Only a

8 very -- I wish I had all my figures with me today, but

9 I left my computer at home by mistake. But I had read

10 very recently the report that was done by the Union of

11 Concerned Scientists on water consumption and energy

12 production and it's available on the Union of

13 Concerned Scientists' website. And most of what I'm

14 going to say is taken from that information.

15 Only a very small fraction of the earth's

16 water is potable. And already huge corporations are

17 buying up water supplies all over the world, which

18 means that before long anybody who can't afford to buy

19 water won't have clean water to drink or may not have

20 water at all because there are water wars going on.

21 We've already had it over the Tennessee River here

22 where Georgia and North Carolina want their share of

23 our water, right?

24 Global warming and climate change, which I

25 see you have noted in your study, are going to affect

1 the supplies of water and threaten -- think about it -

2 - land masses are shrinking; populations are growing.

3 The demand for water will ever be greater,

4 particularly if we are able to continue in what we

5 think of as an advanced civilization.

6 The single largest use of fresh water in

7 the United States is thermal nuclear -- no, I'm sorry

8 -- the thermal energy, either by nuclear or coal. And

9 I have -- the study that I referred to has a pie

10 chart, shows that 41 percent of the water, the largest

11 usage is for these forms of energy production. Now if

12 you look at the chart, here is nuclear, here is coal,

13 oil, gas. Solar thermal uses a lot of water, so does

14 biofuel. However, solar photovoltaic and wind, look,

15 you can't even see a line for how much water is

16 required to produce energy in those ways.

17 Now it's not just on a global scale that

18 we have to think, although I think we need to think

19 that way as well. One of the things that the Union of

20 Concerned Scientists have pointed out is that in the

21 Southeast United States we have a particularly severe

22 problem of water and energy production. That drought

23 and heat have caused many -- and we all know this --

24 many closings, shut-downs of nuclear reactors because

25 the water is too hot or there's not enough of it.

39

1 Same thing has happened in -- and the drought is

2 threatening the nuclear industry in Europe now.

3   Particularly Browns Ferry has been closed,

4 heavens knows, how many times.

5   I was in Texas this summer. The darker

6 the area, the greater the intensity of drought. People

7 are losing their crops. In Fort Worth they're telling

8 you don't use too much water, constantly reminding

9 people not to use too much water. That's this year.

10   If you look at 2007, this is right where

11 TVA is building all these plants, six, going on seven,

12 on the Tennessee River. And TVA wants to put four

13 more at Watts Bar. Now that Tennessee River provides

14 drinking water for the cities of Knoxville,

15 Chattanooga, Huntsville, all the communities in

16 between.

17   The TVA's plan is to become, as Mr.

18 Kilgore said, the foremost producer of nuclear energy

19 in the country. And that means this Watts Bar 2. It

20 also means the plant at Bellefonte, the first one, and

21 then three more.

22   And I propose that this is a threat to our

23 drinking water. It's not what your study says.

24   And I went through and I marked all the

25 sections in which you claim that the impact will be

NEAL R. GROSS
COURT REPORTERS AND TRANSCRIBERS
1323 RHODE ISLAND AVE., N.W.
(202) 234-4433   WASHINGTON, D.C. 20005-3701   (202) 234-4433

40

very small. Section 4, 4221, surface water use impact, based on the NRC staff's independent analysis the staff concludes that because of the small amount -- small volume of water consumed relative to the Tennessee River flow, the impact on surface water use of operating WBN Unit 2 is small.

The same thing it says further on, on ground water use, No. 4-11, Page 4-11, we're told that based on the independent analysis of additional information since the 1978 whatever this is, FES-OL, the NRC staff concludes that the impact on ground water from operating Watts Bar Unit 2 would be small.

Now I asked the question, is this karst topography? And nobody really answered my question. I'm sorry, sir.

MR. MILANO: It is.

MS. FERRIS: Karst topography is limestone. It's got cracks and crevices everywhere. If it gets into -- if radiation gets -- or pollution gets into that, you have got an effect on the ground water.

Now we know about that in Dickson, Tennessee right now, which is also a landfill built on karst topography. There are people who have become terribly ill and they're bringing a lawsuit against

41

1   the county and state for that polluted landfill

2   because of the topography.

3   The same thing is true of the Stones River

4   in Rutherford County. It's built on -- this landfill

5   sits on limestone right over our drinking water

6   supply.

7   If we -- I'm sorry. I think I'm allergic

8   to something here.

9   MR. SAFER: It's the radiation.

10   (Laughter)

11   MS. FERRIS: I hope not. In any event I

12   want to say to you the population of the earth is

13   growing. The water demands are growing. And one way

14   we can make sure that other industry and agriculture

15   and people have enough clean water is to use clean

16   means of producing energy. And we've got them. I

17   couldn't get up here if I didn't know the technology

18   already exists for wind and solar energy production.

19   And I get this argument with people all

20   the time, both at these meetings and elsewhere. Say,

21   oh, they can't produce enough. Well, if we put the

22   billions of dollars into solar and wind energy that we

23   are putting into nuclear energy, we could do it. We

24   went to the moon. We're sending off explorers into

25   space. The technology is there. What's lacking is

NEAL R. GROSS
COURT REPORTERS AND TRANSCRIBERS
1323 RHODE ISLAND AVE., N.W.
(202) 234-4433   WASHINGTON, D.C. 20005-3701   (202) 234-4433

NUREG-0498, Supp 2   E-128   May 2013

42

1   the will.

2   And I would like to -- all of you would be

3   dead without water. Your children will die without

4   water. Other species will die without water. We have

5   got to preserve it. There was a documentary made on

6   water wars called Blue Gold and that's water that

7   they're talking about. We've got to have it and we've

8   got to preserve it while there's still some left to

9   preserve.

10   Thank you.

11   MR. CARPENTER: Other comments?

12   MR. SAFER: I can go again. I wanted to

13   get into some specifics. One thing in this document,

14   and I of course haven't read all of it. It was much

15   harder on the computer. I appreciate having a hard

16   copy now; that does make it a lot easier. I realize

17   it is more costly to the NRC, but this is an important

18   issue.

19   The Watts Bar -- this book says that no

20   other new nuclear facilities within 50 miles are being

21   considered. That's just false. Oak Ridge is within

22   50 miles.

23   TVA is on record as being far beyond

24   considering small modular reactors. They're in

25   communication with the NRC, other branches of the NRC,

NEAL R. GROSS
COURT REPORTERS AND TRANSCRIBERS
1323 RHODE ISLAND AVE., N.W.
(202) 234-4433   WASHINGTON, D.C. 20005-3701   (202) 234-4433

May 2013   E-129   NUREG-0498, Supp 2

43

1 daily. There's meetings; there's been several

2 meetings this month on SMRs and TVA. They plan to

3 build -- first they planned to build six up at the

4 Clinch River site; now I think it's down to two.

5 These are 125 to 150 megawatt reactors

6 that are modular built and they're sunk into a hole

7 150 feet deep in the ground, the same karst geology.

8 So I wish that would be corrected or you check that.

9 I know there was some discussion about whether that's

10 accurate or not.

11 I don't see -- if you speak English,

12 considered means considered. And they're certainly

13 being considered. So that's one thing.

14 The second thing, the highly irradiated

15 used fuel that's often called spent fuel is being

16 stored in fuel pools as long as possible. And that's

17 just the biggest danger that it can be. They need to

18 be moved to a hardened onsite storage. And I know

19 they have to be in the pools for about five years till

20 they cool down. But beyond that point the packing of

21 these pools with more and more rods that way beyond

22 what they were designed for is a real huge risk that

23 the community is taking on and needs to be aware of.

24 And the community needs to support the idea of

25 hardened onsite storage.

NEAL R. GROSS
COURT REPORTERS AND TRANSCRIBERS
1323 RHODE ISLAND AVE., N.W.
(202) 234-4433   WASHINGTON, D.C. 20005-3701   (202) 234-4433

NUREG-0498, Supp 2   E-130   May 2013

44

1   The reason that's not happening is
2   strictly cost. And when you're talking about cost,
3   this whole reactor is nuclear power on the cheap. And
4   I don't know why we're accepting the cheapest possible
5   nuclear power plant. TVA tried to build a new AP
6   1000, two of them at Bellefonte. They found out they
7   was going to be so much more expensive than finishing
8   this reactor and the Bellefonte Unit 1 that they
9   backed off from it.

10   Well, excuse me, but this is not the place
11 to cut costs. If they want to build these things,
12 they have to be state-of-the-art. This is far from
13 state-of-the-art.

14   This ice condenser design is really a joke
15 in the industry. And I mean I talked to the operators
16 at Sequoyah and they just kind of grinned when I asked
17 them about -- the ice condenser design means there's
18 three million pounds of ice, literally three million
19 pounds of frozen water, that's in the reactor within
20 the containment structure. And should they have a
21 loss of coolant, all of that hot gas is supposed to go
22 through that ice room to lessen the pressure. And so
23 they've made the containment less sturdy than the
24 other reactors around the country and around the
25 world.

NEAL R. GROSS
COURT REPORTERS AND TRANSCRIBERS
1323 RHODE ISLAND AVE., N.W.
(202) 234-4433   WASHINGTON, D.C. 20005-3701   (202) 234-4433

May 2013   E-131   NUREG-0498, Supp 2

45

1   Nobody else is building any ice condenser
2   designs ever again. They were built back in the `70s.
3   Sequoyah, ice condenser designs; Watts Bar 1 is an
4   ice condenser design. There's no justification for
5   finishing this thing.
6   I talked about this the last time in 2009.
7   Obviously it was not heard. But just so everybody
8   knows, it's a Rube Goldberg contraption. If you don't
9   know Rube Goldberg, look him up on the internet
10   because he was a fascinating guy. But, you know, I
11   could go on about that.
12   But the other specifics about this
13   particular Environmental Impact Statement, on Page H-
14   3, Appendix H is the Severe Accident Mitigation Design
15   Alternatives. I'll quote, "TVA did not include the
16   contribution from external events in the Watts Bar
17   Nuclear Plant risk estimates."
18   External events. Now whether those is a
19   tornado like the one that almost hit Browns Ferry, a
20   F-5 tornado. The very same day, April 27th, tornadoes
21   came through here and they got pretty close to Watts
22   Bar. They went right across Bellefonte, where that
23   reactor was going to be built as soon as they finish
24   Watts Bar. And it's just kind of -- to me it's kind
25   of eerie, but in the same day all of TVA's nuclear

46

1 power plants were affected by tornadoes. If that's

2 not a message from on high, I don't know what is, to

3 be honest, folks.

4   So those kinds of things, plus terrorist

5 attacks, that's an external event. They just don't

6 even factor that into this. And I'm sorry that we

7 live in a world where that has to be factored in, but

8 we all know that it does. And that these things are -

9 - they're the biggest target for a terrorist that you

10 can imagine and the effects would be -- they'd put

11 9/11 into a footnote of history almost.

12   So these things are huge target for

13 external events and that really needs to be factored

14 into the design. And believe me, in 1970, they

15 weren't factoring in the possibility of a terrorist

16 attack on U.S. soil in the design of the containment

17 structure which, as I've said, is already thinner than

18 other nuclear power plants of that era.

19   I just have to make note that the

20 definition of risk in this document is it's the

21 product of frequency and the consequences of an

22 accident. Work on that one for a while. I don't

23 know.

24   As I said, it's in 6.2.4, where the staff

25 concludes that the environmental consequences of a

47

1   severe accidents are small, 6.2.4. I don't know how

2   you get that.

3   Severe accident mitigation alternatives in

4   6.3, there's a quote in here. I'll read it directly

5   because I didn't copy it all down. They eliminate the

6   severe accident scenarios that were "excessively

7   costly to implement such that the estimated cost would

8   exceed the dollar value associated with completely

9   eliminating all severe accident risks at WBN 2."

10   I take that to mean that some of the risks

11   it was just too costly to mitigate those risks, so we

12   just threw them out because, if it costs too much, we

13   couldn't possibly deal with it. So we'll just deal

14   with the risks. That's what that says to me.

15   And by the way that whole severe accident

16   thing was required of the NRC by the Third Circuit

17   Court's opinion in Limerick Ecology Action, Inc.

18   versus the NRC in 1989. It was a court ordered thing

19   for the NRC to have to take into account these risks.

20   It took a federal court to require that in 1989.

21   If you read this document, you'll see -- I

22   don't think you'll see the word Fukushima or even

23   Chernobyl in there. As I said earlier, they say it's

24   factored into the computer models, but that sure is

25   sanitizing the realities.

48

1   Page 6-15, "It is noted that the risks

2 from deliberate aircraft impacts were explicitly

3 excluded since this was being considered in other

4 forms along with other sources of sabotage." I don't

5 know where the other form is. I asked an individual,

6 a couple of individuals, with the NRC here. They were

7 going to get back to me on that. But again deliberate

8 aircraft impacts were explicitly excluded from this

9 document.

10   And I think that concludes my comments.

11 Thank you.

12   MR. CARPENTER: Any other questions?

13 Comments?

14   MS. HARRIS: I'm not going to ask anything

15 else because you don't get an answer. You get fluffed

16 off and I'm not interested in that. So I'll just put

17 it online in writing. It's become a task to deal with

18 the NRC whenever you get kindergarten answers to

19 chemistry questions. And I'm kind of over it. I just

20 want the time so that I can do the writing. I want

21 the extension that should be granted.

22   MR. CARPENTER: Anything else?

23   MS. FERRIS: I would like to add something

24 to what Donnie said about terrorists. There was a

25 Newsweek article either last week or the week before

NEAL R. GROSS
COURT REPORTERS AND TRANSCRIBERS
1323 RHODE ISLAND AVE., N.W.
(202) 234-4433   WASHINGTON, D.C. 20005-3701   (202) 234-4433

May 2013   E-135   NUREG-0498, Supp 2

49

1 of an FBI undercover agent who was working with

2 extremist groups in the South. And some of the people

3 he encountered were planning an attack on Browns

4 Ferry. So this isn't some hypothetical. I mean there

5 are people out there who would love to attack these

6 reactors and probably many of them don't have any idea

7 of what the real consequences would be.

8   MR. SAFER: I'm sorry you have to listen

9 to me again. I think unfortunately a lot of people

10 that believe the same thing that Kathy and Ann and I

11 do have gotten so discouraged from this process that

12 they just don't show up at these meetings anymore. So

13 I feel like I have to speak for many, many people.

14   I think that one of the things that the

15 community needs to realize is that decommissioning of

16 these reactors, both Unit 1 and Unit 2, is going to be

17 a huge task on down the road. Now the decommissioning

18 fund, TVA and the other utilities put money into it

19 all the time, but they invest that money just like

20 anybody that has a little money tries to invest it in

21 the stock market or wherever to do the best they can.

22 Well, when the stock market took a big hit, that

23 decommissioning fund took the same hit. And there's

24 just simply not as much money as you need to

25 decommission these reactors.

NEAL R. GROSS
COURT REPORTERS AND TRANSCRIBERS
1323 RHODE ISLAND AVE., N.W.
(202) 234-4433   WASHINGTON, D.C. 20005-3701   (202) 234-4433

50

1   And we've all been through this thing

2   recently of the government shutting down and the

3   government saying no more loans, no more deficit. We

4   don't know who is going to be charge of our government

5   in 10, 20, 30 years. But the reality when it comes to

6   these nuclear materials that are being manufactured at

7   these sites is that they're going to have to be dealt

8   with.

9   But if the federal government refuses to

10   do it, it's going to be the community's problem just

11   to safeguard that plant from now until kingdom come

12   again, till eternity. And I think we've all been sort

13   of shaken in our confidence of the federal government

14   being able to continue its obligations into the

15   future.

16   And these materials, these radioactive

17   materials, need to be kept out of the environment for

18   half a million years. Now if you think the federal

19   government is going to be here half a million years

20   from now, I'd like to see your information.

21   But I'm just trying to point out that all

22   of these things are built on a best-case scenario,

23   that everything is going to go perfectly and the world

24   is going to operate in a way that we wish it would.

25   And I think we've all seen that the world does not

51

1   operate in that way.

2   And whether it's Chernobyl or Fukushima or

3   the fire at Browns Ferry or Three Mile Island, these

4   nuclear reactors are just as prone to accidents,

5   mistakes, failures, human error, whatever, terrorism,

6   or whatever it might be as any other human enterprise

7   and any other human activity. They will have

8   problems. They will have worse-case scenarios. It

9   may not happen very often, but that doesn't mean it

10 can't happen tomorrow here. And if you have two

11 reactors, it just doubles the chance.

12   And again the whole clean-up thing is a

13 whole nother issue that whether that money is really

14 going to be there. The same is true for the high

15 level waste that's in those fuel pools. The federal

16 government is trying to figure out how to take care of

17 its responsibilities on that and they're struggling

18 with that. They have struggled with it for over 50

19 years.

20   There's not a single deep repository for

21 radioactive waste that I know of in the world.

22 There's some talk in France about they maybe started

23 to experiment with one and maybe Finland has done a

24 little, but Finland doesn't have very many nuclear

25 power plants. So talking about the amount of high

NEAL R. GROSS
COURT REPORTERS AND TRANSCRIBERS
1323 RHODE ISLAND AVE., N.W.
(202) 234-4433   WASHINGTON, D.C. 20005-3701   (202) 234-4433

NUREG-0498, Supp 2   E-138   May 2013

52

1 level radioactive waste we've been generating in the

2 United States, it's a lot of material and we simply

3 don't have any place to put it.

4   The Blue Ribbon Commission has been

5 working on it for over a year. They're due to release

6 their report which is highly controversial in my

7 opinion. But you can't take it on face value that

8 these materials are going to be able to be handled the

9 way they need to have been.

10   MR. CARPENTER: Thank you. Any other

11 comments? Questions?

12   Anything from the staff?

13   In that case I would like to thank

14 everyone for participating. We did have some very

15 good comments and questions today. We will get back

16 to those who have given us questions as quickly as

17 possible. I do know that we have at least one written

18 set of questions already. If there are any other

19 written questions that you'd like to provide, you can

20 do it both by submitting from here or leaving it with

21 us before you leave.

22   I would like to remind everyone that again

23 there is a meeting comment sheet. We do like to get

24 those. It tells us how to do these meetings a little

25 bit better each time. So thank you for doing that.

NEAL R. GROSS
COURT REPORTERS AND TRANSCRIBERS
1323 RHODE ISLAND AVE., N.W.
(202) 234-4433   WASHINGTON, D.C. 20005-3701   (202) 234-4433

53

1   We will finish up now and reconvene for

2   the second part of this at 6:30 tonight. Anybody that

3   would like to come back, you're more than welcome.

4   We'd like to have you. It will be in this room again.

5   And with that I quit and close the

6   meeting. Thank you again for coming. Thank you.

7   (Whereupon, this portion of the meeting

8   was concluded at 3:20 p.m. to reconvene for the second

9   portion at 6:30 p.m.)

10

11

12

13

14

15

16

17

18

19

20

21

22

23

24

25

Official Transcript of Proceedings

NUCLEAR REGULATORY COMMISSION

Title: Watts Bar Nuclear Plant EIS
                 Public Meeting: Evening Session

Docket Number: (n/a)

Location: Sweetwater, Tennessee

Date: Thursday, December 8, 2011

Work Order No.:   NRC-1313   Pages 1-34

NEAL R. GROSS AND CO., INC.
Court Reporters and Transcribers
1323 Rhode Island Avenue, N.W.
Washington, D.C. 20005
(202) 234-4433

```
 1                UNITED STATES OF AMERICA

 2            NUCLEAR REGULATORY COMMISSION

 3                     + + + + +

 4              PUBLIC MEETING TO DISCUSS

 5               DRAFT SUPPLEMENT TO THE

 6            FINAL ENVIRONMENTAL STATEMENT

 7         FOR WATTS BAR NUCLEAR PLANT, UNIT 2

 8                    + + + + +

 9             Thursday, December 8, 2011

10                  Magnuson Hotel

11             1421 Murrays Chapel Road

12              Sweetwater, Tennessee

13                    + + + + +

14                  Evening Session

15             6:30 p.m. to 8:30 p.m.

16

17

18

19

20

21

22

23

24

25
```

NEAL R. GROSS
COURT REPORTERS AND TRANSCRIBERS
1323 RHODE ISLAND AVE., N.W.
(202) 234-4433   WASHINGTON, D.C. 20005-3701   www.nealrgross.com

20

1 regulations.gov. And that docket ID is what's
2 particularly important there. When you go to
3 regulations.gov, you can put that docket ID and it
4 will bring up the web page for this particular
5 document. And there's a pretty easy form where you
6 can -- just a blank space for you to submit your
7 comments. And you press the Submit button and it
8 heads out to the NRC. And the last way if you so
9 choose, we do also have a fax number for comments.

10   It's like I said the current due date for
11 comments is December 27th. But we are going to look
12 at potentially extending that based on public request.

13   So that concludes my do. And I'll turn it
14 back over to Gene.

15   MR. CARPENTER: Thank you. All right, now
16 this is the real meat and potatoes part of the
17 meeting. Well, the reason that we're here today, and
18 that is to get your questions and comments to the
19 staff. Now the first part of this that we're going to
20 go into is the question and answer portion of this
21 meeting. And specifically if you have any questions
22 that the staff here they can answer, that we will
23 write down and take back and respond back to you in
24 writing. This is the portion that we would like to
25 have you go ahead and ask your questions.

NEAL R. GROSS
COURT REPORTERS AND TRANSCRIBERS
1323 RHODE ISLAND AVE., N.W.
(202) 234-4433   WASHINGTON, D.C. 20005-3701   www.nealrgross.com

May 2013   E-143   NUREG-0498, Supp 2

21

1   As soon as we're finished with all the

2   questions that we may have from the floor, then we

3   will go into the comment period so that any comments

4   that you have that you'd like us to consider then we

5   will take those. All right?

6   Now I'd ask at the beginning of the

7   meeting anybody who was interested in having to be

8   asked first for questions, comments to sign in and to

9   mark it that they would like to do so. With that

10   there is nobody who has done so, so I'm just going to

11   throw it open to the audience. Does anybody have any

12   questions of the staff?

13   Because we are having this transcribed

14   when I bring the mic over to you, please stand up and

15   tell your name when you do so. Or you can sit.

16   MS. HARRIS: Earlier today we talked about

17   --

18   MR. CARPENTER: Give your name.

19   MS. HARRIS: He knows. Ann Harris.

20   MR. CARPENTER: Thank you.

21   MS. HARRIS: Earlier today we talked about

22   some of the documents that you used to make your

23   judgment in here and some of them refer to 40-year-old

24   documents. Now I realize some things haven't changed,

25   but a lot more has changed than has not.

22

1   And I'm wondering on these documents where

2   you used TVA's documents when you did use them and did

3   you just accept TVA's documents without going back and

4   checking to verify in those old documents and did you

5   go and look for new information concerning those same

6   documents because I'm not finding consistency between

7   what you've put in and some things that I personally

8   know about? And I'll put those in my comments. But

9   I'd like to know how you made those determinations.

10   MR. SUSCO: It's kind of a combination of

11   all of those things. So it really kind of depends on

12   the issue. For example, one issue that there's not

13   going to be a lot of new information is going to be on

14   the geology and soils of the area. Nothing has really

15   changed as far as what this -- what Watts Bar is built

16   on in the last 40 years. So we really could use some

17   of the older studies.

18   But for something like aquatic impacts,

19   now we might start -- as a starting point we might

20   look at something that was from when the plant was

21   initially built, but then we're -- I guarantee that

22   we're going to look at new information for that type

23   of impact, in particular because we really want to see

24   what has changed. And we already got the documents

25   from 40 years ago. We can see the delta and what sort

NEAL R. GROSS
COURT REPORTERS AND TRANSCRIBERS
1323 RHODE ISLAND AVE., N.W.
(202) 234-4433   WASHINGTON, D.C. 20005-3701   www.nealrgross.com

23

1 of impact the plant may have had on that particular

2 type of issue.

3   But as far as documents submitted by TVA,

4 yes, we do put a certain amount of trust in our

5 licensees that what they are submitting to us is

6 truthful. But that's not the only place that we look.

7 There's a lot if you look in the referencing section

8 for each of the chapters, you'll see 10 pages for

9 every chapter and all the references that we looked.

10 And only a small portion of those are TVA's. There's

11 a lot of expert studies that we look at from all sorts

12 of different sources. So it's kind of a combination

13 of all those things.

14   MR. CARPENTER: Any other questions for

15 the staff?

16   MR. RIDEN: David Riden from Riceville,

17 Tennessee. I signed the sign-in thing earlier. Had

18 no intent to ask a question, okay? What you said two

19 things that raise two questions in my mind.

20   The first one relates that I've lived

21 three years in Minot, North Dakota prior to 1968, when

22 the Mouse River flooded. And then recently I was a

23 contractor at Fort Calhoun on the Missouri River.

24   NRC was concerned about the data that the

25 utility had used for Fort Calhoun as far as the flood

24

1   projections were for the Missouri River. Fort Calhoun

2   had used data from the Corps of Engineers. And NRC

3   has a process to calculate each utility to calculate

4   that without utilizing the data from the Corps of

5   Engineers. And the reason why I was there was to look

6   back over their information that they were going to

7   present to NRC. And they did extensive updates.

8   Don't need to go into that.

9   But the bottom line is the information

10 provided by the Corps of Engineers was faulty. And

11 they made great improvements at Fort Calhoun and

12 they're still working on it. If you watch the news,

13 if they hadn't prepared for it, they'd be in a lot of

14 trouble, lot worse trouble. And I attribute NRC

15 pushing them to correct what they had there and it

16 made a bad situation a lot better.

17   And my question is has TVA depended on the

18 Corps of Engineers data for anything related to the

19 Tennessee River at Watts Bar? And if they have, will

20 NRC then go back to the Tennessee Valley Authority and

21 ask them the same prudent questions they asked the

22 utility owner on the Missouri River to do it in

23 accordance with the federal regulations and not depend

24 on the Corps of Engineers? So that's the first

25 question.

NEAL R. GROSS
COURT REPORTERS AND TRANSCRIBERS
1323 RHODE ISLAND AVE., N.W.
(202) 234-4433   WASHINGTON, D.C. 20005-3701   www.nealrgross.com

May 2013   E-147   NUREG-0498, Supp 2

1   And if there's any response, you've got my

2   email address. If there's any response back, I would

3   like to know because I'm an old-time resident of East

4   Tennessee, okay?

5   And that leads into the second question.

6   You want me to go ahead with the second question?

7   So that was the first question. I don't

8   know the answer to it even though I'm a former NRC

9   employee and a former TVA employee. I was here in

10 1978, when we were originally trying to start up Watts

11 Bar Unit 1. And I was the Nuclear Assurance Engineer

12 at the time.

13   And TVA upper management gave up on Watts

14 Bar and said, "David, we brought you here to start up

15 Watts Bar. We want you now to go to Sequoyah because

16 we changed our mind. We want to put our effort in

17 Sequoyah." So I went to Sequoyah and I was there

18 helping them start up the Sequoyah units.

19   And for whatever reason they wanted to --

20 it doesn't hurt my feelings -- but TVA ripped me along

21 with everybody else that were supplement people that

22 they moved over to Sequoyah. And when they released

23 me, they said, "Oh, by the way, David, we like what

24 you do. You're going to be back cause we've got a lot

25 more in TVA to do especially at Browns Ferry." So in

26

1   the last 20 years I've spent over half of that time as

2   a contractor for TVA.

3   Before all that I mentioned I spent three

4   years in Minot, North Dakota. That's because I was

5   with the United Stated Air Force. I'm a Nuclear

6   Weapons Specialist. I was there to start up the

7   Minute Man Three Multiple Independent -- where you had

8   three vehicles.

9   So I preface this question with I am pro-

10   nuclear. I'm an environmentalist. And there's

11   nothing with a more green environmental impact than

12   nuclear power. Its carbon footprint is zero. So I

13   come from a pro stance, okay?

14   But the question is -- and I guess I

15   should preface it. Having come from the nuclear

16   weapons industry and then going to the University of

17   Tennessee and getting a degree in nuclear engineering

18   and then getting into the nuclear power part and

19   working for NRC, people would beg me, "David, counting

20   on you to make sure that what you're doing out there

21   is safe. We don't know enough to even ask the

22   question."

23   And this is general. I get this

24   everywhere I go across the United States for the last

25   35 plus years. Will that thing blow up?

27

1   And my answer has always been no, there is

2   no nuclear special grades material in our commercial

3   reactors in the United States that is used to make

4   nuclear weapons grade material. That is all made at

5   Savannah River. Don't even concern yourself.

6   Well, now I have to eat crow because the

7   Department of Energy got into an agreement with NRC

8   and TVA to shut down Savannah River and make their

9   special nuclear grade material at Watts Bar. Doesn't

10   hurt my feelings one bit at all. It's closer to home.

11   That's fine with me.

12   But I happen to know back in 1978 we

13   didn't even consider having a special core in Watts

14   Bar Unit 1 or 2. It wouldn't in the Environmental

15   Impact Statement to even look at that question of that

16   special nuclear grade material in the rods to go into

17   the core for the Department of Energy then to

18   reprocess.

19   So what I would expect or what NRC taught

20   me to do back in 1978, come up with some good words.

21   And so even though I haven't had a chance to read

22   this, one of the good things was you are here to find

23   out if you missed an issue or has something changed.

24   Well, something has changed. And it could be for the

25   good. I'm all for it; I'm not against it.

NEAL R. GROSS
COURT REPORTERS AND TRANSCRIBERS
1323 RHODE ISLAND AVE., N.W.
(202) 234-4433   WASHINGTON, D.C. 20005-3701   www.nealrgross.com

28

1   What I'd like to see are words added to
2   the Final Environmental Impact Statement that address
3   the issue of the relationship with the Department of
4   Energy and the core that's being used that
5   everything's okay. That's what I'd like to see. Good
6   words to address it. And I don't know, maybe it's
7   addressed in here. But if it's not addressed at all,
8   then I think that leaves the door open for somebody
9   later on to raise an issue and I don't like issues
10 that fall under the area of assumptions or unverified
11 assumptions. I would like to know that NRC addressed
12 it and everything is okay.
13   That's the two questions I have.
14   MR. MILANO: Again it's Pat Milano. Since
15 my organization, the organization I'm with, has done
16 the Safety Evaluation Report, in Section 2 of the FSAR
17 and the final Safety Analysis Report as provided by
18 TVA and in Section 2 also of the staff's Safety
19 Evaluation Report, we address issues that relate to
20 site characteristics. And one of which is hydrology.
21   And that information has been significantly
22 supplemented in several of the last supplements to the
23 Safety Evaluation Report that the NRC has done.
24   And let me -- I'll try to give you a
25 little bit of a brief history of what's gone on. The

NEAL R. GROSS
COURT REPORTERS AND TRANSCRIBERS
1323 RHODE ISLAND AVE., N.W.
(202) 234-4433   WASHINGTON, D.C. 20005-3701   www.nealrgross.com

May 2013   E-151   NUREG-0498, Supp 2

1 NRC staff was reviewing the code, the computer code,

2 and the input assumptions that were being used by TVA

3 to assess the overall operation of, you know, in the

4 Tennessee Valley, you know, the river operations. And

5 because of that the staff had a number of questions

6 and comments related to how well TVA had managed that

7 code.

8   And based on the staff's -- based on the

9 issues that the staff raised, TVA spent an over a year

10 effort to upgrade the computer code itself and

11 reassess all the input assumptions that go into

12 utilization of that code. And they did that for the

13 whole river system that's under their control. And so

14 that was recently done and you'll see that documented

15 in the staff's review as such.

16   So as you had originally indicated when

17 you were talking about the Corps of Engineers, it's

18 not -- the Corps of Engineers did not do the studies

19 for flood height and stuff like that. That was done

20 by TVA and utilizing these codes. And the staff, the

21 NRC staff, put a significant effort into reviewing

22 what TVA did in terms of upgrading the code and

23 rerunning it. So that's been documented in Chapter 2.

24   So does that answer your question?

25   MR. RIDEN: Yes.

30

MR. MILANO: Okay.

MR. SUSCO: So on to the second half of that. To be perfectly honest I don't know if this particular environmental statement we're talking about now mentions or discusses that particular issue about --

MR. MILANO: First of all, I'm sorry to take it away from you, Jeremy, but I'll turn it back to him anyway. What you're talking about is the production of tritium. And TVA in their negotiations and their Memorandum of Understanding with the Department of Energy they -- TVA is currently only producing tritium with Unit 1 and there are no current plans at least before us right now for them in the core for Unit 2 to produce tritium.

And they're not -- I can talk to you later a little bit. Some of this is somewhat security sensitive and stuff.

But it's not something that's done throughout the core. There's specific rods that produce the tritium and sequester it and stuff. And again it's currently only planned for Unit 1.

TVA may ultimately elect to do that for Unit 2, but if they do do that, they're going to have to come in and request an amendment to the operating

31

1   license for Unit 2. And right now we're not

2   evaluating tritium production for Unit 2 because TVA

3   has not requested that.

4   And currently the core that TVA has

5   purchased from Westinghouse does not have that

6   capability with it. There are no tritium producing

7   rods in the current core.

8   When DOE and TVA requested to do this for

9   Unit 1, that came in as a specific request. It was

10 evaluated by the NRC staff and as part of that

11 evaluation the NRC staff is required to do an

12 environmental assessment of that. And so that

13 environmental assessment was done for Unit 1, not at

14 the original licensing, but as an amendment to the

15 operating license several years ago.

16   And if you want to know the specific

17 amendment, I'll have to get back with you because I

18 don't generally deal with Watts Bar Unit 1. But you

19 can find that type of information that you're looking

20 for in the information that supported that amendment

21 to the operating license for Unit 1.

22   MS. HARRIS: It's not hard to find in

23 ADAMS since TVA is the only one that makes it.

24   MR. SUSCO: Just to add a little bit.

25 When Pat was talking there -- Becky if you want to

NEAL R. GROSS
COURT REPORTERS AND TRANSCRIBERS
1323 RHODE ISLAND AVE., N.W.
(202) 234-4433   WASHINGTON, D.C. 20005-3701   www.nealrgross.com

NUREG-0498, Supp 2   E-154   May 2013

32

1 raise your hand, she -- it actually is in Chapter 4.

2 And Becky is one of the engineers that worked on

3 putting together different parts of this Environmental

4 Impact Statement.

5   And so I do encourage you to do those.

6 Read Chapter 4. And if somehow we didn't properly

7 characterize it or if there's pieces of information

8 that we're missing, let us know in your public

9 comments and we'll take a look at that.

10   MR. CARPENTER: Any other questions?

11   MS. HARRIS: One of the things that you

12 should know about the tritium is that DOE worked on

13 the evaluations from Sandia when they were done out in

14 New Mexico for a long period of time, like 20 years.

15 And they've had to back off of the amount that they're

16 producing at Watts Bar because it was not designed for

17 that. There's a lot of information out there that you

18 can find.

19   MR. CARPENTER: Thank you. Any other

20 questions?

21   And again if you have questions and you'd

22 like to give it to us in writing, we have some 3x5

23 cards here on the table, or you can send it in via the

24 regulations.gov website, mail it to the Chief, Rules,

25 Announcements and Directives Branch, or fax it in to

33

1   us. We'll always take your questions at any time.

2   If there's no further questions, I'd like

3   to go ahead and open it up for comments. Do we have

4   any comments?

5   And we will take written comments also if

6   there's no public ones here.

7   None?

8   Pat, Jeremy, any last comments?

9   MR. SUSCO: No.

10   MR. MILANO: All I want to mention is I

11   don't want to take thunder away from Gene here, but

12   we're not -- the NRC we're not going to leave here

13   right away. If you've got anything else that's not

14   related maybe to the Environmental Statement, you want

15   to ask some general questions about and stuff like

16   that, we'll be here for a while. And if you want to

17   ask some more questions, we'll do what we can to

18   answer them for you.

19   MR. CARPENTER: Any other?

20   Well, thank you all for coming. We do

21   appreciate your taking the time and effort to come

22   here. I hope that this was informative for you. And

23   I do appreciate the questions and the comments, both

24   in this meeting and in the earlier ones.

25   We will of course try to get the answers

NEAL R. GROSS
COURT REPORTERS AND TRANSCRIBERS
1323 RHODE ISLAND AVE , N.W.
(202) 234-4433   WASHINGTON, D.C. 20005-3701   www.nealrgross.com

NUREG-0498, Supp 2   E-156   May 2013

34

back to you if there's any other questions. We did have written questions in the afternoon session. If there's any further written questions, we will get back to you on those.

So this is, as we put it out earlier, is being transcribed. The transcription will be available later once that is completed.

So at this point unless questions, comments? Thank you all very much for coming. We do appreciate your time. Thank you.

(Whereupon, this meeting was concluded at 7:18 p.m.)

Appendix E

Choctaw Nation of Oklahoma

P.O. Box 1210 • Durant, OK 74702-1210 • (580) 924-8280

Gregory E. Pyle
Chief

Gary Batton
Assistant Chief

November 21, 2011

U. S. NRC
Attention: Mr. Pat Milano
11555 Rockville Pike
Rockville, MD 20852-2746

Dear Pat Milano:

We have reviewed the following proposed project (s) as to its effect regarding religious and/or cultural significance to historic properties that may be affected by an undertaking of the projects area of potential effect.

RE: Draft supplement to the Final Environmental Statement for Watts Bar Nuclear Plant Unit 2

Comments: After further review of the above mentioned project (s), and based on the information provided it has come to our attention that the project *is out of the Choctaw Nation of Oklahoma areas of interest*. A list of states and counties has been provided. If we can further assistance please contacted our office at 1-800-6170 ext. 2216.

Sincerely,

Ian Thompson PhD, RPA
Director Historic Preservation Department
Tribal Archaeologist, NAGRPA Specialist
Choctaw Nation of Oklahoma

By
Caren A. Johnson
Administrative Assistant

Choctaws...growing with pride, hope and success!

UNITED STATES ENVIRONMENTAL PROTECTION AGENCY
REGION 4
ATLANTA FEDERAL CENTER
61 FORSYTH STREET
ATLANTA, GEORGIA 30303-8960

12/15/2011

Stephen J. Campbell, Chief
Watts Bar Special Projects Branch
Division of Operating Reactor Licensing
Office of Nuclear Reactor Regulation
U.S. Nuclear Regulatory Commission
Washington, D.C. 20555-0001

Subject: EPA's Comments on the "Draft Final Environmental Statement, Related to the Operation of Watts Bar Nuclear Plant, Unit 2." Supplement, NUREG-0498, July 2011.

Dear Mr. Campbell,

Pursuant to Section 309 of the Clean Air Act (CAA) and Section 102(2)(C) of the National Environmental Policy Act (NEPA), the U.S. Environmental Protection Agency (EPA) Region 4 has reviewed the "Draft Final Environmental Statement, Related to the Operation of Watts Bar Nuclear Plant, Unit 2. Supplement, NUREG-0498, July 2011.

This Draft Final Environmental Impact Statement (DFEIS) is the results of The Tennessee Valley Authority (TVA), submitting to the U.S. Nuclear Regulatory Commission (NRC) On March 4, 2009, a request to reactivate its application for a license to operate a second light-water nuclear reactor (Unit 2) at the Watts Bar Nuclear (WBN) Plant in Rhea County, TN. EPA understands that the proposed action is NRC issuance of a 40-year facility operating license for WBN Unit 2. WBN Unit 2 is a pressurized-water reactor that could produce up to 3,425 megawatts thermal. This reactor-generated heat would be used to produce steam to drive steam turbines, by providing 1,160 megawatts electric of net electrical power capacity to the region.

For renewal of a license, EPA understands that Title 10 of the Code of Federal Regulations (10 CFR 51.95(c)) states that the NRC shall prepare a Supplemental DFEIS which is a Supplement 2, (NUREG-0498) to previously conducted Environmental Impact Statements, The current Draft FEIS that EPA reviewed serves to meet this requirement. EPA finds that this document appropriately includes an analysis that evaluates the environmental impacts of the proposed action of relicensing WBN Unit 2.

The environmental impacts from the proposed action are appropriately classified as SMALL, MODERATE, or LARGE. As set forth in the GEIS (generic environmental impact statement), Category 1 issues are those defined as meeting all of the following criteria:

- The environmental impacts associated with the issue are determined to apply either to all plants or, for some issues, to plants having a specific type of cooling system or other specified plant or site characteristics.
- A single significance level (i.e., SMALL, MODERATE, or LARGE) has been assigned to the impacts, except for collective offsite radiological impacts from the fuel cycle and from high-level waste and spent fuel disposal.
- Mitigation of adverse impacts associated with the issue is considered in the analysis, and it has been determined that additional plant-specific mitigation measures are likely not to be sufficiently beneficial to warrant implementation.

In summary, EPA notes the following assumptions and conclusions of the Draft FSEIS:

1. The NRC did not note any issues for air quality impacts, nor did the Staff find any new and significant information during the environmental review.

2. The NRC evaluated the direct and indirect impacts due to groundwater use during the license renewal term and concluded that the impacts would be SMALL.

3. The NRC did not find any new and significant surface water issues during the environmental review. The NRC evaluated the direct and indirect impacts of entrainment, impingement, and heat shock from continued operations during the license renewal term on fish and shellfish. After an extensive review including new information, NRC staff found that the adverse effects of entrainment and impingement would be small and would not destabilize or noticeably alter the aquatic biota of the Chickamauga Reservoir. EPA agrees that mitigation measures and the requirements of the NPDES permits would minimize the physical and thermal effects of the heated discharge on aquatic resources.

4. With regards to Solid Radioactive Waste, and Spent Fuel Storage, NRC has not fully determined long-term storage location for Classes B and C Low Level Waste (LLW). Based on the NRC staff's independent review of information since the 1978 FES-OL, NRC concluded that the environmental impacts of radioactive waste storage and disposal associated with WBN Unit 2 would be small.

5. With respect to environmental justice, the NRC also finds that no disproportionately high and adverse human health impacts would be expected in special pathway receptor populations in the region as a result of the consumption of water, local food, fish, and wildlife.

6. The Draft FSEIS however, does not mention the condition of the WBN Unit 2 facility. EPA recommends more discussion on the condition of the WBN Unit 2 physical condition relative to relicensing. NRC should discuss any historical maintenance activities that will demonstrate the condition and structural integrity of Unit 2. The identified additional information (data, analyses, and/or discussions) should be included (or referenced as appropriate) in the Final SEIS.

EPA's review of NRC's Draft FSEIS received an "EC-2" rating, meaning that there are environmental concerns with additional information requested in the FSEIS. Specifically, as outlined in EPA's comment letter dated May 14, 2007, referenced subject, TVA's Draft Supplement Environmental Impact Statement Watts Bar Nuclear Plant Unit 2. We are also, request additional clarifying information on the on-going structural safety analysis and repairs, upgrades and/or retrofits to Watts Bar Unit 2, be mentioned in the FSEIS.

In conclusion, the Draft FSEIS is clearly written and provides useful information for assessment of the proposal to finish and operate Watts Bar Unit 2. If you wish to discuss EPA's comments, please contact me at 404/562-9611 (mueller.heinz@epa.gov) or Larry Gissentanna, of my staff at 404-562-8248 (Gissentanna.larry@epa.gov).

Sincerely,

Heinz J. Mueller, Chief
NEPA Program Office
Office of Policy and Management

Reference Memo dated May 14 2007

UNITED STATES ENVIRONMENTAL PROTECTION AGENCY
REGION 4
ATLANTA FEDERAL CENTER
61 FORSYTH STREET
ATLANTA, GEORGIA 30303-8960

May 14, 2007

Ms. Ruth M. Horton
Tennessee Valley Authority
400 W. Summit Hill Drive, WT 11D-K
Knoxville, TN 37902

RE: EPA Review and Comments on
Draft Supplemental Environmental Impact Statement (DSEIS)
Completion and Operation of
Watts Bar Nuclear Plant Unit 2
CEQ No. 20070113

Dear Ms. Horton:

The U. S. Environmental Protection Agency (EPA), Region 4, reviewed the Draft Supplemental Environmental Impact Statement (DSEIS), pursuant to Section 309 of the Clean Air Act and Section 102 (2)(C) of the National Environmental Policy Act (NEPA). The purpose of this letter is to provide the Tennessee Valley Authority (TVA) with EPA's comments regarding potential impacts of the completion and operation of the Watts Bar Nuclear Plant Unit 2.

The proposed action of completing and operating the Watts Bar Nuclear plant Unit 2 would provide additional baseload capacity, and maximize the use of existing assets. The facility uses intakes from the Tennessee River for plant cooling, and discharges wastewater via three outfalls to the Tennessee River.

Based on EPA's review of the DSEIS, the project received an "EC-1" rating, meaning that environmental concerns exist. Specifically, protecting the environment involves the continuing need for appropriate storage and ultimate disposition of radioactive wastes generated on-site, as well as continuing measures to limit bioentrainment and other impacts to aquatic species from surface water withdrawals and discharges, and compliance with the NPDES Permit.

The National Pollutant Discharge Elimination System (NPDES) Permit Program authorizes the discharge of pollutants from certain facilities to waters of the United States. Administration of the NPDES permit program in Tennessee is delegated by EPA to the Tennessee Division of Water Pollution Control. The Watts Bar Nuclear Plant has an NPDES Permit issued by the Division of Water Pollution Control. The NPDES Permit limits specific pollutant discharges from the plant, requires monitoring of discharges, and regulates the flow and thermal impacts of discharges. The NPDES permittee has operated and is operating in compliance with the NPDES permit requirements.

The DSEIS acknowledges that continuing radiological monitoring of all plant effluents and appropriate storage of spent fuel assemblies and radioactive wastes on-site is required for this project. Ultimately, long-term radioactive waste disposition will require transportation of wastes to a permitted repository site. In particular, please address the following concerns in the FSEIS:

- Solid Radioactive Wastes (page 81): The shipping arrangements for Unit 2 after 2008 appear uncertain with Barnwell's closing. Please provide more information on the availability and disposal costs options for Clive, Utah facility, Sequoyah Nuclear Plant or other disposition options under consideration.

- Spent Fuel Storage (page 83): Clarify whether the referenced dry cask facility is being processed as a definite project with funding to construct it. Is Unit 2 operation contingent on this facility being constructed? Clarify where the current Unit 1 spent fuel is being stored. Does the capacity for this new facility consider the contingency of Yucca Mountain being indefinitely postponed? Is the data in Table 3-24 in addition to the data given for Unit 1, or the cumulative dimensions, capacity, etc.?

In conclusion, the DSEIS is clearly written and provides useful information for assessment of the proposal to finish and operate Unit 2. However, clarification is needed regarding radioactive waste disposition after 2008 and TVA's proposed Dry Cask storage plans. Thank you for the opportunity to comment on this document. We look forward to reviewing the FSEIS. If we can be of further assistance, please contact Ramona McConney of my staff at (404) 562-9615.

Sincerely,

Heinz J. Mueller, Chief
NEPA Program Office

Appendix E

United States Department of the Interior

OFFICE OF THE SECRETARY
Office of Environmental Policy and Compliance
Richard B. Russell Federal Building
75 Spring Street, S.W.
Atlanta, Georgia 30303

TAKE PRIDE®
IN AMERICA

ER 11/1021
9043.1

December 19, 2011

Cindy Bladey, Chief
Rules, Announcements, and Directives Branch (RADB)
Office of Administration
Mail Stop: TWB-05-B01M
U.S. Nuclear Regulatory Commission
Washington, DC 20555-0001

11/10/2011
76 FR 70169 (2)

Re: Comments on the Draft Supplemental Environmental Impact Statement (SEIS) Related to
        the Operation of the Watts Bar Nuclear Plant, Unit 2, Supplement 2. Rhea County, TN
        NUREG-0498, Docket Number 30-391

Dear Ms. Bladey,

The Department of the Interior (Department) has reviewed the Draft Supplemental
Environmental Impact Statement (SEIS) Related to the Operation of the Watts Bar Nuclear
Plant. We have no comments at this time. I can be reached on (404) 331-4524 or via email at
joyce_stanley@ios.doi.gov.

Sincerely,

Joyce Stanley, MPA
Regional Environmental Protection Assistant

for

Gregory Hogue
Regional Environmental Officer

cc: Jerry Ziewitz – FWS
        Brenda Johnson - USGS
        Anita Barnett – NPS
        Li-Tai Sikiu Bilbao – OSM
        OEPC – WASH

SUNSI Review Complete
Template = ADM-013

E-RIDS = ADM-03
Add = B. Clayton (boca)
J. Giacinto (JJga)
C. Fells (Cxf5)

STATE OF TENNESSEE
DEPARTMENT OF TRANSPORTATION
ENVIRONMENTAL DIVISION
SUITE 900 - JAMES K. POLK BUILDING
505 DEADERICK STREET
NASHVILLE, TENNESSEE 37243-0334
(615) 741-3655

December 20, 2011

Mr. Stephen J. Campbell
Watts Bar Special Projects Branch
Division of Operating Reactor Licensing
Office of Nuclear Reactor Regulation
Washington, DC 20555-0001

RE: Environmental Comment Review
      Watts Bar Nuclear Plant, Unit 2, Rhea County, TN

Dear Mr. Campbell:

The Director of the Environment Division has forwarded your recent letter for response. I have reviewed your letter and attachments concerning the Watts Bar Nuclear Plant, Unit 2, Rhea County, Tennessee.

At this time, the Tennessee Department of Transportation is unaware of any conflicts with your proposed project. Please feel free to contact me in the future if other questions arise.

Thank you for the opportunity to review this project.

Sincerely,

Ann Andrews
Environmental Documentation Office

ec: Mr. N.E. Christianson
     Ms Suzanne Herron

Appendix E

UNITED STATES
NUCLEAR REGULATORY COMMISSION
WASHINGTON, D.C. 20555-0001

November 2, 2011

Mr. Joe Carpenter, Chief
Environment and Planning Bureau
Tennessee Department of Transportation
James K. Polk Bldg
505 Deaderick Street, Suite 700
Nashville, TN 37243

SUBJECT   NOTICE OF AVAILABILITY OF THE DRAFT SUPPLEMENT TO THE FINAL
           ENVIRONMENTAL STATEMENT FOR WATTS BAR NUCLEAR PLANT, UNIT 2
           FOR PUBLIC COMMENT (

Dear Mr. Carpenter:

The U.S. Nuclear Regulatory Commission (NRC) staff has completed draft Supplement 2 to
NUREG-0498, "Final Environmental Statement Related to the Operation of Watts Bar Nuclear
Plant, Unit 2." Enclosed is a copy of the draft supplement and the associated Notice of
Availability that will be published in the *Federal Register*. This notice advises the public that the
draft supplement is available for public inspection at the NRC Public Document Room or from
the Publicly Available Records component of NRC's Agencywide Documents Access and
Management System (ADAMS). ADAMS is accessible from the NRC Website at
http://www.nrc.gov/reading-rm/adams.html (online in the NRC Library) and directly from the
NRC Website at www.nrc.gov.

The notice also informs the public that a public meeting on the draft supplement will be held at
the Magnuson Hotel at 1421 Murrays Chapel Road in Sweetwater, Tennessee, on Thursday,
December 8, 2011. As discussed in the enclosed notice, the staff is providing the public with
the opportunity to provide comments. The meeting will consist of two sessions, which will cover
the same subjects. The sessions will convene at 2:00 p.m. and 6:30 p.m. and will continue until
4:00 p.m. and 8:30 p.m., as necessary. The meeting will be transcribed and will include
(1) a presentation of the contents of the draft supplement to the environmental statement, and
(2) the opportunity for interested government agencies, organizations, and individuals to provide
comments on the draft report. Additionally, the NRC staff will host informal discussions 1 hour
before the start of each meeting session.

Please note that the public comment period for the draft Supplement 2 to NUREG-0498 ends on
December 27, 2011.

J. Carpenter   - 2 -

A separate Notice of Availability of the draft Supplement 2 to NUREG-0498 will be placed in the *Federal Register* through the U.S. Environmental Protection Agency. If you have any questions regarding this matter or wish to comment on the draft Supplement 2 to NUREG-0498, please contact Mr. Pat Milano at 301-415-1457 or by email at Patrick.Milano@nrc.gov.

Sincerely,

Stephen J. Campbell, Chief
Watts Bar Special Projects Branch
Division of Operating Reactor Licensing
Office of Nuclear Reactor Regulation

Docket No. 50-391

Enclosure
Draft NUREG-0498, Supplement 2

cc w/o encl. Listserv

ML12005A082

PUBLIC SUBMISSION

As of: January 03, 2012
Received: December 28, 2011
Status: Pending Post
Tracking No. 80f8a7b6
Comments Due: January 24, 2012
Submission Type: Web

Docket: NRC-2008-0369
Environmental Assessment and Finding of No Significant Impact: Watts Bar Nuclear Plant, Unit 2

Comment On: NRC-2008-0369-0016
Draft Supplement 2 to Final Environmental Impact Statement Related to the Operation of Watts Bar Nuclear Plant, Unit 2; Tennessee Valley Authority

Document: NRC-2008-0369-DRAFT-0006
Comment on FR Doc # 2011-32909

Submitter Information

Name: Donna Budnick
Address:
41 Brown Road West
Winchester, TN, 37398

General Comment

It is my understanding that TVA's nuclear power plants cannot withstand a 7 or great earthquake. Please rethink continuing to build/refurbish such a power plant until such time as one can be built to withstand a severe earthquake. Watts Bar is not so very far away from the New Madrid fault line

PUBLIC SUBMISSION

As of: January 10, 2012
Received: January 04, 2012
Status: Pending_Post
Tracking No. 80f8d8b0
Comments Due: January 24, 2012
Submission Type: Web

Docket: NRC-2008-0369
Environmental Assessment and Finding of No Significant Impact: Watts Bar Nuclear Plant, Unit 2

Comment On: NRC-2008-0369-0016
Draft Supplement 2 to Final Environmental Impact Statement Related to the Operation of Watts Bar Nuclear Plant, Unit 2; Tennessee Valley Authority

Document: NRC-2008-0369-DRAFT-0007
Comment on FR Doc # 2011-32909

11/10/2011
76 FR 70169   (3)

Submitter Information

General Comment

SAMA is incomplete. It is based on a PRA that does not comply with applicable RG or ASME/ANS standard.

SUNSI Review Complete
Template = ADM-013

E-RIDS = ADM-03
(add = B. Clayton (bac2)
J. Giacinto (jfg2)
C. Fells (cxf5)

https://fdms.erulemaking.net/fdms-web-agency/component/contentstreamer?objectId=0900006480f8d8b...   01/10/2012

the **Chickasaw Nation** HEADQUARTERS

Arlington at Mississippi / Box 1548 / Ada, OK 74821-1548 / (580) 436-2603

Bill Anoatubby
Governor

Jefferson Keel
Lieutenant
Governor

January 19, 2012

Mr. Patrick Milano
Project Manager
US Nuclear Regulatory Commission
Washington, D. C. 20555-0001

Dear Mr. Milano:

    Thank you for your letter regarding the Notice of Availability of the draft supplement to the Final Environmental Statement for Watts Bar Nuclear Plant Unit 2 for public comment.

    Through your notification the Chickasaw Nation is made aware that the US Nuclear Regulatory Commission is reviewing an application submitted by the Tennessee Valley Authority for an operating license for Watts Bar Nuclear Plant (WBN), Unit 2 located in Rhea County, Tennessee. We are in agreement with the NRC to review the permit.

    After reviewing your information, we are in agreement with the assessment and have no objections to the proposed undertaking. We concur with your finding of no adverse effect to historic properties and we accept the special conditions set forth in this report. We do not presently know of any specific historic properties or properties of significant religious or sacred value. In the event your agency becomes aware of the need to enforce other statutes we request to be notified under NEPA, NAGPRA, AIRFA and ARPA.

    If you have any questions, please contact Ms. LaDonna Brown, historic preservation officer at (580)272-5593, Ladonna.brown@chickasaw.net.

Sincerely,

Jefferson Keel, Lt. Governor
The Chickasaw Nation

United States Department of the Interior

OFFICE OF THE SECRETARY
Office of Environmental Policy and Compliance
Richard B. Russell Federal Building
75 Spring Street, S.W.
Atlanta, Georgia 30303

ER 11/1021
9043.1

January 23, 2012

Justin Poole
U.S. Nuclear Regulatory Commission
Office of Administration
Mail Stop: TWB-05-B01M
Washington, DC 20555-0001

Re: Comments and Recommendations on the Draft Supplemental Environmental Impact
        Statement (SEIS) Related to the Operation of Watts Bar Nuclear Plant, Unit 2,
        Supplement 2 in Rhea County, TN

Dear Mr. Poole:

The United States Department of the Interior (Department) has reviewed the Draft Supplemental
Environmental Impact Statement (DSEIS) for the Operation of Watts Bar Nuclear Plant and
offers the following comments.

General

The electronic version of many of the figures in chapters 2, 3, and 4 are difficult to interpret. In
particular, figures 2-4 and 2-5 (pgs. 2-15 and 2-16) are inadequate to allow the reviewer to
identify the important water related features. We suggest that the final version include legible
versions of these figures; this will allow the public to better assess the adequacy of the
information provided in the text.

Additionally, we have reviewed the referenced Environmental Review Distribution Transmittal
and the NRC DSEIS, including a Biological Assessment for the federally endangered gray bat
(Myotis grisescens), pink mucket *Lampsilis abrupta),* rough pigtoe *(Pleuroberna plenum),*
fanshell *(Cyprogenia Ste garia),* dromedary pearly mussel *(Dromus dromas),* orangefoot
pimpleback *(Plethobasus cooperianus),* and the threatened snail darter *(Percina tanasi).* Since
the preparation of the DSEIS, the laurel dace *(Chrosomus saylori)* was listed as endangered (76
FR 48722 4874 1) on September 8, 2011, and is known to occur with the project assessment
area. The sheepnose mussel *(Plethobasus* cyphyus) is proposed for listing as endangered (76 FR
3392 3420) and occurs in the project assessment area. The Tennessee Ecological Services Field

Watts Bar Nuclear Plant – ER 11/1021

Office has completed section 7 consultation with the NRC on the project and has no other substantive comments on the DSEIS to offer at this time.

Thank you for the opportunity to review and comment on the DSEIS. If you have any questions concerning our comments, please contact Steve Alexander at (931) 528-6481 (ext. 210) or via email at steven_alexander@fws.gov. I can be reached on (404) 331-4524 or via email at joyce_stanley@ios.doi.gov.

Sincerely,

Joyce Stanley, MPA
Regional Environmental Protection Assistant

for

Gregory Hogue
Regional Environmental Officer

cc: Jerry Ziewitz – FWS
Brenda Johnson - USGS
Anita Barnett – NPS
Li-Tai Sikiu Bilbao – OSM
OEPC – WASH

2

Harmon, Curran, Spielberg + Eisenberg LLP

(H)(C)(S•E)

1726 M Street NW, Suite 600
Washington DC 20036-4523

202.328.3500 | office
202.328.6918 | fax
HarmonCurran.com

January 24, 2012

Cindy Bladey, Chief
Rules, Announcements, and Directives Branch
Office of Administration, Mail Stop TWB-05-B01M
U.S. Nuclear Regulatory Commission
Washington, D.C. 20555-0001
Posted on: www.regulations.gov

SUBJECT: *SACE Comments on DEIS for Watts Bar Unit 2, Docket ID NRC-228-0396*

Dear Ms. Bladey:

On behalf of Southern Alliance for Clean Energy ("SACE"), I am writing to submit comments on the Draft Environmental Impact Statement ("DEIS") for the proposed Watts Bar Unit 2 ("WBN2") nuclear power plant. SACE has been admitted as an intervenor in the operating license proceeding for WBN2, where it has raised two important environmental issues that the Tennessee Valley Authority's ("TVA's") Final Supplemental Environmental Impact Statement ("FSEIS") for WBN2 has failed to address or resolve: impacts to the aquatic environment and environmental concerns raised by the Fukushima Accident and the Fukushima Task Force. Because the DEIS does not correct these failures, we believe it violates the National Environmental Policy Act ("NEPA").

Aquatic Impacts. In Contention 7 (which was admitted by the Atomic Safety and Licensing Board ("ASLB") in *Tennessee Valley Authority* (Watts Bar nuclear Plant, Unit 2), LBP-09-26, 70 NRC 939, 981-90 (2009)), SACE has challenged the adequacy of TVA's FSEIS for WBN2 to address the impacts of WBN2 on aquatic organisms. A copy of Contention 7, including the supporting declaration of Dr. Shawn Paul Young, is attached for your consideration as Exhibit 1. Although TVA conducted additional environmental studies that were intended to address our concerns, they are not sufficient to support TVA's claim that the aquatic environmental impacts of WBN2 are insignificant. TVA also distorts the aquatic impacts of WBN2 by characterizing the baseline condition of the Tennessee River as a reservoir rather than a free-flowing river that has been adversely affected by dams and industrialization.

The DEIS has not resolved the issues raised in Contention 7 because it merely adopts the analysis and conclusions of TVA's FSEIS with respect to aquatic impacts. Our continuing concerns about the inadequacy of TVA's environmental analysis are documented in Contention 7 and the attached response by SACE to a recent summary disposition motion by TVA: Southern Alliance for Clean Energy's Opposition to Tennessee Valley Authority's Motion for Summary Disposition of Contention 7 Regarding Aquatic Impacts of Watts Bar Unit 2 and attached Declaration of Shawn Paul Young, Ph.D. A copy of SACE's response to TVA's summary disposition motion is attached for your consideration as Exhibit 2.

Environmental Implications of Fukushima Accident. SACE has also submitted a contention challenging TVA's failure to amend the FSEIS for WBN2 to address the environmental

Harmon, Curran, Spielberg + Eisenberg LLP

implications of the Fukushima Task Force Report issued by the NRC in July 2011: Motion to Admit new Contention Regarding the Safety and Environmental Implications of the Nuclear Regulatory Commission Task Force Report on the Fukushima Dai-ichi Accident (August 11, 2011). A copy of our contention, including the supporting declaration of Dr. Arjun Makhijani, is attached for your consideration as Exhibit 3. We are awaiting a ruling on the admissibility of the contention by the ASLB. The concerns raised by the Fukushima Task Force and our contention remain unaddressed by the DEIS, which does not even mention the Fukushima accident.

We believe the deficiencies in outlined in Exhibits 1 through 3 are very grave. By TVA's own admission, the Tennessee River *"is the most diverse temperate freshwater ecosystem in the world."* Programmatic EIS for Reservoir Operations Study, § 4.7.[1] Neither TVA nor the NRC Staff has grappled with the significance of the impacts of WBN2 to aquatic organisms, and thus they have given no serious consideration to mitigation measures that could protect the fragile and extraordinarily important ecosystem of the Tennessee River.

The NRC's failure to address the environmental implications of the Fukushima Task Force is also extremely grave, given that the Task Force has called for a complete upgrade of the NRC's program for mandatory safety regulations and has targeted WBN2 for specific recommendations. For these reasons, we believe the DEIS fails to satisfy NEPA in significant respects.

Sincerely,

/s/
Diane Curran
Counsel to SACE

---

[1] http://www.tva.gov/environment/reports/ros_eis/.

July 13, 2009

UNITED STATES OF AMERICA
NUCLEAR REGULATORY COMMISSION

BEFORE THE SECRETARY

In the Matter of)	
)	Docket No. 50-391
Tennessee Valley Authority)	
)	
(Watts Bar Unit 2))	

PETITION TO INTERVENE AND REQUEST FOR HEARING

I.   INTRODUCTION

Pursuant to 10 C.F.R. § 2.309 and the notice published by the Nuclear Regulatory

Commission ("NRC" or "Commission") at 74 Fed. Reg. 20,350 (May 1, 2009), Petitioners

Southern Alliance for Clean Energy ("SACE"), Tennessee Environmental Council ("TEC"), We

the People ("WTP"), the Sierra Club, and Blue Ridge Environmental Defense League

("BREDL") hereby request a hearing and petition to intervene in this proceeding regarding the

Tennessee Valley Authority's ("TVA's") updated application for a facility operating license

("OL") for the Watts Bar Nuclear Plant ("WBN") Unit 2. Petitioners' standing to intervene is

described in Section II of this pleading, and Petitioners' contentions are set forth in Section III.

This proceeding is highly unusual in that TVA's updated OL application follows a

lengthy hiatus in the WBN Unit 2 OL proceeding: TVA submitted its Final Environmental

Statement for construction of WBN Units 1 and 2 in 1972 (TVA, Final Environmental

Statement, Watts Bar Nuclear Plant Units 1 and 2 (1972) ("FES")), and was issued construction

permits for both units in January 1973. Final Supplemental Environmental Impact Statement for

the Completion and Operation of Watts Bar Nuclear Plant Unit 2, at 5 (2007) ("FSEIS"). TVA

of mitigative measures must follow an analysis of impacts and be informed by it, otherwise it is meaningless.

TVA's FSEIS and SAMA analysis are thus insufficient to satisfy the requirements of NEPA, because they fail to address the environmental impacts of aircraft attacks on WBN Unit 2. As the Power Reactor Security Rule and Aircraft Impacts Rule clearly show, the Commission regards an aircraft attack on WBN as a reasonably foreseeable event and NEPA therefore requires TVA to present a more complete impact analysis. *San Luis Obispo Mothers for Peace v. NRC*, 449 F.3d 1016, 1031 (9th Cir. 2006) (finding that "the possibility of a terrorist attack [on a nuclear power plant] is not so 'remote and highly speculative' as to be beyond NEPA's requirements.")[17]

Contention 7:   **Inadequate Consideration of Aquatic Impacts**

TVA claims that the cumulative impacts of WBN Unit 2 on aquatic ecology will be insignificant (FSEIS Table S-1 at page. S-2, and Table 2-1 at page. 30). TVA's conclusion is not reasonable or adequately supported, and therefore it fails to satisfy 10 C.F.R. § 51.53(b) and NEPA.

TVA's discussion of aquatic impacts is deficient in three key respects. First; TVA mischaracterizes the current health of the ecosystem as good, and therefore fails to evaluate the impacts of WBN2 in light of the fragility of the host environment. Second, TVA relies on outdated and inadequate data to predict thermal impacts and the impacts of entrainment and impingement of aquatic organisms in the plant's cooling system. Third, TVA fails completely to

---

[17] While Petitioners recognize that the Commission has refused to apply the *Mothers for Peace* decision as precedent in circuits other than the U.S Ninth Circuit, Petitioners believe that this position is inconsistent with NEPA and that the decision should, therefore, be applied in all reactor licensing decisions.

31

analyze the cumulative effects of WBN2 when taken together with the impacts of other industrial facilities and the effects of the many dams on the Tennessee River.

Basis and Discussion

This contention is supported by the expert declaration of Dr. Shawn Paul Young (July 11, 2009) (Attachment 6) ("Young Declaration").

WBN's cooling system has two sets of cooling water intakes, located at different points along the Tennessee River, and one set of outfalls. The original cooling system for WBN was a closed-cycle cooling system, with intakes and outfalls located on the upper end of Lake Chickamauga. In 1998, when the closed cycle cooling system proved insufficient, TVA supplemented WBN's intake capacity by converting the intakes from an unused fossil fuel plant to a Supplemental Condenser Cooling Water ("SCCW") system for WBN Unit 1. In effect, the SCCW system is a once-through cooling system. FSEIS at 24. The intake for the SCCW is at the lower end of Watts Bar Reservoir, which lies upstream of the WBN plant. TVA continued to use the original outfall from the unused fossil fuel plant on Lake Chickamauga, however. WBN thus currently withdraws water from intake structures at two different locations, and it also discharges thermal effluent through two different outfalls.

TVA's finding that WBN Unit 2 will have no significant impacts on aquatic life in the Tennessee River is inadequately supported in the following respects:

1.   TVA's conclusion that cumulative impacts will be insignificant is based on the faulty premise that the aquatic ecosystem that will be affected by WBN Unit 2 is currently in a good state of health. In fact, data in TVA's own environmental studies, as well as available literature, show that the health of the Tennessee River ecosystem, including Lake Chickamauga where WBN Units 1 and 2 are located, is damaged, fragile, and quite vulnerable to the additional

32

impacts that would be posed by WBN Unit 2's cooling water system. Young Declaration at ¶ III.A.1.

The Tennessee River is an extraordinarily diverse and unique ecosystem that supports over 200 fish species, including twenty species that are found only in the Tennessee River. Young Declaration at ¶ III.B.1. Yet the ecosystem also harbors the highest number of imperiled species of any large river basin in North America. *Id.* at ¶ III.B.2. TVA incorrectly portrays the ecosystem as healthy, when its health and diversity are actually in steep decline. *Id.* at ¶¶ III.C.1-9. TVA asserts, for example, that the freshwater mussel communities are in "excellent" health because their population is "constant." But, in fact, the mussel population is only constant because it is not reproducing, which is a sign of poor health. *Id.*

By characterizing the health of fish and benthic organisms as "good" or "excellent," TVA rationalizes its failure to take a hard look at the reasons why these species are declining. While dams may be the primary cause of these ill effects, they are not the only contributor. *Id.* at ¶ III.C.10. TVA has not taken the necessary steps to evaluate how the effluent from WBN Units 1 and 2 may contribute to the stresses on the fragile health of fish communities, or how these facilities may interfere with mussel reproduction. *Id.* at ¶¶ III.C.6,9.

2.   TVA relies on outdated and inadequate data to predict the effects of WBN Unit 2's cooling system on fish, mussels, and other aquatic organisms. In particular, the FSEIS understates the potential impacts of the coolant intake system (*i.e.,* entrainment and impingement) and the thermal impacts of the coolant discharge system on fish and benthic organisms, by relying on poor or outdated data, distorted interpretations of data, and assumptions and extrapolations in lieu of recent monitoring studies. Young Declaration at ¶ III.A.2.

33

Given their lack of mobility, fish eggs and most fish larvae cannot escape the intake flow velocity and are sucked into the intake canal and cooling system. Phytoplankton and zooplankton, which constitute important food sources for fish, mussels, and aquatic insects, may also be entrained due to their lack of mobility. Fish and other organisms pass through the plant's cooling system, suffering injury or death though physical contact, rapid pressure or temperature change, and chemical poisoning from biocides and other chemicals introduced into the water. *Id.* at ¶ III.D.5.

. Knowledge of the ichthyoplankton population distribution in relation to intakes across time and space is very important to an understanding of entrainment impacts, because ichthyoplankton tend to be patchy (high numbers clumped into a specific portion of the water column). This patchy distribution creates a high level of vulnerability to entrainment mortality if the organisms are located near intakes, because they cannot simply avoid the intakes. But TVA has not collected sufficient data to understand the distribution of icthyoplankton populations or how they are affected by the Watts Bar intakes. That is because TVA has not taken direct measurements of entrainment, even though direct measurements are recommended by the U.S. Environmental Protection Agency. Instead, it has extrapolated entrainment estimates from outdated and inadequate data. *Id.* at ¶¶ III.D.7-10.

TVA's conclusion that entrainment impacts are insignificant is based upon an unsupported assumption: that population densities are uniform across the river channel and from the surface to the bottom of the river. The data do not support this assumption, however, because the numbers are all relative, expressed in percentages. It is therefore impossible to determine what the actual populations of organisms are. *Id.* at ¶¶ III.D.11-13. TVA also does not provide

34

any data for fish eggs, which may be found in high abundance during different times of the year and are very vulnerable to entrainment. *Id.* at ¶¶ III.D.14-15.

TVA's impingement data are likewise inadequate to support the FSEIS' finding of no significant impact. For instance, TVA failed to follow-up on a survey conducted at the SCCW intake that found an increased level of impingement in comparison to earlier surveys. *Id.* at ¶ III.D.16. TVA also failed to update the thirty-five-year-old data on which it relied for its conclusions about impingement impacts at the WBN Unit 1 intake. Additionally, TVA inappropriately treats its impingement data for the Lake Chickamauga and Watts Bar Reservoir intakes as if they were the same. The vicinities of the two intakes, however, have very different habitat characteristics and are therefore likely to support very different populations of aquatic organisms. *Id.* at ¶ III.D.17.

TVA further states that thermal impacts will be insignificant, even though TVA's conclusions are contradicted by its own acknowledgement of the need to relocate mussels in the vicinity of the SCCW discharge to avoid mortality from elevated temperatures. *Id.* at ¶ III.E.2. And TVA provides no evidence, such as scientific studies or field observations, to justify its conclusion. *Id.* For instance, TVA is missing a number of basic data sets with respect to thermal impacts, including data on overall drift communities, and data on spatial and temporal distribution of ichthyoplankton in relation to thermal mixing zones. *Id.* at ¶ III.E.3.a. Other factors neglected by TVA (which must be understood in order to properly assess thermal impacts on aquatic life), include characteristics of the thermal plume; variations in the size and temperature profile of the mixing zone; the temperatures in the core of the thermal plume (rather than at the edge) and whether they have an effect on aquatic organisms; and the effects of high

35

temperatures on fish eggs and larvae, which are highly vulnerable to elevated and rapidly changing temperature. *Id.* at ¶¶ III.E.3.b-f.

Finally, TVA fails to show that it has accounted for the impacts of overflow from the holding ponds, where excess cooling water may be stored at very high temperatures. *Id.* at ¶ III.E.4.

3.   TVA does not adequately address the cumulative impacts of WBN Unit 2 in conjunction with the impacts of the numerous water impoundments on the Tennessee River, or with other industrial facilities such as the ten fossil fuel-burning plants, the six operating nuclear reactors, and the five additional reactors for which TVA has sought operating licenses. Each of these facilities affects the Tennessee River continuum. That is, each facility not only affects the immediate environment, but those changes are then felt throughout the river as a domino effect.

The portion of the Tennessee River in the vicinity of WBN is an important part of the river continuum, as are all other segments of the river. Each segment has its own complex ecological balance that is required to support a diverse population of fish and other organisms, providing different habitats needed at different life history stages that must match available food and habitat needs in time and space. Each new industrial facility that is added to the environment will compound the existing disruptions to these interrelated aquatic ecosystems, and further remove the Tennessee River from any semblance of the natural state which would be necessary to restore or even halt the deterioration of the hundreds of declining, threatened, and endangered aquatic species in the Tennessee River Basin. Young Declaration at ¶ III.A.3.

The FSEIS is thus inadequate because it does not contain a discussion of these cumulative industrial impacts or the degree to which WBN Unit 2 will contribute to them.

36

IV. CONCLUSION

For the foregoing reasons, Petitioners have demonstrated that they have standing and that

their contentions are admissible. Therefore, they are entitled to a hearing on their contentions.

Respectfully submitted,

Diane Curran
Matthew D. Fraser
HARMON, CURRAN, SPIELBERG, & EISENBERG, L.L.P.
1726 M Street N.W., Suite 600
Washington, D.C. 20036
202-328-3500
Fax: 202-328-6918
e-mail: dcurran@harmoncurran.com
      mfraser@harmoncurran.com

July 13, 2009

37

December 20, 2011

UNITED STATES OF AMERICA
NUCLEAR REGULATORY COMMISSION
BEFORE THE ATOMIC SAFETY AND LICENSING BOARD

In the Matter of)	
)	Docket No. 50-391-OL
Tennessee Valley Authority)	
)	
(Watts Bar Unit 2))	

SOUTHERN ALLIANCE FOR CLEAN ENERGY'S OPPOSITION
TO TENNESSEE VALLEY AUTHORITY'S MOTION
FOR SUMMARY DISPOSITION OF CONTENTION 7 REGARDING
AQUATIC IMPACTS OF WATTS BAR UNIT 2

I.   INTRODUCTION

Pursuant to 10 C.F.R. §§ 2.1205 and the Atomic Safety and Licensing Board's

("ASLB's") orders of May 26, 2010 and December 1, 2011, Southern Alliance for Clean

Energy ("SACE") hereby responds to Tennessee Valley Authority's ("TVA's") Motion

for Summary Disposition of Contention 7 (Nov. 21, 2011). This response is supported by

the attached Statement of Disputed Material Facts and Declaration of Dr. Shawn Young

(Dec. 20, 2011) ("Young Declaration").

As discussed below and as demonstrated in the Statement of Disputed Material

Facts and Young Declaration, TVA fails to demonstrate that the concerns raised in

Contention 7 have been resolved by recent studies conducted by TVA. To the contrary,

as discussed in Dr. Young's declaration, although the new data is incomplete and

inaccurately analyzed, it shows that Watts Bar Unit 1 ("WBN1") has a significant impact

on the environment that will be exacerbated by the operation of Watts Bar Unit 2

("WBN2"). Therefore there is a genuine dispute of material fact between the parties and summary disposition should be denied.

II.   STANDARD FOR SUMMARY DISPOSITION

NRC regulations at 10 C.F.R. § 2.1205 govern summary disposition motions and direct Licensing Boards to "apply the standards for summary disposition set forth in Subpart G."[1] Under Subpart G, summary disposition is appropriate if the filings in the proceedings, statements of the parties and affidavits, if any, "show that there is no genuine issue as to any material fact and that the moving party is entitled to a decision as a matter of law."[2] In a motion for summary disposition, the moving party bears the burden to demonstrate the absence of a genuine issue as to any material fact.[3] Any doubt as to the existence of a genuine issue of material fact is resolved against the moving party.[4] "Because the burden is on the moving party, the Board must examine the record in the light most favorable to the non-moving party and give the non-moving party the benefit of all favorable inferences that can be drawn from the evidence."[5]

A party opposing a motion for summary disposition need not show a likelihood of success on the merits, but rather, only that there is a genuine issue of fact to be evaluated

---

[1] 10 C.F.R. § 2.1205(c).
[2] *Id.* § 2.710(d)(2).
[3] *Id.* § 2.325; *Advanced Med. Sys., Inc.* (One Factory Row, Geneva, Ohio, 44041), CLI-93-22, 38 NRC 98, 102 (1993); *Entergy Nuclear Vermont Yankee LLC* (Vermont Yankee Nuclear Power Station), LBP-06-5, 63 NRC 116, 121 (2006) (quoting *Private Fuel Storage, LLC* (Independent Spent Fuel Storage Installation), LBP-01-39, 54 NRC 497 (2001).
[4] *Entergy Nuclear Vermont Yankee LLC* (Vermont Yankee Nuclear Power Station), LBP-06-5, 63 NRC 116, 121 (2006) (citing *Advanced Med. Sys., Inc.* (One Factory Row, Geneva, Ohio, 44041), CLI-93-22, 38 NRC 98, 102 (1993)).
[5] *Id.*

2

at the evidentiary hearing.[6] Indeed, summary disposition "is not a tool for trying to convince a Licensing Board to decide, on written submissions, genuine issues of material fact that warrant resolution at a hearing."[7] A licensing board should not conduct a "trial on affidavits," but rather "determine whether there is a genuine issue for [hearing]."[8] In making this determination, "the evidence of the non-movant is to be believed, and all justifiable inferences are to be drawn in his favor."[9]

Moreover, summary disposition is rarely appropriate when conflicting expert opinions are involved.[10] Indeed, "competing expert opinions present the 'classic battle of the experts' and it [is] up to [the finder of fact] to evaluate what weight and credibility each expert opinion deserves."[11] At the summary disposition stage, "[r]egardless of the level of the dispute . . . it is not proper for a Board" to choose which expert has the better of the argument.[12]

---

[6] *Advanced Med. Sys., Inc.* (One Factory Row, Geneva, Ohio, 44041), CLI-93-22, 38 NRC 98, 102 (1993)

[7] *Entergy Nuclear Vermont Yankee LLC* (Vermont Yankee Nuclear Power Station), LBP-06-5, 63 NRC 116 , 121 (2006) (*quoting Private Fuel Storage, L.L.C.* (Independent Spent Fuel Storage Installation), LBP-01-39, 54 N.R.C. 497, 509 (2001)).

[8] *Entergy Nuclear Generation Co. and Entergy Nuclear Operations, Inc.* (Pilgrim Nuclear Power Station), CLI-10-11, 71 NRC 287, 297 (2010).

[9] *Id.* (quoting *Anderson v. Liberty Lobby,* 477 U.S. 242, 247-48 (1986).

[10] *Entergy Nuclear Vermont Yankee LLC* (Vermont Yankee Nuclear Power Station), LBP-06-5, 63 NRC 116 , 122 (2006) (citing *Phillips v. Cohen,* 400 F.3d 388, 399 (6th Cir. 2005)).

[11] *Id.*

[12] *Entergy Nuclear Vermont Yankee LLC* (Vermont Yankee Nuclear Power Station), LBP-06-5, 63 NRC 116, 121 (2006) (citing Private Fuel Storage, L.L.C. (Independent Spent Fuel Storage Installation), LBP-01-39, 54 NRC 497, 510 (2001)).

3

III.   ARGUMENT

A.   **TVA Mischaracterizes the Requirements of NEPA.**

With respect to two issues TVA claims that it is entitled to summary disposition as a matter of law under the National Environmental Policy Act ("NEPA"). But TVA misinterprets NEPA and ignores the salient facts of this case. First, TVA asserts that it is permissible under NEPA to use the "current" *i.e.,* degraded condition of the aquatic ecosystem in the Tennessee River near WBN1 and WBN2 as a "baseline" to evaluate the cumulative impacts of WBN2. TVA Motion at 21. In support of this proposition, TVA cites *Calvert Cliffs Unit 3 Nuclear Project, LLC* (Combined License Application for Calvert Cliffs Unit 3), LBP-09-04, 69 NRC 170, 201 (2009), in which the Atomic Safety and Licensing Board ("ASLB") approved the use of a baseline that effectively constituted a "snapshot" of the current condition of the aquatic environment. But, as the ASLB recognized in *Calvert Cliffs,* NEPA sets no hard and fast rule regarding appropriate baseline conditions and instead calls for application of a "rule of reason." *Id.* at 203. In each case, the appropriate scope of the baseline for a project is a "functional project: an applicant must provide enough information and in sufficient detail to allow for an evaluation of important impacts." *Southern Nuclear Operating Co.* (Early Site Permit for Vogtle ESP Site), LBP-07-3, 65 NRC 237, 256 (2007) (citing Office of Nuclear Reactor Regulation, "Standard Review Plans for Environmental Review for Nuclear Power Plants," NUREG-155 at 4.3.2-1 to -2 (Oct. 1999); office of Nuclear Regulatory Research, General Site Suitability Criteria for Nuclear Power Stations, Regulatory Guide 4.7 at 4.7-14 to -15 (rev. 2, April 1998).

4

In *Calvert Cliffs*, the ASLB did not rule out the possibility that past conditions could be relevant to a baseline analysis. 69 NRC at 203. Instead, it found that the petitioners had "not justified requiring individual examination of the environmental effects of reactors located at a substantial distance from the Calvert Cliffs site."

In this case, SACE has made a very strong case that historic conditions in the Tennessee River are uniquely relevant to the cumulative impacts of WBN2. As discussed in Contention 7, the Tennessee River "is an extraordinarily diverse and unique ecosystem that supports over 200 fish species, including twenty species that are found only in the Tennessee River." Petition to Intervene at 33. Moreover, the river "harbors the highest number of imperiled species of any large river basin in North America." *Id.* TVA does not dispute these assertions. In fact, TVA itself has recognized that the Tennessee River is a unique environmental resource from not just a national perspective but a global one:

> Aquatic resources occurring in the TVA region are important from local, national, and global perspectives. Tennessee has approximately 319 fish species, including native and introduced species, and 129 freshwater mussels (Etnier and Starnes 1993), Parmalee and Bogan 1998). The Tennessee-Cumberland Rivers have the highest number of endemic fish, mussel, and crayfish species in North America (Schilling and Williams 2002). *This is the most diverse temperate freshwater ecosystem in the world.*

Programmatic EIS for Reservoir Operations Study, § 4.7.[13] Clearly, any impacts to an ecosystem that is unique in the entire *planet* for its diversity are "important." *Southern,* 65 NRC at 256. To accept TVA's assertion that for purpose of an EIS affecting this unique ecosystem, current deteriorated condition could be considered appropriate for purposes of evaluating impacts and alternatives would be equivalent to pounding nails

---

[13] (http://www.tva.gov/environment/reports/ros_eis/).

5

into its coffin. If the narrow species diversity of a reservoir is considered the baseline for the WBN2 environmental analysis, then any hope of mitigation measures to sustain or restore the vestiges of diversity that remain will be effectively extinguished by the environmental analysis whose purpose is to protect the environment.

For instance, as Dr. Young discusses in his Declaration in Section F, TVA operates the dams and the power plants on the Tennessee River as a single system. This system includes ten different tributaries with a high number of fish and mussel species. By failing to use a baseline that takes into account the fragile health of these tributaries, TVA effectively writes off any mitigation measures that could aid their survival and consigns them to oblivion. This is the type of blindered and harmful decision-making that NEPA was intended to avoid. *Robertson,* 490 U.S. at 349. Such a result would be all the more egregious in light of the fact that the indefinite existence of dams on major rivers is no longer a foregone conclusion. As reported on the American Rivers website, over 600 dams in the U.S. have been removed over the past 50 years.[14]

The licensing and operation of WBN2 is just one of many industrial projects that will affect the aquatic health of the Tennessee River. If TVA and the NRC are allowed to ignore the true baseline condition of the river in the EIS for Watts Bar, then not only is any opportunity for mitigation of the effects of WBN2 lost, but future decisions will be affected by the bad assumptions of these EISs. That outcome is not consistent with the purposes of NEPA.

---

[14] http://www.americanrivers.org/our-work/restoring-rivers/dams/projects/2011-dam-removal-resource-guide.html.

6

TVA also asks the ASLB to dismiss SACE's claim that TVA fails to show that it accounted for the hydrothermal impacts of overflow from the holding pond, on the ground that the holding pond has never been used. TVA Motion at 19. According to TVA, this shows that use of the holding pond is a "worst case scenario" which need not be addressed under Supreme Court precedent. *Id.* (citing *Robertson v. Methow Valley Citizens Council*, 490 U.S. 332, 354-55 (1989)). This argument is absurd. Clearly, TVA anticipated that a holding pond might be needed, otherwise TVA would never have included a holding pond in its design. Thus, the potential need for the holding pond can hardly be characterized as "speculative." TVA Motion at 19. TVA's argument therefore should be rejected.

B.   TVA Has Failed To Demonstrate that Facts Material to the Claims Of Contention 7 are Undisputed.

1.   Claims of Contention 7

Contention 7 challenges the adequacy of TVA's Final Supplemental Environmental Impact Statement ("FSEIS") for the proposed Watts Bar Unit 2 nuclear power plant ("WBN2"). Contention 7 disputes the reasonableness of the FSEIS' conclusion that the cumulative impacts of WBN2 on the aquatic ecology of the Tennessee River are insignificant in three respects. First, TVA mischaracterizes the current health of the ecosystem as good, and therefore fails to evaluate the impacts of WBN2 in light of the fragility of the host environment. Second, TVA relies on outdated and inadequate data to predict thermal impacts and the impacts of entrainment and impingement of aquatic organisms in the plant's cooling system. Third, TVA fails completely to analyze the cumulative effects of WBN2 when taken together with the

7

impacts of other industrial facilities and the effects of the many dams on the Tennessee River.

2. Summary disposition of Contention 7 is inappropriate because material facts are in dispute.

TVA claims that it has conducted studies that resolve the three major deficiencies identified in Contention 7. As discussed in Dr. Young's attached Declaration, this is not correct. With respect to the inadequacy of TVA's previous data and analyses, TVA has made some progress by collecting new data on entrainment, impingement, freshwater mussels, and thermal impacts during 2010. But TVA has only started to catch up with its failure to collect the appropriate data that would be reasonably sufficient to evaluate impacts on aquatic resources by collecting only one year of data for entrainment, impingement, freshwater mussels, and thermal impacts over the preceding years. TVA still has not collected an amount of data that is reasonably necessary to evaluate the effects of WBN1 on aquatic organisms in the Tennessee River, and therefore it does not have enough information to extrapolate the impacts of WBN2. Young Declaration, par. II-2.

In addition, there are still big gaps in the information that TVA has collected. For example, TVA collected entrainment data for the Condenser Cooling Water ("CCW") system only and did not include the Supplemental Condenser Cooling Water ("SCCW") system. In addition, TVA did not collect impingement data for all key locations. And TVA's Hydrothermal Study does not address important parameters such as Outfall 101 or the amount of time that fish larvae remain in the thermal plume. Young Declaration, par. II-3.

8

Finally, TVA's description of its method of analyzing aquatic impacts indicates a troubling lack of care or competence. For example, by adding widely divergent diurnal and nocturnal entrainment measurement, TVA violates guidance of the U.S. Environmental Protection Agency ("EPA") and grossly overstates the size and diversity of the fish population. Some of the studies relied on by TVA had to be revised after they were released, indicating that TVA has significant problems ensuring the quality of its measurements and analyses. It is reasonable to expect that the results from TVA's biological studies will be accurate in order to support TVA's conclusions. In too many instances, however, TVA makes significant mistakes. Young Declaration, par. II-4.

With respect to TVA's mischaracterization of the health of the aquatic environment as good, TVA has done nothing to alleviate the concerns raised by Contention 7. Although as discussed above, TVA's data collection is insufficient to present a reasonable picture of the health of the Tennessee River, the data that TVA has collected do not indicate, as TVA claims, that WBN1's impacts on the aquatic ecosystem have been insignificant. Rather, they point to already-significant aquatic impacts by WBN1 that are likely to be significantly exacerbated by the operation of WBN2. Young Declaration, par. II-5.

Further, despite alarming evidence of significant decline in the diversity and numbers of aquatic organisms in the Tennessee River in the vicinity of WBN, TVA continues to assert that the aquatic health of the river is good. The only way that TVA can present such a clean bill of health is to mischaracterize the baseline condition of the Tennessee River as a large reservoir where one would expect to see a limited number of

9

species of aquatic organisms. In reality, the Tennessee Rive is a fragile and rapidly deteriorating riverine ecosystem with remnants of the greatest species diversity of any river in the United States. By falsely painting a rosy picture of aquatic health in the river, TVA understates the significance of the impacts of WBN1 and WBN2, and thus minimizes the benefits that could be achieved by implementing alternatives that would reduce the impacts of the cooling system on organisms in the river. Young Declaration, par. II-6.

Finally, TVA still does not address the cumulative impacts of WBN2 in conjunction with the impacts of the numerous water impoundments on the Tennessee River, or with other industrial facilities such as the ten fossil fuel-burning plants, the six nuclear reactors that are already in operation, and the five additional reactors for which TVA has sought operating licenses. The combined operation of WBN1 and WBN2, by itself, may cause changes in how Watts Bar Dam is operated. TVA and the NRC Staff both acknowledge that in order to stay within thermal discharge limits stated in the NPDES that requests for additional discharge from Watts Bar Dam may be needed. Thus, operating WBN alone would change reservoir operations in the middle- Tennessee Basin that would be supported by water releases or hydrological adjustments in upper-Tennessee River Basin. The effects of more alterations to the hydrological cycle of the basin on aquatic organisms, especially the already declining native fish and freshwater mussel species, must be addressed. Given the extensive portfolio of energy and industrial facilities that the Tennessee River supports and that the management agencies must

10

maintain adequate water for all these facilities, this is an extremely important omission. Young Declaration, par. II-7.

TVA's claims are also contradicted by the NRC Staff's Draft Supplemental Environmental Statement ("DSES") for WBN2. TVA claims, for instance, that WBN Unit 1 was originally designed to operate only in a closed cycle cooling mode via the Condenser Cooling Water ("CCW") system. After TVA began operation of Unit 1, it determined that a supplemental cooling system would increase the efficiency of the plant. Accordingly, TVA began to use a Supplemental Condenser Cooling Water ("SCCW") system in 1998. Disputed as to the reason TVA began to use the SCCW. The original cooling system was under-designed and would have prevented WB1 from achieving rated power output on hot summer days. Some form of cooling tower enhancement or supplemental cooling was/is necessary for WB1 to achieve rated output on hot summer days (when the highest annual demand is experienced on the TVA system). This is supported by the NRC's Draft SFEIS at page 3-4, which states:

> Evaporation of cooling-water system water from the cooling-tower increases the concentration of dissolved solids in the cooling-water system. In most closed-cycle wet cooling systems, a portion of the cooling water is removed and replaced with makeup water from the source (for WBN, the Tennessee River) to limit the concentration of dissolved solids in the cooling system and in the discharge to the receiving water body.
>
> *Because the WBN cooling tower cannot remove the desired amount of heat from the circulating water during certain times of the year, TVA added the Supplemental Condenser Cooling Water (SCCW) system to the cooling system for the WBN reactors* (TVA 1998). The SCCW draws water from behind Watts Bar Dam and delivers it, by gravity flow, to the cooling-tower basins to supplement cooling of WBN Unit 1. This cooling system would also be used for Unit 2. The temperature of this water is usually lower than the temperature of the water in the cooling-tower basin and, as a result, lowers the temperature of the water being used to cool the steam in the condensers. Slightly less water enters the cooling-

11

tower basins through the SCCW intake than leaves the cooling-tower basins and is discharged to the Tennessee River through the SCCW discharge structure (TVA, 2010). *Since the SCCW has been operating, elevated total dissolved solids in blowdown water have not been a concern because a large volume of water continually enters and leaves the cooling-tower basins* (PNNL 2009).

(emphasis added). Had TVA more robust cooling system in the first place, the SCCW would never have been considered necessary by TVA and TVA would not now be proposing to operate WBN2 with the SCCW.

Accordingly, all of the allegedly undisputed material facts alleged by TVA are disputed by SACE. Therefore summary disposition of Contention 7 is inappropriate.

B.   TVA's Studies Have Not Mooted Contention 7.

TVA argues that Contention 7 is "fundamentally a contention of omission."

Motion at 1. Therefore, according to TVA, the ASLB should dismiss Contention 7 because it has now performed the studies demanded by the contention. *Id.* TVA contradicts its own argument, however, by conceding that the contention claims that TVA's aquatic studies were "inadequate and outdated." *Id.* Indeed, the contention itself repeatedly refers to the inadequacy of TVA's discussion of aquatic impacts:

> TVA claims that the cumulative impacts of WBN Unit 2 on aquatic ecology will be insignificant (FSEIS Table S-1 at page. S2, and Table 2-1 at page. 30). *TVA's conclusion is not reasonable or adequately supported*, and therefore fails to satisfy 10 C.F.R. § 51.53(b) and NEPA. *TVA's discussion of aquatic impacts is deficient* in three key respects. First; TVA mischaracterizes the current health of the ecosystem as good, and therefore fails to evaluate the impacts of WBN2 in light of the fragility of the host environment. Second, *TVA relies on outdated and inadequate data* to predict thermal impacts and the impacts of entrainment and impingement of aquatic organisms in the plant's cooling system. Third, TVA fails completely to analyze the cumulative effects of WBN2 when taken together with the impacts of other industrial facilities and the effects of the many dams on the Tennessee River.

12

Petition to Intervene and Request for Hearing at 31-32 (July 13, 2009) (emphasis added). A discussion of the inadequacy and inaccuracy of TVA's studies also runs throughout the basis for the contention. *See id.* at 32 ("TVA's finding that WBN Unit 2 will have no significant impacts on aquatic life in the Tennessee River is inadequately supported"); *id.* at 33 ("TVA incorrectly portrays the ecosystem as healthy"); *id.* at 33 ("TVA relies on outdated and inadequate data to predict the effects of WBN Unit 2's cooling system on fish, mussels and other aquatic organisms"); *id.* (TVA relies on "poor and out outdated data, distorted interpretations of data, and assumptions and extrapolations in lieu of recent monitoring studies"); *id.* at 34 ("TVA has not collected sufficient data to understand the distribution of ichtyoplankton populations or how they are affected by Watts Bar intakes") *id.* (TVA's conclusion that entrainment impacts are insignificant is based upon an unsupported assumption"); *id.* at 36 ("TVA does not adequately address the cumulative impacts of WBN Unit 2 in conjunction with the impacts of the numerous water impoundments on the Tennessee River). Thus, both the plain language and the context of the contention show that it is a contention of adequacy, not omission. *Southern Nuclear Operating Co.* (Early Site Permit for Vogtle ESP Site), LBP-08-2, 67 NRC 54, 65 (2008). Accordingly, contrary to TVA's arguments, SACE was not required to amend the contention to address each study that TVA prepared in order to maintain the viability of Contention 7.

IV. CONCLUSION

For the foregoing reasons, the ASLB should deny TVA's Motion for Summary Disposition of Contention 7.

13

Respectfully submitted,

Electronically signed by
Diane Curran
HARMON, CURRAN, SPIELBERG, & EISENBERG, L.L.P.
1726 M Street N.W., Suite 600
Washington, D.C. 20036
202-328-3500
Fax: 202-328-6918
e-mail: dcurran@harmoncurran.com

December 20, 2011

14

December 20, 2011

UNITED STATES OF AMERICA
NUCLEAR REGULATORY COMMISSION
BEFORE THE ATOMIC SAFETY AND LICENSING BOARD

In the Matter of	)
	)
Tennessee Valley Authority	) Docket No. 50-391-OL
	)
(Watts Bar Unit 2)	)
	)

SOUTHERN ALLIANCE FOR CLEAN ENERGY'S
STATEMENT OF DISPUTED MATERIAL FACTS

Southern Alliance for Clean Energy ("SACE") respectfully submits the following statement of disputed material facts in response to Tennessee Valley Authority's ("TVA's") Statement of Material Facts on Which No Genuine Issue Exists (Nov. 21, 2011). SACE responds as follows:

I. **Procedural Background**

A. Licensing History for Watts Bar Nuclear Plant

1. On May 14, 1971, TVA applied for a Construction Permit ("CP") for the Watts Bar Nuclear Plant ("WBN"). The NRC issued CPs for WBN Units 1 and 2 on January 23, 1973, and construction began. TVA substantially completed construction of Unit 1 in 1985. **Undisputed.**

2. On June 30, 1976, TVA first filed an application for an operating license ("OL") for WBN Units 1 and 2. On February 7, 1996, the NRC issued an OL for Unit 1 that authorized operation at 100% power. **Undisputed.**

3. Between 1973 and 2008, the NRC extended the CP for Unit 2 on several occasions. During this time, TVA maintained WBN Unit 2 in deferred plant status, in accordance with the NRC's "Policy Statement on Deferred Plants." **Undisputed.**

4. On August 3, 2007, TVA informed the NRC Staff its intention to resume and complete construction of WBN Unit 2. TVA updated its original OL application for WBN Unit 2 on March 4, 2009, prompting the NRC to publish a notice of hearing in the Federal Register on May 1, 2009. **Undisputed.**

5. Throughout this time, TVA and the NRC completed a number of environmental reviews of WBN. On November 9, 1972, TVA issued a Final Environmental Statement for WBN Units 1 and 2 ("TVA 1972 FES"). On December 1, 1978, the NRC issued its Final Environmental Statement evaluating the operation of Units 1 and 2 ("NRC 1978 FES"). The NRC supplemented its 1978 FES on April 1, 1995 ("NRC 1995b"), in order to re-examine environmental considerations before issuing an OL for WBN Unit 1. **Undisputed.**

6.   When TVA reactivated construction of WBN Unit 2, it also submitted its Final Supplemental Environmental Impact Statement ("2007 FSEIS") to the NRC on February 15, 2008. The NRC published its draft supplement to the final environmental statement ("Draft SFES") on October 31, 2011. **Undisputed.**

B.   Intervention in Current Proceeding

7.   After TVA updated its OL application for WBN Unit 2 and the NRC issued a Notice of Opportunity for Hearing on May 1, 2009, five organizations (Southern Alliance for Clean Energy ("SACE"), Tennessee Environmental Council, We the People, the Sierra Club, and Blue Ridge Environmental Defense League) jointly filed a Petition to Intervene and Request for Hearing, which included seven contentions. Among those, Contention 7 challenged TVA's analysis of the impact of operation of WBN Unit 2 on the aquatic environment. In Contention 7, SACE alleged:

> TVA claims that the cumulative impacts of WBN Unit 2 on aquatic ecology will be insignificant (FSEIS Table S-1 at page. S2, and Table 2-1 at page. 30). [sic] TVA's conclusion is not reasonable or adequately supported, and therefore fails to satisfy 10 C.F.R. § 51.53(b) and NEPA.

> TVA's discussion of aquatic impacts is deficient in three key respects. First; TVA mischaracterizes the current health of the ecosystem as good, and therefore fails to evaluate the impacts of WBN2 in light of the fragility of the host environment. Second, TVA relies on outdated and inadquate data to predict thermal impacts and the impacts of entrainment and impingement of aquatic organisms in the plant's cooling system. Third, TVA fails completely to analyze the cumulative effects of WBN2 when taken together with the impacts of other industrial facilities and the effects of the many dams on the Tennessee River. **Undisputed.**

8.   The NRC Staff and TVA subsequently filed answers addressing the Petition. On September 3, 2009, SACE filed a Motion for Leave to Amend Contention 7, along with an Amended Contention 7. Both TVA and the NRC Staff filed responses opposing SACE's Motion and Answers to the Amended Contention. SACE thereafter filed a reply to the Answers to the Amended Contention on October 5, 2009. **Undisputed.**

9.   On November 19, 2009, this Board granted the Petition to Intervene on behalf of SACE, admitting two contentions. The Board denied SACE's Motion to Amend Contention 7, instead admitting Contention 7 as originally presented. Although the Board admitted Contention 1 along with Contention 7, TVA moved to dismiss Contention 1 as moot on April 19, 2010. The Intervenors did not oppose that motion, and the Board granted TVA's unopposed Motion and dismissed Contention 1 accordingly. As a result, only Contention 7 remains to be resolved. **Undisputed.**

C.   New Information on the Record – TVA's Aquatic Studies and NRC's Draft SFES

10.   In direct response to the issues raised by SACE in Contention 7, TVA collected extensive new data on the current health of the aquatic environment and the impact of operation

of WBN Unit 1 on that environment, prepared numerous updated and expanded aquatics-related analyses, documented the analyses in published reports and studies, and disclosed these reports and studies to the NRC Staff and SACE. **Undisputed except with respect to TVA's characterization of the data as "extensive." As discussed throughout Dr. Young's Declaration, there are significant gaps and inadequacies in the data.**

A complete list of those studies, including the dates that TVA disclosed each to SACE and the NRC Staff, follows:

> a.   Comparison of Fish Species Occurrence and Trends in Reservoir Fish Assemblage Index Results in Chickamauga Reservoir Before and After [WBN] Unit 1 Operation (June 2010) ("RFAI Study"), which TVA disclosed to SACE and the NRC Staff on July 15, 2010;
>
> b.   Analysis of Fish Species Occurrences in Chickamauga Reservoir – A Comparison of Historic and Recent Data (Oct. 2010) ("Fish Species Order (Granting TVA's Unopposed Motion to Dismiss SACE Contention 1) (June 2, 2010) (unpublished). Occurrences Study"), which TVA disclosed to SACE and the NRC Staff on November 15, 2010;
>
> c.   Mollusk Survey of the Tennessee River Near [WBN] (Rhea County, Tennessee) (Nov. 2010) ("Mollusk Survey"), which TVA disclosed to SACE and the NRC Staff on January 18, 2011;
>
> d.   Discussion of the Results of the 2010 Mollusk Survey of the Tennessee River Near [WBN] (Rhea County, Tennessee) (Mar. 2011) ("Discussion of Mollusk Survey"), which TVA disclosed to SACE and the NRC Staff on March 15, 2011;
>
> e.   Aquatic Environmental Conditions in the Vicinity of [WBN] During Two Years of Operation, 1996-1997 (June 1998, Revised June 2010) ("Revised Aquatics Study"), which TVA disclosed to SACE and the NRC Staff on July 15, 2010;
>
> f.   Comparison of 2010 Peak Spawning Seasonal Densities of Ichthyoplankton at [WBN] at Tennessee River Mile 528 with Historical Densities during 1996 and 1997 (Apr. 2011, Revised Nov. 2011) ("Peak Spawning Entrainment Study"), which TVA disclosed to SACE and the NRC Staff on April 15, 2011;
>
> g.   Fish Impingement at [WBN] Intake Pumping Station Cooling Water Intake Structure during March 2010 through March 2011 (Mar. 2011, Revised Apr. 2011) ("Impingement Study"), which TVA disclosed to SACE and the NRC Staff on May 16, 2011; and
>
> h.   Hydrothermal Effects on the Ichthyoplankton from the [WBN] Supplemental Condenser Cooling Water Outfall in Upper Chickamauga Reservoir (Jan. 2011) ("Hydrothermal Study"), which TVA disclosed to SACE and the NRC Staff on February 15, 2011. **Undisputed.**

11.   SACE has not raised any concerns with respect to these studies with the NRC or this Board. **Undisputed.**

12.   The NRC Staff's Draft SFES, dated October 31, 2011, concurs with TVA's findings in its aquatics studies. Section IV, below, discusses the specific conclusions drawn by the Staff that are relevant to TVA's aquatic studies. **Undisputed.**

II.   **Description of the Proposed Project**

A.   General Information

13.   The WBN site is located is located in Rhea County, Tennessee, on the west bank of the Tennessee River, in the upper Chickamauga Reservoir at Tennessee River Mile ("TRM") 528. **Undisputed.**

14.   The Tennessee River System is approximately 650 miles long and is comprised of riverine and lacustrine environments, created by numerous dams and locks on the system, most of which have been in place since the 1940s. Chickamauga Dam, completed in 1940 at TRM 471, impounds Chickamauga Reservoir downstream of WBN. Watts Bar Hydroelectric Dam impounds the Watts Bar Reservoir 1.9 miles upstream of WBN. **Undisputed.**

15.   The Tennessee River is also host to numerous industrial facilities. For example, WBN is located approximately one mile downstream of the decommissioned Watts Bar Fossil Plant. **Undisputed.**

16.   TVA is the licensee and operator of the existing WBN Unit 1, a Westinghouse pressurized water reactor that began full commercial operation on May 27, 1996. **Undisputed.**

17.   WBN Unit 1 was originally designed to operate only in a closed cycle cooling mode via the Condenser Cooling Water ("CCW") system. After TVA began operation of Unit 1, it determined that a supplemental cooling system would increase the efficiency of the plant. Accordingly, TVA began to use a Supplemental Condenser Cooling Water ("SCCW") system in 1998. **Disputed as to the reason TVA began to use the SCCW. The original cooling system was under-designed and would have prevented WB1 from achieving rated power output on hot summer days. Some form of cooling tower enhancement or supplemental cooling was/is necessary for WB1 to achieve rated output on hot summer days (when the highest annual demand is experienced on the TVA system). This is supported by the NRC's Draft SFEIS at page 3-4, which states:**

> Evaporation of cooling-water system water from the cooling-tower increases the concentration of dissolved solids in the cooling-water system. In most closed-cycle wet cooling systems, a portion of the cooling water is removed and replaced with makeup water from the source (for WBN, the Tennessee River) to limit the concentration of dissolved solids in the cooling system and in the discharge to the receiving water body.

> *Because the WBN cooling tower cannot remove the desired amount of heat from the circulating water during certain times of the year, TVA added the Supplemental Condenser Cooling Water (SCCW) system to the cooling system for the WBN reactors (TVA 1998). The SCCW draws water from behind Watts Bar Dam and delivers it, by gravity flow, to the cooling-tower basins to supplement cooling of WBN Unit 1. This cooling system would also be used for Unit 2. The temperature of this water is usually lower than the temperature of the water in the cooling-tower basin and, as a result, lowers the temperature of the water being used to cool the steam in the*

condensers. **Slightly less water enters the cooling-tower basins through the SCCW intake than leaves the cooling-tower basins and is discharged to the Tennessee River through the SCCW discharge structure (TVA, 2010).** *Since the SCCW has been operating, elevated total dissolved solids in blowdown water have not been a concern because a large volume of water continually enters and leaves the cooling-tower basins* (PNNL 2009).

(emphasis added). Had TVA more robust cooling system in the first place, the SCCW would never have been considered necessary by TVA and TVA would not now be proposing to operate WBN2 with the SCCW.

18.   The present proceeding pertains to the OL for WBN Unit 2. The added operation of WBN Unit 2 may result in minimal increased demands on that aquatic environment both for cooling water intake and cooling water discharge. **Disputed as to the term "minimal." As discussed in Dr. Young's Declaration throughout, the already-stressed Tennessee River aquatic environment will be further stressed by additional CCW intake and discharge and increased SCCW discharge to accommodate the operation of both WB1 and WB2 cooling towers and the increased cumulative cooling tower blowdown discharge to the Tennessee River as a result of WB2 operation. The combined operation of two units will have substantial impacts on the Tennessee River.**

B.   WBN Cooling System Intake

19.   WBN Unit 2 shares intake channels with Unit 1. Operation of Unit 1 withdraws cooling water from CCW and SCCW intake channels. Under dual unit operation, WBN will continue to draw cooling water from the CCW and SCCW intake channels. **Undisputed.**

20.   The SCCW system is gravity driven. As a result, intake flow and velocity for the SCCW depends on the water level behind the Watts Bar Dam. **Undisputed.**

21.   Flow through the CCW is driven by the IPS, rather than gravity. The IPS will draw more water at a higher flow rate under dual unit operation than for operation of Unit 1 alone. CCW maximum intake velocities will not increase under dual unit operation because the intake will draw water through additional openings. **Undisputed.**

22.   Studies show that the hydraulic entrainment from dual unit operation will result in an additional entrained amount of 0.2% of the flow in the Chickamauga Reservoir. The resulting total hydraulic entrainment represents approximately 0.5% of the flow in the Chickamauga Reservoir. This increased hydraulic entrainment will result in a proportionate increase in entrainment of the ichthyoplankton present in the water column. **Disputed as to this calculation is only partly correct, and only accurate at a very specific river flow past WBN Plant. As discussed in Dr. Young's Declaration at par. III-A.13-14, the 0.2% hydraulic entrainment for WB1 is based upon TVA using a long term average river flow past WBN of 27,000 cfs. Using 3,500 cfs, which is the minimum amount of flow from Watts Bar Dam that permits TVA to discharge thermal and chemical effluent through Outfall 101, the hydraulic entrainment increases to 2.1% (10 times higher). Then, with the addition of Unit 2 almost doubling hydraulic entrainment, the hydraulic entrainment at a flow of 3,500 cfs further increases to approximately 4.0% (20 times higher). Also, only data collected by field studies**

in combination with proper methods for calculation may accurately characterize ichthyoplankton entrainment under any level of hydraulic entrainment.

23. Studies show that CCW flow rates resulting from dual unit operation will average 134 cubic feet per second ("cfs") at summer pool levels and 113 cfs at winter pool levels, an increase from those rates observed under operation of Unit 1 alone: 73 cfs and 68 cfs, respectively. (The maximum intake velocities will not change under dual unit operation because of the additional IPS openings available to accommodate increased flow.) The increased flow rates in the CCW intake channel resulting from dual unit operation will result in a proportionate increase in the rates of fish impingement. **Disputed. It is important to note that TVA identifies the makeup flow through the IPS as 174 fps, double the withdrawal from the Tennessee River that would occur with only WBN1 online, and an increase in warm blowdown discharge to the Tennessee River from 135 cfs to 170 cfs, a 26 percent increase. These are substantial increases, independent of the role of the SCCW. See Table 3-1 of the DFES, at page 3-9:**

	WB1&WB2	WB1 only
Blowdown rate when diffusers are discharging from cooling towers and YHP	4.81 m³/s (170 cfs)[b]	3.82 m³/s (135 cfs)[a]
IPS makeup flow	4.93 m³/s (174 cfs)[c]	2.5 m³/s (88 cfs)
SCCW		
Intake flow rate	7.1 m³/s (250 cfs)	7.31 m³/s (258 cfs)
Discharge flow rate	8.46 m³/s (299 cfs)	7.48 m³/s (264 cfs)

The rates of fish impingement may exponentially increase. Similar to the issue of hydraulic versus ichthyoplankton entrainment, only field monitoring will accurately determine impingement rates. *See* **Young Declaration, ¶¶ III-A.13-14.**

C. WBN Cooling System Output

24. WBN Unit 2 shares cooling water discharge outfalls with Unit 1. **Undisputed.**

25. The thermal discharge from WBN operation is bound by thermal limits established by TVA's NPDES permit. The NPDES system establishes legally enforceable, aquatic health-based limits on hydrothermal discharges, in accordance with state and federal statutes. The Tennessee Water Pollution Control Division ("TDEC") issued a new NPDES permit for the operation of WBN Units 1 and 2 on June 30, 2011, most recently revised on August 31, 2011. **Undisputed, except for the facts that the existence of a legal limit does not ensure there will be no significant impacts to aquatic organisms and is not a guarantee that the operation will stay within the limit.**

26. TVA's NPDES permit sets discharge limits for each of the WBN outfall points under operation of WBN Units 1 and 2 that are unchanged from the limits set for Unit 1 operation. **Undisputed.**

27. For Outfall 101, the discharge point for blowdown water from the CCW system, the NPDES permit for operation of WBN Units 1 and 2 allows discharge only when the release from Watts Bar Dam is at least 3500 cfs, and specifies a discharge temperature limit of 35°C.

These requirements are unchanged from those set in TVA's NPDES permit for operation of Unit 1 alone. **Undisputed.**

28. For Outfall 102, the discharge point for the CCW holding ponds, the NPDES permit for dual unit operation allows discharge only under emergency situations. Even then, the NPDES permit limits the temperature of discharged water to 35°C and requires that TVA make every effort to use this outfall only when the flow of the receiving waters meets or exceeds 3500 cfs. This condition is unchanged from that in the NPDES permit for WBN Unit 1. **Undisputed.**

29. For Outfall 113, the discharge point for the SCCW system, the NPDES permit for operation of Units 1 and 2 specifies a discharge temperature limit based on the receiving water. For example, the NPDES permit requires that the temperature rise at the edge of the mixing zone shall not exceed 3°C relative to an upstream control point. The limits that apply to Outfall 113 in the current NPDES permit are unchanged from those established in the NPDES permit for WBN Unit 1 operation. **Undisputed.**

30. Because the thermal discharge limits established by TVA's NPDES permit for dual unit operation are unchanged from those for Unit 1 operation, thermal impacts on the aquatic environment resulting from WBN operation will not be materially different under dual unit operation than they are for operation of Unit 1 alone. **Disputed. There will be substantial increases in discharge from the CCW and SCCW. See Table 3-1 of the DFES, at page 3-9:**

	WB1&WB2	WB1
Blowdown rate when diffusers are discharging from cooling towers and YHP	4.81 m³/s (170 cfs)[b]	3.82 m³/s (135 cfs)[a]
IPS makeup flow	4.93 m³/s (174 cfs)[c]	2.5 m³/s (88 cfs)
SCCW		
Intake flow rate	7.1 m³/s (250 cfs)	7.31 m³/s (258 cfs)
Discharge flow rate	8.46 m³/s (299 cfs)	7.48 m³/s (264 cfs)

Also, as discussed in Dr. Young's Declaration at pars. III-C.1-11, the already-stressed Tennessee River aquatic environment will be further stressed by additional CCW and SCCW thermal discharge from cumulative cooling tower blowdown discharge to the Tennessee River as a result of WB2 operation.

II. Description of TVA's Aquatics Studies

31. As noted in ¶ 10 above, TVA conducted a number of aquatics studies in direct response to the assertions made by SACE and its expert, Dr. Young, in Contention 7. Those studies, which are described in more detail below, collectively provide data on fish and mussel populations in the WBN vicinity, and the entrainment, impingement, and hydrothermal impacts on those species that result from operation of WBN Unit 1. In addition, TVA conducted some of the studies to resolve alleged errors in TVA's original studies identified by SACE and Dr. Young. **Undisputed that TVA conducted the studies described in pars. (A) through (G) below. Disputed that the studies resolve Dr. Young's concerns, as discussed throughout his Declaration.**

A.   Comparison of Fish Species Occurrence and Trends in Reservoir Fish Assemblage Index Results in Chickamauga Reservoir Before and After WBN Unit 1 Operation (June 2010) ("RFAI Study")

32.   In Contention 7, SACE and Dr. Young claimed that TVA relies on poor and outdated data about the health of the aquatic community in the WBN vicinity in lieu of recent monitoring studies. Dr. Young challenged TVA's characterization of the health of the fish community in the WBN vicinity, which TVA based in part on measured RFAI data. In response to those allegations, TVA conducted this new study to explain RFAI methodology and evaluate the aquatic community in the WBN vicinity using that methodology. **Undisputed.**

33.   First, this study provides a detailed explanation of TVA's RFAI methodology. TVA created the RFAI methodology based on industry standards for biological indices, including those approved by TDEC and the U.S. Environmental Protection Agency ("EPA"), for use in its Vital Signs monitoring program. TVA has conducted fish sampling in the Chickamauga Reservoir every year since 1993, in support of this program. **Undisputed as to the conduct of the RFAI study every year since 1993. Disputed as to the consistency, accuracy, and usefulness of the study to portray aquatic health in the Tennessee River near WBN1.** *See* **Young Declaration, ¶¶ III-E.1-20.**

34.   RFAI methodology uses twelve fish community metrics from four general categories: Species Richness and Composition; Trophic Composition; Abundance; and Fish Health. For each metric, scores are given on a scale from 1 to 5, with a score of 5 indicating optimum health. The resulting scores range from 12-60, broken down as follows: 12-21 ("Very Poor"), 22-31 ("Poor"), 32-40 ("Fair"), 41-50 ("Good"), or 51-60 ("Excellent"). RFAI scores have an intrinsic variability of ±3 points. **Undisputed as to the description of the RFAI methodology. Disputed as to the consistency, accuracy, and usefulness of the RFAI methodology to portray aquatic health in the Tennessee River near WBN1.** *See* **Young Declaration, ¶¶ III-E.3-17.**

35.   RFAI methodology addresses all five attributes or characteristics of a Balanced Indigenous Population ("BIP"), which is required by the Clean Water Act. If an RFAI score reaches 70% of the highest attainable score of 60 (i.e., 42), or if fewer than half of the RFAI metrics receive a low (1) or moderate (3) score, then normal community structure and function are considered to be present, indicating that BIP is maintained. **Undisputed that this is a description of TVA's methodology for compliance with the BIP requirement. Disputed as to the fact that RFAI methodology only addresses four not five attributes, and to the consistency, accuracy, and usefulness of this methodology to portray aquatic health in the Tennessee River near WBN1.** *See* **Young Declaration, ¶¶ III-E.1-20.**

36.   Second, this study evaluates the health of the aquatic environment in the WBN vicinity based on recent fish surveys and the RFAI methodology. The study found that RFAI scores from the site downstream of the WBN intake and thermal discharge have averaged 44 from 1996 to 2008 (i.e., during operation of WBN Unit 1), indicating that the aquatic health of that area is "good" even during WBN operation. **Undisputed that this is a description of TVA's RFAI study, results, and conclusions. Disputed as to the consistency, accuracy, and usefulness of this methodology to portray aquatic health in the Tennessee River near**

WBN1 and the concluding scores to properly correlate with the true health of the fish community. *See* **Young Declaration, ¶¶ III-E.1-20.**

37.   Third, this study compares the health of that environment as reflected in RFAI scores from before and after WBN operation. Scores from every sample year (1993-2008) were at least 42, i.e., 70% of the highest attainable score of 60. As a result, the study concluded that both before and after WBN operation, BIP has been maintained. **Undisputed that this is a description TVA's RFAI study, results, and conclusions. Disputed as to the consistency, accuracy, and usefulness of this methodology to portray aquatic health in the Tennessee River near WBN1.** *See* **Young Declaration, ¶¶ III-E.1-20.**

38.   SACE has not challenged the methodology or findings of this study with this Board. **Undisputed that SACE has not challenged the most recent iteration of the RFAI study before the Board. Contention 7, however, criticizes the methodology and results of previous RFAI studies, which have not changed in any significant respect.** *See* **Young Declaration, ¶ E-III.1.**

B.   Analysis of Fish Species Occurrences in Chickamauga Reservoir – A Comparison of Historic and Recent Data (Oct. 2010) ("Fish Species Occurrences Study")

39.   SACE claimed in Contention 7 that TVA relies on inadequate and outdated data to form its conclusion that fish populations in the WBN vicinity are in good health, and has not taken steps necessary to evaluate how effluent from WBN may affect fish communities. In direct response, TVA conducted this study to analyze extensive historic and recent fish survey data from the WBN vicinity, and compare the current prevalence of fish species to historic (i.e., pre-operational) values. **Undisputed except with respect to TVA's characterization of the data as "extensive." As discussed in Dr. Young's Declaration at pars III-E.2-3, there are significant inadequacies in the analyses found in this report.**

40.   This study uses the extensive fish survey data available for the WBN vicinity, dating back to 1947. Because it also provides recent survey data for the fish populations in the WBN vicinity, this study inherently reflects the impact of the current operation of WBN Unit 1 on those populations. **Undisputed to the extent that TVA states it used fish survey data back to 1947 and provides recent survey data. Disputed with respect to TVA's characterization of the data as "extensive" and TVA's conclusion that this study alone inherently reflects the impact of the current operation of WBN1 on fish populations. See Dr. Young's Declaration throughout.**

41.   In analyzing the collective historical fish survey data for the Chickamauga Reservoir, this study takes into consideration the variations in survey methods employed over the past 60 years. Variations in survey methodology preclude direct comparisons between historical and recent surveys. This study also compared the results of fish sampling efforts in various Tennessee River reservoirs subject to similar conditions to understand widespread patterns and behavior of species in reservoir environments. **Disputed. While the study may acknowledge the variations in survey methods employed over the years, it does not cure the mistakes of the past, and instead perpetuates them. TVA either has an "extensive" fish species survey/study for historical comparison, which shows significant decline of fish species overtime, including since operation of Unit 1, or TVA has an unreliable, outdated, and**

inadequate means to properly evaluate impacts from WBN. **The different sampling methods do not detract from the fact that there has been a decline in fish species pre- and post-WBN operation, which is evidence that the health of the fish community is poor** *See* Dr. Young's Declaration at pars. III-E.1-20.

42.   This study found that species occurrence and abundance in the Chickamauga Reservoir has changed from 1947 to 2009. Many of these changes took place before operation of WBN Unit 1 began. **Undisputed to the extent that TVA asserts that many of the changes in species occurrence and abundance in the Chickamauga Reservoir took place before the operation of WBN1 began. Disputed to the extent that TVA implies that changes after WBN1 operation began are insignificant.** *See* Young Declaration, ¶¶ III-E.1-20.

43.   One major cause of this change is impoundment of the Tennessee River, which began in the 1930s and has altered habitats required for various life stages of aquatic species. Some of the species not found in recent surveys require unimpounded, free flowing riverine environments. **Undisputed to the extent that impoundment of the Tennessee River is a major cause of the decline in species occurrence and abundance. Disputed to the extent that TVA implies that changes after WBN1 operation began are insignificant.** *See* **Young Declaration, ¶¶ III-E.1-20 and III-D.4-7, III-D.10, and III-A.6-9.**

44.   The study found that another reason for the change in species diversity and abundance is that most species that have not been collected in recent times have historically never been caught frequently or in large numbers in Chickamauga Reservoir. **Undisputed that this is a conclusion of the study. Disputed as a rationale for the decline of indigenous species present and decline of indigenous species abundance. The fact that species have not been caught in the reservoir is a meaningful indication of the decline of indigenous fish species.** *See* Young Declaration, ¶¶ III-E.1-20.

45.   Finally, the study found that changes in fish survey methods account for some of the changes in findings of species occurrence and abundance. Certain survey methods, such as hoop nets, trap nets, and cove rotenone sampling, that were effective for targeting certain species, are no longer in use. **Undisputed in that this is a conclusion of the study. Disputed as being used as rationale for the decline of the fish community. Even with TVA's many changes in methods, a clear pattern of declining indigenous fish species and their abundance pre- and post-WBN operation is clear.** *See* Young Declaration, ¶¶ III-E.1-20 and III-D.4-7, III-D.10, and III-A.6-9.

46.   As a result, this study concluded that there is no basis to support a finding that operation of WBN Unit 1 caused the observed changes in fish species and occurrence in the Chickamauga Reservoir. **Undisputed as to the study's stated conclusion. Disputed as to whether the conclusion is accurate that there is no basis to support a finding that operation of WBN1 caused the observed changes in fish species and occurrence.** *See* Young Declaration, ¶¶ III-A.1-14, III-B.1-5, III-C.1-12, and III-E.1-20.

47.   SACE has not challenged the methodology or findings of this study with this Board. **Undisputed.**

C.   Mollusk Survey of the Tennessee River Near [WBN] (Rhea County, Tennessee) (Oct. 28, 2010, Revised Nov. 24, 2010) ("Mollusk Survey"), and Discussion of the Results of the 2010 Mollusk Survey of the Tennessee River Near [WBN] (Rhea County, Tennessee) (Mar. 2011) ("Discussion of Mollusk Survey")

48.   In Contention 7, SACE claimed that TVA relies on inadequate and outdated data to estimate the effects of WBN operation on mussels in the WBN vicinity. In support, Dr. Young alleged that the mussel community in the WBN vicinity is not in good health, and that TVA has not given sufficient consideration of the impact of WBN operation on that community. **Undisputed.**

49.   To remedy those alleged deficiencies, TVA engaged an outside consultant to conduct a survey of the mussel community in the WBN vicinity in 2010. The consultant conducted semi-quantitative and quantitative mollusk sampling in three sample areas at which TVA has previously conducted pre-operational and operational mollusk surveys. **Undisputed.**

50.   Because WBN Unit 1 was in operation in 2010 and had been in operation for more than a decade, this survey inherently reflects the impact of the operation of WBN Unit 1 on the mussel community in the WBN vicinity. **Disputed as to a one year survey capturing the population trend of a mussel community. It was reasonable for TVA to have contracted for a multi-year study when it was decided to apply for the operating license.**

51.   The consultant provided the results in the Mollusk Survey. TVA subsequently produced Discussion of Mollusk Survey, analyzing the results of the Mollusk Survey and comparing those results to preoperational (1983 to 1994) and operational (1996 to 1997) monitoring of the mollusk communities at WBN. **Undisputed.**

52.   These studies agree that the Chickamauga Reservoir in the WBN vicinity is not the ideal habitat for mussels. Still, the 2010 survey found that the mussel community in the WBN vicinity is in substantially similar condition as it was near the end of the previous operational monitoring period (1996 to 1997), in both species composition and the number of mussels collected. In addition, the 2010 survey collected juveniles of at least five mussel species, evidencing reproduction of mollusks in the WBN vicinity. **Undisputed as to the agreement a reservoir may not be ideal habitat for mussels. Disputed as to what results the consultant produced versus what conclusions TVA drew from that data. Disputed as to the mussel community in the WBN vicinity being in substantially similar condition as it was near the end of the previous operational monitoring period and the significance of the collection of five juvenile mussel species. *See* Young Declaration, ¶¶ III-D.1-7.**

53.   As a result, this study concluded that there is no basis to support a finding that the relatively low densities of mussels in the WBN vicinity are the result of operation of WBN Unit 1. **Undisputed that this is the conclusion stated. Disputed as to the accuracy and reasonableness of the conclusion. *See* Young Declaration, ¶ III-D.4-7.**

54.   SACE has not challenged the methodology or findings of this study with this Board. **Undisputed.**

D.   Aquatic Environmental Conditions in the Vicinity of Watts Bar Nuclear Plant During Two Years of Operation, 1996-1997 (June 1998, Revised June 2010) ("Revised Aquatics Study")

55.   TVA completed the initial Aquatics Study in 1998, comparing pre-operational (1973 to 1979, 1982 to 1985) and operational (1996 to 1997) aquatic monitoring in the WBN vicinity. The original study focused on the effects of WBN operation on fish (juveniles and adults), benthic macroinvertebrates, and water quality. As part of the analysis of the effects on fish, the study estimated entrainment of ichthyoplankton and impingement of fish resulting from operation of WBN Unit 1. **Undisputed.**

56.   The original study concluded that ichthyoplankton were present in relatively low densities in the vicinity of the WBN intake, and that those that were present had passed through the turbines of the Watts Bar Dam. The study also found that most spawning that occurs in Chickamauga Reservoir occurs downstream of the WBN intake. In other words, relatively few ichthyoplankton were available to be entrained at the WBN intake. The original study concluded that the percent of ichthyoplankton entrained was very low, and that WBN entrainment has no impact on the fish populations in the WBN vicinity. **Undisputed with respect to TVA's description of the study. Disputed in Contention 7. Disputed as to accuracy of results and conclusions.** *See* **Young Declaration, ¶ III-A.2 and III-A.12.**

57.   TVA revised this study in direct response to concerns raised by SACE in Contention 7, and by Dr. Young in support of Contention 7, that TVA's methods for estimating entrainment were flawed. Dr. Young claimed that TVA erroneously assumed that distribution of ichthyoplankton across the reservoir is uniform, and did not take into account variations in seasonal abundance of ichthyoplankton. Dr. Young also alleged that TVA should estimate entrainment using actual intake water demand and river flow values. **Undisputed as to stated information. Disputed as to the Aquatics Study was also revised after Dr. Young identified major clerical and mathematical errors that had gone unnoticed for over a decade.**

58.   In response to Dr. Young's concerns, TVA revised the entrainment analysis to account for seasonality of ichthyoplankton occurrence and reservoir releases from Watts Bar Dam. TVA also used actual intake water demand and reservoir flow values. **Undisputed that TVA revised its entrainment analysis to account for seasonality of ichthyoplankton occurrence and reservoir releases and that TVA used actual intake water demand and reservoir flow values. Disputed as to whether TVA did, in fact, account for seasonality of ichthyoplankton occurrence prior to the Peak Entrainment Study in 2010.** *See* **Young Declaration, ¶ III-A.2.**

59.   After conducting the revised entrainment estimates, TVA found that its overall conclusions regarding entrainment were unchanged. Estimated entrainment rates remained very low. For samples collected in 1996, percent entrainment in the revised analysis was estimated to be 0.29% for fish eggs and 0.57% for fish larvae. For samples collected in 1997, percent entrainment in the revised analysis was estimated to be 0.02% for fish eggs and 0.22% for fish larvae. **Undisputed that TVA has describe the results of the study. Disputed is the accuracy and validity of these results.** *See* **Young Declaration, ¶ III-A.2.**

60.   TVA's experts concluded that these rates are "low" and therefore there is no impact to the ichthyoplankton populations of Chickamauga Reservoir as a result of operation of WBN Unit 1. **Undisputed as to the description of the conclusion by TVA's experts. Disputed as to the reasonableness of the conclusion. The data were not sufficient to support the conclusion as this study was only for a 3-month period during only 2 years, one of which Unit 1 was not even operational or only at partial-capacity for a majority of time. The Revised Aquatics Study has the same shortcomings and still arrives at the same conclusions that are disputed in Contention 7.** *See* **Young Declaration, ¶ III-A.2 and III-A.12.**

61.   SACE has not challenged the methodology or findings of this study with this Board. **Undisputed.**

E.   Comparison of 2010 Peak Spawning Seasonal Densities of Ichthyoplankton at [WBN] at Tennessee River Mile 528 with Historical Densities During 1996 and 1997 (Apr. 2011, Revised Nov. 2011) ("Peak Spawning Entrainment Study")

62.   TVA conducted this study to respond to SACE and Dr. Young's concerns that TVA's methods for estimating entrainment were flawed, and that TVA should have taken direct measurements of entrainment. **Undisputed.** TVA collected raw data on actual entrainment at WBN during Unit 1 operation from March 2010 through March 2011, to ensure that all of SACE and Dr. Young's concerns regarding entrainment estimates were addressed, and in direct response to requests from SACE and Dr. Young for recent actual entrainment monitoring at WBN during operation of WBN Unit 1. **Undisputed with respect to the assertion that TVA collected raw data on actual entrainment at WBN1 in 2010-11. Disputed as to whether the data collected were sufficient to resolve Dr. Young's concerns.** *See* **Young Declaration, ¶ III-A.4.**

63.   This study reports entrainment resulting from operation of WBN Unit 1, as measured during the peak spawning period of April through June, 2010. TVA used this timeframe to address SACE and Dr. Young's concern that TVA account for the spawning patterns of fish species in the Chickamauga Reservoir and the high abundance of ichthyoplankton during certain times of year. **Disputed with respect to the assertion that the study reports entrainment from operation of WBN1 as measured through the peak spawning period in 2010. This study only reports entrainment at the CCW, and does not report entrainment by the SCCW. Thus, the cumulative entrainment due to operation of WBN Unit1 is not known. Disputed with respect to whether the data collected were sufficient to resolve Dr. Young's concerns.** *See* **Young Declaration, ¶ III-A.5.**

64.   This study concluded that measured entrainment rates at the WBN in 2010 were below one half of one percent of the ichthyoplankton population in the WBN vicinity, and consistent with those calculated for the same period during the first two years of operation of Unit 1, 1996 to 1997, when consistent calculation methods were applied. Specifically, the study found that the percent of entrained eggs in 2010 (0.12%) was within the range for 1996 (0.2%) and 1997 (0.2%). Likewise, the study found that the percent of entrained larvae in 2010 (0.40%) was within the range for 1996 (0.88%) and 1997 (0.22%). **Undisputed that TVA correctly describes the study's results. Disputed with respect to the accuracy of the results.** *See* **Young Declaration, ¶¶ III-A.2, III-A.5, and III-A.10-11.**

65.   TVA's experts concluded that these entrainment rates are "very low," and are not adversely affecting the fish population in the WBN vicinity. **Undisputed that this is the conclusion by TVA's experts. Disputed as to the accuracy and reasonableness of the conclusion.** *See* **Young Declaration, ¶ III-A.1-12.**

66.   The increased water intake demand for the CCW caused by dual unit operation will result in an estimated increase in hydraulic entrainment of approximately 0.2%. This study found that ichthyoplankton entrainment will increase proportionately with hydraulic entrainment. This increase will result in entrainment percentages that are still less than 1% of the ichthyoplankton population. This study concluded that, as a result, dual unit operation will not result in a material change in entrainment impacts. **Disputed as to the accuracy and reasonableness of this conclusion, and the rationale/methodology to arrive at this conclusion.** *See* **Young Declaration, ¶ III-A.13-14.**

67.   SACE has not challenged the methodology or findings of this study with this Board. **Undisputed.**

F.   Fish Impingement at [WBN] Intake Pumping Station Cooling Water Intake Structure During March 2010 through March 2011 (Mar. 2011, Revised Apr. 2011) ("Impingement Study")

68.   This study analyzes raw impingement data collected at the CCW intake during operation of WBN Unit 1 from March 2010 through March 2011. **Undisputed.** TVA used this data, in combination with the existing recent SCCW impingement data, to estimate the annual impingement mortality of fish in the vicinity of WBN as the result of operation of WBN Unit 1, and to predict the impact from operation of Unit 2. **Disputed as to the fact that TVA did not update the SCCW impingement in conjunction with the CCW impingement in this study.** TVA conducted this study in response to allegations by SACE and Dr. Young that TVA's analysis of the effects of WBN operation on the aquatic community was deficient because TVA had not conducted recent studies of actual impingement at the CCW intake. **Undisputed with respect to the assertion that TVA conducted the study. Disputed as to whether the study was sufficient to resolve Dr. Young's concerns.** *See* **Young Declaration, ¶ III-B.1-5.**

69.   This study found that total impingement values in 1996 to 1997 (161) were less than those measured in 2010 to 2011 (13,573). This study also found, however, that mortality resulting from a cold shock event dominated impingement mortality at WBN in 2010 to 2011. Shad in the Southeastern United States, including the Chickamauga Reservoir, are susceptible to cold shock. When temperatures fall below 50°F, they become lethargic and more susceptible to impingement. The study found that the most significant impingement events observed at WBN in 2010 to 2011 were the result of cold shock. **Undisputed as to the accuracy of TVA's description of the study's conclusions. Disputed as to the implication that cold shock, not the operation of WBN1, is the most significant cause of impingement mortality.** *See* **Young Declaration, ¶ III-B.1-4.**

70.   Excluding the cold shock event, this study found that fewer fish and number of species were impinged in 2010 to 2011, than in 1996 to 1997. The EPA endorses an impingement modeling approach that excludes the effects of extreme environmental conditions.

The EPA also acknowledges the effects of cold shocks on shad. **Disputed as to the cause of mortality. The mortality was caused by impingement against a man-made structure due to intake flow velocities not just the physiological consequences of cold temperatures.** *See* **Young Declaration, ¶ III-B.1-4.**

71.   This study concludes that low numbers of impinged fish in both 1996-97 and 2010-11 indicate that impingement resulting from operation of WBN Unit 1 will not materially affect fish populations in the WBN vicinity. **Disputed as to the reasonableness of the study duration being adequate to determine this conclusion, and as to the reasonableness of the conclusion.** *See* **Young Declaration, ¶ III-B.1-5.**

72.   Dual unit operation will result in increased withdrawal of water through the CCW intake channel. Impingement will likewise increase at a rate that is proportional to the increase in flow rate. This study concluded that the impingement increase from dual unit operation would still be very small when compared to the effects of cold shock and winter kills on shad. As a result, TVA's experts concluded that operation of Unit 2 will not result in material increases in impingement at WBN. **Disputed as to this methodology that was also used similarly by TVA to arrive at conclusions of entrainment from the combined operation of Unit 1 and 2.** *See* **Young Declaration, ¶ III-A.13-14.**

73.   SACE has not challenged the methodology or findings of this study with this Board. **Undisputed.**

G.   Hydrothermal Effects of the Ichthyoplankton from the Watts Bar Nuclear Plant Supplemental Condenser Cooling Water Outfall in Upper Chickamauga Reservoir (Jan. 2011) ("Hydrothermal Study")

74.   This study analyzes the hydrothermal impacts of WBN operation, based on in-river testing in the vicinity of the WBN outfall during WBN operation in May and August, 2010. TVA conducted this study in direct response to claims by SACE and Dr. Young that TVA should study the hydrothermal effects of operation of WBN Unit 1 on the aquatic environment in the WBN vicinity. Dr. Young alleged that TVA does not provide data on spatial or temporal distribution of ichthyoplankton in relation to thermal mixing zones, does not evaluate the impact of discharge temperatures on ichthyoplankton, and does not account for impacts of variations in the size or temperature profile of the mixing zone. **Undisputed**

75.   In direct response to these claims, TVA designed this study to document the flow patterns and characteristics of the thermal plume from WBN, and track the thermal plume in conjunction with ichthyoplankton sampling. This allowed TVA to understand the temporal and spatial distribution of ichthyoplankton and exposure rates to thermal discharges. **Disputed as to the fact that TVA failed to study the thermal discharge from Outfall 101 in conjunction with Outfall 113 to encompass the cumulative thermal discharge from WBN, and failed to address exposure rates and the effects of abrupt temperature changes on ichthyoplankton in this study.** *See* **Young Declaration, ¶ III-C.3-4.**

76.   TVA conducted this study in May and August, 2010, because those time frames represented extreme conditions: peak abundance of fish eggs and larvae, near maximum ambient water temperatures, and no release from the upstream Watts Bar Dam. **Undisputed as to**

timeframe of study. Disputed as to whether this would be representative over time as this study only represents a few points in time, not adequately addressing environmental variability. *See* **Young Declaration, ¶ III-C.2.**

77.   This study found that, even under these extreme conditions, water temperatures did not approach the limits established by TVA's NPDES permit for operation of WBN Units 1 and 2. **Disputed as study results directly stated to the contrary.** *See* **Young Declaration, ¶ III-C.11.** Because discharge temperatures did not exceed those set in TVA's NPDES permit, this study concluded that there was no risk of thermal damage to ichthyoplankton from operation of WBN. **Disputed as to accuracy and reasonableness of these conclusions.** *See* **Young Declaration, ¶ III-C.1-11.**

78.   Even if operation of WBN Units 1 and 2 causes effluent temperatures to rise above those measured even under extreme conditions for Unit 1, TVA is bound by its NPDES discharge limits. Accordingly, dual unit operation does not pose any greater risk of thermal damage to the aquatic community in the WBN vicinity than does operation of Unit 1 alone. **Disputed as to accuracy and reasonableness of these conclusions.** *See* **Young Declaration, ¶ III-C.1-11.**

79.   SACE has not challenged the methodology or findings of this study with this Board. **Undisputed.**

IV.   Overview of the Draft SFES Conclusions Regarding TVA's Aquatic Studies

80.   As noted previously, the NRC Staff's Draft SFES concurs with the findings presented in TVA's aquatics studies. **Undisputed.**

81.   Specifically, the Staff concurred with TVA's findings regarding entrainment impacts, concluding in the Draft SFES that hydraulic entrainment would have a very minor impact on the aquatic biota in the vicinity of WBN. The Staff agrees that existing levels of measured entrainment under Unit 1 operation are too low to be readily detected in the aquatic populations in the WBN vicinity, and the additional water withdrawn via the CCW intake will not be noticeable or furthermore destabilizing to the aquatic ecology in the WBN vicinity. Moreover, the Staff concludes that the water withdrawn from the SCCW intake will actually decrease under dual unit operation. In drawing these conclusions, the Staff relies in part on the Revised Aquatics Study and the Peak Spawning Entrainment Study. **Undisputed. It should be noted that the NRC Staff has not conducted any independent studies to support its conclusions.**

82.   The Staff's conclusions regarding impingement impacts are similar. The Staff finds that measured levels of impingement under operation of WBN Unit 1 are low and impingement effects are too minor to be readily detected in aquatic populations in the WBN vicinity. The increased flow rates for the CCW intake under dual unit operation will not alter that conclusion, concludes the Staff, and the decreased flow rates for the SCCW intake will not increase impingement effects. The Staff relied in part on the Impingement Study in drawing these conclusions. **Undisputed. It should be noted that the NRC Staff has not conducted any independent studies to support its conclusions.**

83.   With respect to thermal impacts from operation of WBN Unit 2, the Staff concludes that this effect also will be undetectable and will not destabilize or noticeably alter the aquatic biota in the WBN vicinity. The Staff based this conclusion in part on the Hydrothermal Study, as well as limits set by the NPDES permit. **Undisputed. It should be noted that the NRC Staff has not conducted any independent studies to support its conclusions.**

84.   The Staff concludes in the Draft SFES that although the impoundments and industrial facilities have a significant cumulative impact on the aquatic biota in the WBN vicinity, "the overall impacts on aquatic biota, including Federally listed threatened and endangered species, from impingement and entrainment at the SCCW and IPS [i.e., CCW] intakes and from thermal . . . discharges as a result of operating Unit 2 on the WBN site are SMALL." **Undisputed. It should be noted that the NRC Staff has not conducted any independent studies to support its conclusions.**

Respectfully submitted,

Electronically signed by
Diane Curran
HARMON, CURRAN, SPIELBERG, & EISENBERG, L.L.P.
1726 M Street N.W., Suite 600
Washington, D.C. 20036
202-328-3500
Fax: 202-328-6918
e-mail: dcurran@harmoncurran.com

December 20, 2011

December 20, 2011

UNITED STATES OF AMERICA
NUCLEAR REGULATORY COMMISSION

BEFORE THE ATOMIC SAFETY AND LICENSING BOARD

In the Matter of	)	
Tennessee Valley Authority	)	
Completion and Operation License	)	Docket No. 50-391OL
Watts Bar Nuclear Plant Unit 2	)	
	)	

DECLARATION OF SHAWN PAUL YOUNG, PH.D.

Under penalty of perjury, I, Shawn Paul Young, declare as follows:

I. STATEMENT OF PURPOSE AND PROFESSIONAL QUALIFICATIONS

1.   My name is Shawn Paul Young, Ph.D. I have been retained by Southern Alliance for

Clean Energy ("SACE") as an expert consultant in this matter. I submit this declaration as a

private consultant to SACE in this matter.

2.   I am currently employed as a Fish Biologist for the Kootenai Tribe of Idaho. I also

maintain a private environmental consulting business. My current business address is P.O. Box

507, Bonners Ferry, Idaho 83805.

3.   My professional and educational experience is summarized in the curriculum vitae

attached to this declaration. To summarize, I received a B.S. in Environmental Studies from

Northland College; a M.S. in Aquaculture, Fisheries, and Wildlife Biology from Clemson

University; and a Ph.D. in Fisheries and Wildlife Sciences from Clemson University. I have

fourteen years of experience researching the effects of human activities on fisheries and aquatic

ecosystems. In addition to my professional qualifications, I am an avid outdoorsman. I have

fished, hunted, and enjoyed nature in every manner since my early childhood.

4.   As listed in my curriculum vitae, I have authored and published peer-reviewed articles and reports relevant to fisheries and aquatic ecology. I have been consulted by public, state, federal, and academic sectors in the subject area of fish and aquatic ecology. I have delivered scientific presentations at numerous professional meetings, academic seminars, and citizen fishing association functions.

5.   I am familiar with the application of Tennessee Valley Authority ("Applicant" or "TVA") for an Operating License ("OL") at the Watts Bar Nuclear Plant site and related documents, including TVA's 2007 Final Supplemental Environmental Impact Statement ("FSEIS"); the Draft Final Environmental Statement ("DFES") issued by the NRC Staff in December 2011; and the Joint Affidavit of TVA staff, Dennis Scott Baxter and John Tracy Baxter, and experts, Dr. Charles Coe Coutant and Dr. Paul Neil Hopping, supporting TVA's Motion for Summary Disposition of Contention 7. I have reviewed these documents with particular reference to their description and analysis of the additional unit's expected heat budget, water intake, water consumption, and thermal discharge into the Tennessee River; and the proposed reactor's potential impacts on the aquatic organisms of the Tennessee River.

6.   I am providing this declaration in support of Intervenors' Contention 7 -- Impacts on Aquatic Resources of the Tennessee River. That contention and its supporting declaration expressed my view that TVA's conclusion in the FSEIS that the cumulative impacts of WBN Unit 2 on aquatic ecology will be insignificant is not reasonable or adequately supported. My opinion was based on three fundamental problems with TVA's data and analysis. First, TVA mischaracterizes the current health of the ecosystem as good, and therefore fails to evaluate the impacts of WBN2 in light of the fragility of the host environment. Second, TVA relies on outdated and inadequate data to predict thermal impacts and the impacts of entrainment and

2

impingement of aquatic organisms in the plant's cooling system. Third, TVA fails completely to analyze the cumulative effects of WBN2 when taken together with the impacts of other industrial facilities and the effects of the many dams on the Tennessee River. This declaration explains the basis for my scientific opinion that the concerns I raised in Contention 7 have not been resolved by the studies cited in TVA's Motion for Summary Disposition, the NRC Staff's DFES, or the Joint Affidavit submitted by TVA's experts.

7.   I have arrived at my conclusions dealing with the matters stated herein based upon material fact found within the documents related to Watts Bar Nuclear Units 1 and 2, and within relevant scientific literature produced by other scientists pertaining to this subject, and believe them to be true and correct.   The opinions and conclusions I express in this affidavit are my own and should not be attributed to any other person or entity.

II.   SUMMARY OF MY PROFESSIONAL OPINION REGARDING TVA'S ASSERTIONS.

1.   Relying on several studies that it has conducted in response to Contention 7, as well as the DFES, TVA claims that it has resolved the three major deficiencies identified in Contention 7. But this is not correct.

2.   With respect to the inadequacy of TVA's previous data and analyses, TVA has made some progress by collecting new data on entrainment, impingement, freshwater mussels, and thermal impacts during 2010. But TVA has only started to catch up with its failure to collect the appropriate data that would be reasonably sufficient to evaluate impacts on aquatic resources by collecting only one year of data for entrainment, impingement, freshwater mussels, and thermal impacts over the preceding years. TVA still has not collected an amount of data that is reasonably necessary to evaluate the effects of WBN1 on aquatic organisms in the Tennessee

3

River, and therefore it does not have enough information to extrapolate the impacts of WBN2. See pars. III-A.5, III-B.3-4, and III-C.1-2 below.

3.   In addition, there are still big gaps in the information that TVA has collected. For example, TVA collected entrainment data for the Condenser Cooling Water ("CCW") system only and did not include the Supplemental Condenser Cooling Water ("SCCW") system. See par. III-A.4 below. In addition, TVA did not collect impingement data for all key locations. See par. III-B.5 below. And TVA's Hydrothermal Study does not address important paramaters such as Outfall 101 or the amount of time that fish larvae remain in the thermal plume. See par. III-C.4 below.

4.   Finally, TVA's description of its method of analyzing aquatic impacts indicates a troubling lack of care or competence. For example, by adding widely divergent diurnal and nocturnal entrainment measurement, TVA violates guidance of the U.S. Environmental Protection Agency ("EPA") and grossly overstates the size and diversity of the fish population. See pars. III-C.6 and III-C.10 below. Some of the studies relied on by TVA had to be revised after they were released, indicating that TVA has significant problems ensuring the quality of its measurements and analyses. See pars. III-A.2 and III-A.11, and pars. III-C.6-9 below.   It is reasonable to expect that the results from TVA's biological studies will be accurate in order to support TVA's conclusions. In too many instances, however, TVA makes significant mistakes.

5.   With respect to TVA's mischaracterization of the health of the aquatic environment as good, TVA has done nothing to alleviate my concern. Although as discussed above, TVA's data collection is insufficient to present a reasonable picture of the health of the Tennessee River, the data that TVA has collected do not indicate, as TVA claims, that WBN1's impacts on the aquatic

4

ecosystem have been insignificant. Rather, they point to already-significant aquatic impacts by WBN1 that are likely to be significantly exacerbated by the operation of WBN2.

6.   Further, despite alarming evidence of significant decline in the diversity and numbers of aquatic organisms in the Tennessee River in the vicinity of WBN, TVA continues to assert that the aquatic health of the river is good. The only way that TVA can present such a clean bill of health is to mischaracterize the baseline condition of the Tennessee River as a large reservoir where one would expect to see a limited number of species of aquatic organisms. In reality, the Tennessee Rive is a fragile and rapidly deteriorating riverine ecosystem with remnants of the greatest species diversity of any river in the United States. By falsely painting a rosy picture of aquatic health in the river, TVA understates the significance of the impacts of WBN1 and WBN2, and thus minimizes the benefits that could be achieved by implementing alternatives that would reduce the impacts of the cooling system on organisms in the river.

7.   Finally, TVA still does not address the cumulative impacts of WBN2 in conjunction with the impacts of the numerous water impoundments on the Tennessee River, or with other industrial facilities such as the ten fossil fuel-burning plants, the six nuclear reactors that are already in operation, and the five additional reactors for which TVA has sought operating licenses. The combined operation of WBN1 and WBN2, by itself, may cause changes in how Watts Bar Dam is operated. TVA and the NRC Staff both acknowledge that in order to stay within thermal discharge limits stated in the NPDES that requests for additional discharge from Watts Bar Dam may be needed. Thus, operating WBN alone would change reservoir operations in the middle- Tennessee Basin that would be supported by water releases or hydrological adjustments in upper-Tennessee River Basin. The effects of more alterations to the hydrological cycle of the basin on aquatic organisms, especially the already declining native fish

5

and freshwater mussel species, must be addressed. Given the extensive portfolio of energy and

industrial facilities that the Tennessee River supports and that the management agencies must

maintain adequate water for all these facilities, this is an extremely important omission.

III.   **STATEMENT OF PROFESSIONAL OPINION REGARDING ADEQUACY OF TVA'S RECENT BIOLOGICAL STUDIES TO ADDRESS THE ENVIRONMENTAL IMPACTS OF THE PROPOSED WATTS BAR 2 NUCLEAR POWER PLANT ON AQUATIC ORGANISMS**

Over the past two years, TVA has conducted or revised eight studies which it claims to

resolve the concerns raised by Contention 7. The studies are the following:

- Aquatic Environmental Conditions in the Vicinity of Watts Bar Nuclear Plant During Two Years of Operation, 1996-1997 (June 1998, Revised June 2010) ("Revised Aquatics Study")

- Comparison of 2010 Peak Spawning Seasonal Densities of Ichthyoplankton at [WBN] at Tennessee River Mile 528 with Historical Densities During 1996 and 1997 (Apr. 2011, Revised Nov. 2011) ("Peak Spawning Entrainment Study")

- Fish Impingement at [WBN] Intake Pumping Station Cooling Water Intake Structure During March 2010 through March 2011 (Mar. 2011) ("Impingement Study")

- Hydrothermal Effects of the Ichthyoplankton from the Watts Bar Plant Supplemental Condenser Cooling Water Outfall in Upper Chickamauga Reservoir (Jan. 2011) ("Hydrothermal Study")

- Mollusk Survey of the Tennessee River Near [WBN] (Rhea County, Tennessee) (Oct. 28, 2010, Revised Nov. 24, 2010) ("Mollusk Survey")

- Results and Discussion of the 2010 Mollusk Survey of the Tennessee River Near [WBN] (Rhea County, Tennessee) (Mar. 2011) ("Discussion of Mollusk Survey")

- Comparison of Fish Species Occurrence and Trends in Reservoir Fish Assemblage Index Results in Chickamauga Reservoir Before and After WBN Unit 1 Operation (June 2010) ("RFAI Study")

- Analysis of Fish Species Occurrences in Chickamauga Reservoir – A Comparison of Historic and Recent Data (Oct. 2010) ("Fish Species Occurrences Study")

As discussed below, these studies do not resolve the concerns raised in Contention 7.

A.   **Revised Aquatics Study and Peak Entrainment Study**

6

1.   TVA asserts that it has revised its method for estimating entrainment impacts and has also collected raw data on actual entrainment associated with WBN1 for one year. TVA Motion at 14-15. TVA asserts that these studies show the rate of entrainment is very low. *Id.* In my professional opinion, however, TVA's studies do not provide a reasonable degree of support for the conclusion that the rate of entrainment is low. In fact, they indicate a rate of entrainment that is unacceptable.

2.   The Revised Aquatics Study is a revision of the "Aquatics Study" for which TVA collected ichthyoplankton data in order to estimate entrainment at WBN Unit 1 only during April – June 1996 and 1997, not the entire year, a major shortcoming. The timing of the original Aquatic Study corresponded to the commencement of operation of WBN Unit 1. The study results were published in 1998. TVA concluded that WBN Unit 1 ichthyoplankton entrainment was low and had insignificant impacts on the fish community. In 2009, I identified major errors in this document that had major implications. TVA revised this study, and released a revision in 2010 that did not include an additional level of detail for data presentation and analysis to assess whether the errors were properly rectified. Further, TVA's conclusions remained unchanged. Based upon the original erroneous document, in 1998, TVA convinced the Tennessee Department of Environment and Conservation ("TDEC") to allow termination of the entrainment monitoring program mandated in the original NPDES permit. Therefore, since 1997, TVA had not collected any post-operational entrainment study at Unit 1.

3.   After SACE's contention 7 was admitted, TVA conducted one year of entrainment monitoring during 2010 to compare the results against 1996 and 1997 entrainment data. The Peak Entrainment Study was a survey of the ichthyoplankton drift past the Supplemental Condenser Cooling Water ("SCCW") discharge (Outfall 113) and the Unit 1 water intake

7

pumping structure for the CCW system. The Peak Entrainment study was conducted in conjunction with the "Hydrothermal Study" in order to also determine ichthyoplankton abundance at the SCCW intake, and in the SCCW discharge under two different thermal mixing zone scenarios.

4.   In the Peak Entrainment Study, TVA collected ichthyoplankton along a transect from riverbank to riverbank below the SCCW discharge plume and above the intake pumping structure (IPS) for the CCW. As such, the study provides only a minimal account of the conditions in the Tennessee River. In order to make a reasonable analysis of the impacts of WBN1 on the river and the likely impacts of WBN2, TVA should have been collecting entrainment data regularly since WBN1 went online in 1996. For any reasonable biologist, two measurements taken thirteen years apart would not provide a sufficient basis for an analysis of entrainment impacts. TVA should have collected data for at least three years after WBN1 began operating in order to determine any annual variability of ichthyoplankton abundance. And TVA should have updated those measurements after it decided to pursue an operating license for WBN2, with at least two more years of measurements.

5.   TVA's data collection for the Peak Entrainment Study was incomplete because TVA reported entrainment measurements only for the CCW intake. Even though TVA collected ichthyoplankton samples at the SCCW intake and in Watts Bar forebay, TVA did not present the data or calculate entrainment rates for the SCCW within the Peak Entrainment Study. Instead, TVA only presented data on ichthyoplankton abundance near the SCCW intake within the Hydrothermal Study, and again did not present any entrainment rates. Thus, TVA failed to adequately estimate total entrainment at the WBN1 water intake structures. The omission is significant because Tables 2 and 3 of the "Hydrothermal Study" list the results of

8

ichthyoplankton abundance at and near the SCCW intake in Watts Bar Reservoir forebay. The results listed in the hydrothermal study show that 300% more fish larvae were captured at the SCCW intake on May 11-12, 2010 (Table 3) than were captured in the forebay nearby (Table 2). This indicates that a very high level of entrainment may be occurring at the SCCW intake. TVA, however, failed to recognize this significant material fact.

6.   In any event, the results that TVA reported for the CCW intake show that WBN1 has had significant impacts on the aquatic environment and that operation of WBN2 is also likely to impose significant additional impacts. First, the Peak Entrainment Study shows that ichthyoplankton abundance in the vicinity of WBN has declined significantly since operation of WBN1 commenced. The abundance of ichthyoplankton was substantially lower in 2010 than in post-operational surveys during years 1996 and 1997 as calculated and listed by TVA in the Revised Aquatics Study. As stated in the Peak Entrainment Study at page 3 with respect to fish larvae:

> Average densities (525, 924, 282), peak seasonal densities (1,387; 1,699; 828) and dates of peak densities (06/03, 05/15, 05/16) for larvae during April through June 1996, 1997, and 2010, respectively, are presented in Table 5. *All of these values for samples collected during 2010 were slightly lower than the range of the two previous years (1996 and 1997) of monitoring.*

(emphasis added). TVA and the NRC Staff failed to properly acknowledge the significant decline as a very important material fact in their respective analyses and conclusions.

7.   The Peak Entrainment Study also reported a decline in the number of fish eggs between 1996 and 2010: average densities were reported as 262, 150, and 75 and peak seasonal densities were reported as 1,095, 1,004, and 811 for April through June 1996, 1997, and 2010, respectively. The significance of this decline is not discussed by either TVA in its Motion or the NRC Staff in the DES.

9

8.   Based on the data reported in the Peak Entrainment Study, (Table 7, p. 19), larger than anticipated entrainment events occurred at WBN1. Daily entrainment rates of fish larvae were as high as 8.65% (June 21, 2010) during peak ichthyoplankton abundance. In my professional opinion, such a high rate of entrainment may have adverse impacts on the fish community. This measurement is very significant, given that hydraulic entrainment will double at the IPS for the CCW with the addition of WBN2, likely doubling ichthyoplankton entrainment. Larval fish entrainment events may double from 8.5% to 17%, a rate of entrainment that would certainly have a significant impact on the health of the fish population.

9.   The Peak Entrainment Study also reported in Table 7 that daily entrainment rates of fish eggs were as high as 4.08% (May 16, 2010) during peak ichthyoplankton abundance. In my professional opinion, an egg entrainment rate of 4% is high enough to have a potentially adverse impact on the fish community. This measurement is very significant, given that hydraulic entrainment will double at the IPS for the CCW with the addition of WBN2, likely doubling fish egg entrainment events from 4.0% to 8.0%. At 8%, the impacts would indeed be significant.

10.   I am also concerned about potential errors in the Peak Entrainment Study. At page i, TVA stated that another revision should be released sometime this month, December 2011. This indicates to me that there may be more errors in the study.

11.   Further, I identified errors in methodology TVA used to complete calculations in the "Hydrothermal Study" which may have consequences for the Peak Entrainment Study. Both studies should have used the same formula to calculate the number of ichthyoplankton within 1,000 m^3 of source water from the number of organisms actually captured in the volume of water actually sampled to catch those organisms. Within the Hydrothermal Study, the number of ichthyoplankton density per 1,000 m^3 of water was estimated to determine how many fish eggs

10

and fish larvae were exposed to high water temperatures in the SCCW thermal plume during the day and during the night. To arrive at an estimate of the daily abundance per 1,000 m^3 of water, the day and night estimates should have been averaged, not added together. See pars. III-C. 6-9, below in this declaration. Thus, results for daily ichthyoplankton abundance at the SCCW intake are incorrect; and since the two studies incorporate similar methods to estimate ichthyoplankton densities, similar errors in calculations may have been made in the Peak Entrainment Study also. However, the entrainment study lists results in a different manner that does not allow one to determine this.

12.   In conclusion, the Revised Aquatic Study and the Peak Entrainment Study do not support TVA's conclusion that the environmental impacts from entrainment at the current IPS for the CCW intake with one reactor are insignificant, nor do they support a conclusion that the additional impacts of WBN2 would be insignificant. To the contrary, the data reported shows that the impacts from entrainment from the IPS for the CCW from one reactor unit alone may be large and warrants further investigation. Further, the Hydrothermal Study suggests that entrainment at the current SCCW intake may be also be significant with large impacts to the fish community.

13.   As a general matter, TVA also mischaracterizes the relationship between river flow and entrainment. According to TVA, studies show that the hydraulic entrainment from dual unit operation will result in an additional entrained amount of 0.2% of the flow in the Chickamauga Reservoir. Statement of Material Facts, par. 22. TVA asserts that the resulting total hydraulic entrainment represents approximately 0.5% of the flow in the Chickamauga Reservoir; and that this increased hydraulic entrainment will result in a proportionate increase in entrainment of the ichthyoplankton present in the water column. *Id.*

11

14. TVA's calculation is only partly correct, and only accurate at a very specific river flow past WBN Plant. The 0.2% hydraulic entrainment for WB1 is based upon TVA using "a long term average river flow past WBN of 27,000 cfs." See Footnotes 58-60 and Joint Affidavit par. 37. However, the flow past WBN may vary widely depending on seasonal precipitation levels and daily operations of Watts Bar Dam immediately upstream of WBN. Therefore, hydraulic entrainment will vary depending on amount of water in the Tennessee hydrosystem and how much flow is released from Watts Bar Dam. For instance, using CCW water withdrawal rate of 88 cfs (NRC DFES Table 3-1 at page 3-9) and river flow of 3,500 cfs, which is the minimum amount of flow from Watts Bar Dam that permits TVA to discharge thermal and chemical effluent through Outfall 101, the hydraulic entrainment increases to 2.5% (12.5 times higher). Then, with the addition of Unit 2 doubling hydraulic entrainment, the hydraulic entrainment at a flow of 3,500 cfs further increases to approximately 5.0% (25 times higher). Also, with higher hydraulic entrainment, the probability of entraining more ichthyoplankton increases. However, one cannot assume that ichthyoplankton entrainment will increase proportionately. In fact, ichthyoplankton may increase exponentially. The increase depends on the proximity of ichthyoplankton to water intakes. Only data collected by field studies in combination with proper methods for calculation may accurately characterize ichthyoplankton entrainment under any level of hydraulic entrainment. I note that this is a similar issue in regards to impingement.

B.   **Impingement Study**

1.   TVA claims that impingement data it collected between March 2010 and March 2011 at the CCW intake show that impingement rates under normal conditions were unchanged from those that TVA historically measured at the CCW intake, but that unusually cold weather in the winter of 2011 produced high impingement rates. TVA also cites the DES for the proposition

12

that impingement impacts during operation of both WBN1 and WBN2 would be "too low to noticeably alter the aquatic community". TVA Motion at 16-17.

2.   I disagree with TVA that the Impingement Study provides sufficient data on which to reach a conclusion about impingement impacts of either WBN1 or dual operation of WBN1 and WBN2.

3.   Although WBN1 has been operating since 1996, the last time TVA took an impingement measurement for the CCW was in 1997. Although TVA has planned for some time to finish building and operate WBN2, it made no effort to measure impingement rates until 2010, after Contention 7 was admitted for a hearing. For any reasonable fish biologist, two measurements taken more than ten years apart would not suffice to provide the basis for any analysis of the impingement impacts of WBN1.

4.   The circumstances of the 2010 measurements illustrate my point. In comparison to the 161 fish impinged in March 1996 through 1997, 13,573 were impinged in 2010. See Attachment 15, page 3. TVA attributes this exponential increase to cold weather in 2010. But it is also possible that the through-screen velocity of water flowing into the CCW intake is partially responsible for the high impingement rate. At page 1, the Impingement Study lists the through-screen velocity as 0.67 fps. The EPA recommends that through-screen velocity be kept below 0.5 fps, however, in order to reduce entrainment and impingement. Without more data over a period of several years, the contribution of the cold and plant operating conditions to the rate of impingement can only be guessed at. In short, it is not possible for TVA to make up for years of neglect in only one year.

5.   TVA also failed to take impingement measurements for all key locations. The Impingement Study sampled fish impingement at the IPS for the CCW only, and did not include

13

the SCCW. A study was conducted in 2000 to evaluate impingement at the SCCW intake above Watts Bar Dam; however, this study did not monitor an entire year. This study still showed that impingements may also occur at the SCCW intake (p. 6, Watts Bar Nuclear Plant Supplemental Condenser Cooling System Fish Monitoring Program, January 2001); yet, TVA still did not conduct impingement monitoring at the SCCW during 2010 in conjunction with the CCW study to determine the cumulative impingement by current operations of WBN Unit 1.

C.   Hydrothermal Study

1.   In the Hydrothermal Study, TVA reports the results of monitoring the water temperatures in the thermal plume of the SCCW (Outfall 113) during May and August 2010. TVA recorded water temperatures during the two mixing zone scenarios that occur daily, the active mixing zone when Watts Bar Dam releases water and the passive mixing zone when Watts Bar Dam does not release water. TVA also completed ichthyoplankton sampling at and near the SCCW above Watts Bar Dam, and downriver of Watts Bar Dam below the actual thermal plume during both day and night. TVA asserts that the Hydrothermal Study shows that thermal discharges from WBN1 and WBN2 will not have a significant impact on aquatic organisms. TVA Motion at 18-19.

2.   TVA should have conducted the study over several years to characterize thermal plume water temperatures and ichthyoplankton abundance that may vary across years due to variable climatic conditions, and due to variable operations of Watts Bar Dam caused by variable hydrological conditions in the Tennessee River Basin.

3.   The Hydrothermal Study also failed to address important parameters. For instance, it did not include any data or analysis for Outfall 101 (discharge at the CCW diffuser), which releases heated effluent when the dam discharge exceeds 3,500 cfs. Outfall 101 should have been

14

included, especially in light of the fact that ichthyoplankton may drift through Outfall 113 mixing zone and then into the Outfall 101 mixing zone. This omission is significant.

4.   In addition, contrary to statements in the Motion for Summary Disposition and the DFES, the Hydrothermal Study did not list nor discuss ichthyoplankton exposure rates i.e., the amount of time fish eggs and larvae remain in the thermal plume. The omission of this information is significant because the early life stages of fish, especially eggs and larvae are vulnerable to abrupt temperature change such as those found at Outfall 113 and 101, and exposure to such water temperature changes caused by WBN heat waste discharge may cause high mortality rates. Abrupt temperature changes are detrimental to fish eggs and larvae. Also, abrupt temperature change affects species differently. This is an important omission because a rapid increase of 5 – 10° F can kill fish eggs and fish larvae, and from the data presented, most of the ichthyoplankton likely experienced this as they drifted through the SCCW mixing zone. Further, not only are ichthyoplankton exposed to the SCCW thermal plume, but these same fish eggs and larvae then drift through the CCW diffuser thermal plume below. A second abrupt temperature increase further elevates risk of mortality from the heat discharged from WBN.

5.   The Hydrothermal Study is also deficient because TVA failed to report and discuss the fact that an alarming number of ichthyoplankton were likely entrained by the SCCW and subsequently killed by heat within the SCCW system before being discharged back into the river. This is an extremely important consideration in this matter. Further, the portion of ichthyoplankton in the Watts Bar Reservoir forebay not directly entrained and killed by the SCCW would likely pass through the dam and then still would be subjected and potentially killed by the waste heat in the SCCW and CCW (Outfalls 113 and 101) thermal plumes. The use of the SCCW creates a "double whammy" for fish eggs and larvae, likely causing an

15

alarming level of mortality. TVA does not adequately describe this situation or adequately analyze presented data that shows significant mortality may be occurring via both pathways.

6.   The conclusions of the Hydrothermal Study are also based on incorrect methodology that leads to distorted results. In reporting the results of ichthyoplankton sampling, TVA added the daytime and nighttime measurements rather than averaging them, thus giving a distortedly high population reading. For instance Table 4 on page 25 of the Hydrothermal Survey shows that on May 11-12, 2010, during daytime sampling, TVA estimated 75 organisms per 1,000 m^3 of water at the SCCW outfall. During the nighttime sampling, TVA estimated 8,232 organisms for the same volume of water. TVA then reported the number of organisms per volume of 1,000 m^3 of water for the sampling period as 8,307. In actuality, however, the number of organisms ranged between 75 and 8,232, with an average of approximately 4,153 fish larvae per 1000 m^3 of water during a 24-hour diel cycle.

7.   There is no controversy about what method TVA should have employed – it is listed in the "Materials and Methods" section of TVA's April 2011 "Peak Entrainment Study." For TVA not to notice another significant error in its own reporting raises fundamental questions regarding TVA's methodology for all of its studies.

8.   TVA's methodological error has several implications in the analyses of impacts on the fish community. This error results in the overstatement of the size of the fish population in the river, which in turn will lead to an understatement of the percentage of fish that are affected by entrainment. This has major implications for the validity of the "Entrainment Study" because it results in an incorrect estimate of the percentage of organisms that were entrained at the CCW. If the same error found in the Hydrothermal Study was made during calculations of ichthyoplankton abundance for the Peak Entrainment Study, the results listed in the Peak

16

Entrainment Study are not accurate, and TVA conclusions are not based on accurate material facts. In addition, the original Aquatics Study also had major errors, and one cannot be sure those errors have been remedied in the Revised Aquatics Study. Both documents used to compare post-operation entrainment and the associated impacts have had major errors casting doubt on the validity of TVA's analyses and conclusions.

9.   Another significant error can be found in Tables 5 through 10. Table 10 lists the total ichthyoplankton abundance found at the five different sampling stations across the survey transect. However, the reported total number of ichthyoplankton captured is less than the reported number of ichthyoplankton that were captured at just *one* of the individual sampling stations. This error raises serious questions about the actual results of the study, not to mention TVA's competence and quality assurance procedures for conduct of biological monitoring and anlaysis.

10.   TVA also failed to note the significance of the great discrepancy between the daytime and night-time population measurements, or to analyze how they may be affected by daily variations in thermal plume temperature. In light of the size of the discrepancy, TVA should have undertaken more studies of the differences between daytime and nighttime fish populations. It should also evaluate changes in nighttime operations to reduce the rate of entrainment of aquatic organisms.

11.   The Hydrothermal Study showed that thermal discharge observed for current operation of Unit 1 is already near the limits set in the NPDES permit. TVA's temperature data shows that it is staying within its permit limit of a 5°F daily average change from upriver temperature at the downstream edge of the mixing zone; however, the results from the May and August 2010 tests show that it is operating on the edge of those limits with only Unit 1 operating. As stated at

17

page 5, the maximum difference between ambient and surface temperature reached 5°F during the May night test, 5.34°F during the May day test, and 5.36°F during the August day test. Also, at the point of discharge, the Hydrothermal Study shows that SCCW discharge water is 10°F hotter than the water above the SCCW thermal plume and above Watts Bar Dam. Organisms drifting downriver nearest the point of discharge will likely suffer from this abrupt temperature change, especially fish eggs and larvae. These impacts were not considered by TVA. See above in par. 4.

D.   Mollusk Survey, Discussion of Mollusk Survey, and Revised Aquatics Study

1.   As discussed in Contention 7, TVA's assertion in the FEIS that mussel health is "excellent" because their population is "constant" is contradicted by evidence that mussel populations are declining. Contention 7 at page 33. TVA responded to my criticism by hiring a consultant to conduct a new mussel survey utilizing new and expanded methodology. The study site evaluated mussel beds within transects in the same general areas as previous TVA mussel surveys near WBN Plant. Each mussel was identified by species and age. TVA compared the results from the 2010 study with previous mussel studies, including the post-operational mussel surveys in 1996 and 1997. The results from the 1996 and 1997 post-operational surveys are found within the original "Aquatic Study" published in 1998 and the recent "Revised Aquatics Study".

2.   TVA no longer asserts that mussel health near WBN1 is excellent. Instead, it states that the studies it conducted "agree that the Chickamauga Reservoir in the WBN vicinity is not the ideal habitat for mussels." Statement of Material Fact, page 19. Nevertheless, TVA's experts state that the survey results demonstrated "that the current mussel community adjacent to WBN is stable and that some species are reproducing." Baxter and Coutant, par. 72. They assert that

18

the mussel community in the WBN vicinity is in "substantially similar condition as it was near the end of the previous operational monitoring period (1996 to 1997), in both species composition and the number of mussels collected." In addition, they state that the 2010 survey "collected juveniles of at least five mussel species, evidencing reproduction of mollusks in the WBN vicinity." *Id.* Based on these results, TVA contends that "there is no basis to support a finding that the relatively low densities of mussels in the WBN vicinity are the result of operation of WBN Unit 1."

3.   I disagree with TVA's assertions. The data collected by TVA show that health of the freshwater mussel community around WBN1 is poor and declining. The data also show a connection between the poor health of the mussel community near WBN1 and the operation of WBN1.

4.   There can be no doubt that the health of the mussel community near WBN1 is poor and also declining. The data provided in the Mollusk Survey show that freshwater mussel abundance has declined significantly in the area affected by the SCCW since it began cooling Unit 1 in 1999. TVA failed to address three significant trends reflected in this data. First, the abundance of mussels at the three study sites changed significantly between 1996-97 and 2010. In 1996-97, just before the SCCW went into operation for WBN1 in 1998, 344 mussels were collected from the upper bed located just upriver of WBN. That bed now lies within the SCCW discharge plume (p. 40, Revised Aquatics Study). By 2010 the abundance of mussels at the upper bed had been reduced by approximately half to 175 (p. 4, Mollusk Study). This is a major concern, given that the site is within the mixing zone for the SCCW outfall, which had not been in use for a substantial time prior to or during the 1996-97 surveys.

19

5.   The data also show that mussel abundance in both the middle and lower sites increased since 1996-97 (p. 40, Revised Aquatics Study and p. 4, Mollusk Study. These increases may be due to better sampling techniques employed in 2010, or to better reservoir system management practices implemented at Watts Bar Dam. The Discussion of Mollusk Survey does not explain this development. Quite possibly, the SCCW may be thwarting a rebounding mussel population in the vicinity of WBN.

6.   Second, the experimental boulder field to provide increased mussel habitat as a mitigation measure for the use of the SCCW had very few mussels – only five -- indicating this action was a failure. TVA's experts attribute this failure to the force of the water flowing from Watts Bar Dam. Baxter and Coutant, par. 70. But they do not acknowledge that the boulder field is located near the SCCW. The death of most relocated mussels, and the substantial decline of mussel numbers in the upper bed show the SCCW has and will continue to have substantial adverse impacts on the mussels near WBN.

7.   Finally, the data indicates that a significant number of mussel species are still unable to reproduce and recruit new members to sustain their local populations. The recent survey found the presence of juveniles for four of the 17 species, indicating some reproduction and recruitment is taking place. However, for the other 13 species -- including two endangered species -- no juveniles were present, indicating a lack of reproduction and recruitment capacity, which will lead to eventual local extirpation. In addition, the four reproducing species that were found near WBN1 are just a fraction of the 64 mussel species known to once inhabit the Tennessee River in the vicinity of present day WBN Plant. Thus, only 6% of the indigenous freshwater mussel species remain viable at this time.

20

8.   In paragraph 74 of their affidavit, TVA's experts assert that I erroneously extrapolated TVA's characterization of the Rservor Benthic Macroinvertibrate Index (RBMI") for the benthic macroinvertebrate community in the SBN vicinity, to the freshwater mussel community specifically. They are incorrect. My opinion is based on a passage in TVA's FSEIS on page 55 which states:

> Another aspect of the Vital Signs Monitoring Program is the benthic index, which assesses the quality of benthic communities in the reservoirs (including upstream inflow areas such as that around WBN). The tailwaters of Watts Bar Dam support a variety of benthic organisms *including several large mussel beds*. One of these beds has been documented along the right-descending shoreline immediately downstream from the mouth of Yellow Creek. To protect these beds, the state has established a mussel sanctuary extending 10 miles from TRM 520 to TRM 529.9. Since the institution of the Vital Signs Monitoring Program, the quality of the benthic community in the vicinity of the WBN site has remained relatively constant. The riverine tailwater reach downstream of Watts Bar Dam and WBN rated "good" in 2001 and the rating has increased to "excellent" in 2003-2005 (Appendix C, Tables C-4 and C-5).

(emphasis added). This paragraph specifically discusses freshwater mussels as part of the benthic community evaluated under TVA's Vital Signs Monitoring program. Mussels are benthic macroinvertebrates, and are represented in Metric 2 – "Long-lived Organisms" of the Reservoir Benthic Index (Table 6. Biological Monitoring of the Tennessee River near Watts Bar Nuclear Discharge, 2008). Therefore I did not misinterpret the passage stated in the FSEIS in expressing my opinion that when only four out of 64 (i.e., 6% of) freshwater mussel species once found in the vicinity of WBN remain reproductively viable, in no way can any aspect of the aquatic community be rated in "excellent" health.

9.   I do not believe TVA has a reasonable basis for placing the blame for mussel decline solely on river impoundment. While it is clear that river impoundment has severely impacted the mussel community, the results of the 2010 surveys show an alarming decline of mussels in the vicinity of the SCCW. This is evidence that current WBN operations have had a large impact on

21

mussel health and that adding another reactor unit will increase and perpetuate these negative impacts.

10.　　Another factor which indicates that the health of macroinvertebrates in general is declining is the dominance of only four species including the Asiatic clam, a non-native, invasive species. As shown in the "Revised Aquatics Study" at page 34, during operational monitoring in 1996-1997, only four of 104 aquatic invertebrate species found made up 87.5%. Further, the average density of aquatic macroinvertebrates per square meter actually declined by more than 50% from 1997 to 2008 in the vicinity of WBN. In 1997, 424 organisms per square meter were reported (Appendix C. Aquatic Ecological Health Determinations for TVA Reservoirs – 1997). In 2008, only 187 organisms per square meter were reported (Table 8. Biological Monitoring of the Tennessee River near Watts Bar Nuclear Discharge, 2008). In 2007 and 2008, even TVA's Reservoir Benthic Index (RBI) score used to monitor the macroinvertebrate community fell to the "fair" category.

E.　　RFAI Study and Fish Species Occurrences Study

1.　　TVA uses Reservoir Fish Assemblage Index ("RFAI") "scores" to provide general ratings of the fish community within TVA reservoirs. As discussed by TVA's experts in par. 55, TVA uses the RFAI to determine whether a "Balanced Indigenous Population" is being maintained as required by the EPA under the Clean Water Act. As discussed in Contention 7 and my supporting declaration, I believe TVA's RFAI scores are biased and misleading, and do not properly reflect the true state of the Tennessee River's aquatic resources. TVA's RFAI Study and Fish Species Occurrence Study do not resolve my concerns.

2.　　In the Fish Species Occurrence Study, TVA analyzed and scored new and historical fish survey data to determine the current presence of fish species, and compared the presence of

22

species before and after operation of WBN Unit 1. TVA claims that a comparison of scores between 1993 and 2008 shows that both before and after operation of WBN1, TVA has maintained a "balanced indigenous population" ("BPI"). Statement of Material Facts at pp. 14-15. In the RFAI Study, TVA also concludes that "long-term data trends suggest that the ecological health of the fish community in Chickamauga Reservoir inflow has been maintained." See page 13 of Attachment 9. Furthermore, TVA states that: "The species composition of the fish assemblage of Chickamauga Reservoir has changed somewhat, but not markedly, over the decades of sampling by TVA." See page 19 of Attachment 10. Neither study remedies my concerns in Contention 7.

3. In my professional opinion, the RFAI and Fish Species Occurrence studies does not present a reliable or reasonably accurate picture of the health of aquatic organisms near WBN1, for several reasons. First, TVA's method for conducting RFAI studies has changed over the years, making the scores difficult to compare. And the history of the RFAI program indicates that the older scores are unreliable because the methodology for deriving those scores was questioned by EPA and others. In an EPA guidance document, for example, EPA includes improvement of the RFAI in a list of "Research Needs:"

> *Research Needs* – TVA has been actively developing assessment tools for its reservoirs for several years. The move to a multimetric approach for reservoir fish began in 1990. Successive steps in this development process have brought continued improvement to the RFAI. Potential improvements in the fish indices include using a simple random sampling design rather than a fixed station design to enhance statistical validity with little increase in variability. Use of the index in reservoirs or other river systems is necessary to test its performance under a wider range of conditions than is available in the Tennessee river. *Correlation with known human-induced impacts remains a critical need before general acceptance of the fish index as a reliable method to address reservoir environmental quality.*

EPA 841-B-98-007 - Lake and Reservoir Bioassessment and Biocriteria: Technical Guidance Document, Appendix D: Biological Assemblages, Section D.5 Fish, pp. 176-177 (Undated).

23

(http://water.epa.gov/type/lakes/assessmonitor/bioassessment/upload/lakereservoirbioassess-

biocrit-app-d.pdf) (emphasis added). Second, TVA's summation of data in the Fish Species

Occurrence study is biased, and TVA attempts to portray sampling gear changes as the reason for

the decline of fish species near WBN and Chickamauga Reservoir in general to mask the reality

that the fish community has experienced significant decline pre- and post-WBN operation from

cumulative man-made impacts to the aquatic ecosystem.

4.   The scientific community has also criticized the RFAI's inability to correlate with

environmental degradation or accurately reflect true patterns in environmental health within and

among reservoirs:

> More recently, a second TVA reservoir version of the IBI [Index of Biotic Integrity] has
> been developed, termed the Reservoir Fish Assemblage Index (RFAI, Jennings, Karr, and
> Fore, personal communication). The RFAI has a somewhat different set of 12 metrics
> (Table 4), with the changes in metrics designed to improve sensitivity to environmental
> degradation and to increase adaptability to different types of reservoirs. *However, results
> from applications of both the original TVA version and the newer RFAI have often not
> accurately reflected what are believed to be the true patterns in environmental health
> within and among reservoirs,* and additional modifications will probably be necessary to
> develop better versions of the IBI for impoundments (Jennings, personal
> communication).

Davis, W. S., and T. S. Simon, Biological Assessment and Criteria: Tools for Water Resource

Planning and Decision Making, pp. 260-261 (Lewis Publishers: 1995) (emphasis added).

5.   However, even the biased RFAI scores declined post-operation, thus undermining TVA's

claim that the RFAI scores show that the "good health" of aquatic organisms near WBN1 has not

declined. TVA Joint Affidavit, par. 57.

6.   Some of the problems with TVA's RFAI methodology can be seen in the 12 metrics

described in Paragraph 52 of TVA Joint Affidavit for assessing four general categories of fish

health characteristics: Species Richness and Composition, Trophic Composition, Abundance,

24

and Fish Health. For each metric, scores are given on a scale from 1 to 5, with a score of 5 indicating optimum health.

7.   TVA's RFAI scores are predominantly biased by inappropriate assessments of the first category "Species Richness and Composition" and its 8 metrics (i - viii), and the lack of appropriate metrics within the third category "Abundance" (metric xi).

8.   Species Richness and Composition – Metric (i) is described as:

> i. Total number of indigenous species: Greater numbers of indigenous species are considered representative of healthier aquatic ecosystems. As conditions degrade, numbers of species at an area decline.

Metric (i) is misleading because it reports only the mere presence of a species, and does not account for its actual abundance, reproductive viability, and future existence within the fish community under evaluation. There is no metric to account for this within the "Abundance" category. A threatened or endangered species would register positively under this metric even though its future existence is doubtful. Several indigenous species were present in only one or two years within a decade sampling period. Again, there is no metric to account for these important trends of indigenous fish decline within the "Abundance" category. Further, the percent of native species is biased by hatchery stockings of species that may otherwise have disappeared from Chickamauga Reservoir.

9.   Appendix 1 of Attachment 9 to TVA's Motion illustrates my point. Appendix 1 shows that only one Largescale stoneroller was captured in 2004 and 2008 and zero were captured in all other years from 1999-2009. Yet, these two individuals that were collected during a 10-year sampling period represent species presence in Tables 2 and 3. Similarly, River redhorse (two individuals) and Smallmouth redhorse (one individual), which are Catostomids or suckers, show population trends near WBN similar to the Largescale Stoneroller. Thus, while one or two

25

individual fish could not reasonably be characterized as a healthy or even viable population, the RFAI considers its presence as a positive attribute. Further, several intolerant species were found during 2009 in the following numbers: Chestnut Lamprey (0), Steelcolor shiner (4), Emerald Shiner (1), Black redhorse (5), Golden redhorse (3), Northern Hogsucker (0). In comparison, several tolerant species were found during 2009 in the following numbers: Bluegill (471), Gizzard shad (131), and Largemouth bass (61). Nevertheless, in 2009, TVA gave this metric a score of 5 (see Attachment 9, p. 144, Appendix 2-A). In my view, given the extremely low abundance of indigenous fish species and the high abundance of tolerant species, this metric should receive a score of 1, or an equivalent metric should be incorporated into the "Abundance" category to properly represent the extremely low abundance of numerous indigenous species.

10. Metric (ii) in the category of "Species Richness and Composition" is described as:

ii. Number of centrarchid species: Sunfish species (excluding black basses) are invertivores and a high diversity of this group is indicative of reduced siltation and suitable sediment quality in littoral areas.

Metric (ii) yields misleading results because it uses only one of several families of fishes that are commonly used to assess the status of a fish community, and because Centrarchids are not representative of the most vulnerable indigenous fish species. TVA neglected to use other families more representative of the Tennessee River such as Percidae (which includes darters), Catostomidae (i.e.,suckers), and Cyprinidae (i.e., minnows). These families were highly diverse and plentiful historically; are intolerant to human disturbance and pollution; and all have suffered severe decline in the Tennessee River. TVA gave this metric a 5, the highest score. The only attribute this reflects is that Centrarchids, which thrive in reservoirs, are well-represented. If one of the other three families were used, this metric would be scored a 1.

26

11.   Metric (iii) in the category of "Species Richness and Composition" is described

as:

> iii. Number of benthic invertivore species: Due to the special dietary
> requirements of this species group and the limitations of their food source in
> degraded environments, numbers of benthic invertivore species increase with
> better environmental quality.

As with metric (i), metric (iii) evaluates only the presence of a species, and does not account for

its actual abundance, reproductive viability, and future existence in the environment under

evaluation. Again, there is no similar metric in the "Abundance" category to measure the actual

numbers of a species. If those factors were taken into account, TVA could not have given this

metric a score of 3. Given the steep decline of benthic invertivores as described in par. 9, the

score should be 1.

12.   Metric (iv) in the category of "Species Richness and Composition" is described

as:

> iv. Number of intolerant species: This group is made up of species that are particularly
> intolerant of physical, chemical, and thermal habitat degradation. Higher numbers of
> intolerant species suggest the presence of fewer environmental stressors. The higher
> number of these species would be a positive indicator

Metric (iv) should account for status of suckers, minnows, and darters as well as locally

endangered or extirpated species such as sturgeon and paddlefish because these fish are

intolerant and in decline. As with metrics (i) and (iii), metric (iv) evaluates only the presence of

a species, and does not account for its actual abundance, reproductive viability, and future

existence in the environment under evaluation. Again, there is no similar metric in the

"Abundance" category to measure the actual numbers of a species. If those factors were taken

into account, TVA could not have given this metric a score of 5. This metric suffers from the

27

same bias as Metric (i). TVA gave this metric a score of 5, but it should have received a score of

1.

13.   Metric (v) and Metric (vi) in the category of "Species Richness and Composition"

are described as:

> v. Percentage of tolerant individuals (excluding Young-of-Year): This metric
> signifies poorer water quality with increasing proportions of individuals tolerant of
> degraded conditions.

> vi. Percent dominance by one species: Ecological quality is considered reduced if one
> species inordinately dominates the resident fish community.

Metric (v) should identify a fish species community that is dominated by species tolerant of

disturbance and poor water quality. Metric (vi) should identify a fish species community that is

unbalanced and dominated by only one or few species. These are negative attributes whose

scores should be inversely proportional to the degree they exist. TVA's RFAI sampling shows a

high percentage of tolerant species such as bluegills. See par. 19 below. Further, the fish

community is currently dominated by bluegills (See par. 19); thus, the score should be a 1.

TVA, however, gave Metric (v) a score of 3. TVA correctly gave Metric (vi) a score of 1, which

is evidence that the fish community no longer supports a balanced indigenous population.

14.   Metric (vii) in the category of "Species Richness and Composition" is described

as:

> vii. Percentage of non-indigenous species: This metric is based on the assumption that
> non-indigenous species reduce the quality of resident fish communities.

Like metrics (v) and (vi), this is a negative attribute, whose score should be inversely

proportional to the degree it exists. Metric #7 should identify a fish species community that has

a significant number of non-indigenous species, i.e. species that are not indigenous to the

Tennessee River whether intentionally or unintentionally stocked. TVA sampling shows several

28

non-indigenous species present; and, that the percent of native species is biased by hatchery stockings of species that may otherwise have disappeared from Chickamauga Reservoir. TVA properly scored this metric with a 1, again indicating that the fish community no longer supports a balanced indigenous population.

15.   Metric (viii) in the category of "Species Richness and Composition" is described as:

> viii. Number of top carnivore species: Higher diversity of piscivores is indicative of the availability of diverse and plentiful forage species and the presence of suitable habitat.

Metric (viii) should identify a fish species community that is in proper balance with an adequate carnivore population, or fish that eat other fish and serve as the upper food chain predators. However, this metric may also be biased by hatchery stockings that are used to support a sport fishery. Often hatchery supplementation is used to artificially support a fish population for recreational purposes when the aquatic system no longer supports natural reproduction. Recreational fisheries often target these predatory fish species such as striped bass, sauger, and walleye, all of which are stocked by the State of Tennessee into Chickamauga Reservoir because of lack of natural reproduction to support fishing. The lack of reproduction is due to the alterations of the Tennessee River and the resulting poor ecological health. While TVA scored this metric at 5, the score should be a 3.

16.   The category "Abundance" is as equally important as "Species Richness and Composition"; yet, "Abundance" is only represented by one metric (metric xi) as compared to "Species Richness and Composition" which is represented by eight metrics. This is a major omission that leads to the inappropriately high RFAI scores that overstates the health of the fish community. Metric xi is described as:

> xi. Average number per run (number of individuals): This metric is based upon the assumption that high quality fish assemblages support large numbers of individuals.

29

Metric (xi) is highly biased by the ever-increasing numbers of bluegills and other species that thrive in a man-made environment and now dominate the fish community. The increase of bluegills masks the low number of other native species in decline. TVA, scoring this metric based upon the definition, gave it a 5. However, if this category incorporated similar metrics as "Species Richness and Composition" based upon actual abundance, or number of individuals captured, all of the metrics designed to monitor indigenous fish species would receive RFAI scores of 1, the lowest possible.

17.   Paragraph 53 of the Joint Affidavit describes the method for evaluating total RFAI scores as follows:

> Because there are 12 metrics, RFAI scores range from 12 to 60. The aquatic community health is indicated by the following ranges of scores: 12-21 ("Very Poor"), 22-31 ("Poor"), 32-40 ("Fair"), 41-50 ("Good"), or 51-60 ("Excellent").

TVA's final 2009 RFAI score for the area near WBN Plant was a 44 in the "Good" category. Correcting for the bias of the RFAI would lead to a score of 28, or a "Poor" rating of the health of the fish community. I believe the "poor" rating, which is a significantly different picture of the fish community in the vicinity of WBN than that of TVA's analyses, more accurately represents the status of the fish community of the Tennessee River in the vicinity of WBN Plant.

18.   The score that I estimated is also consistent with other data which show a decrease in the level of diversity and the size of existing populations since WBN1 began operating. For instance, a comparison of the NRC's 1978 Final Environmental Impact Statement (FEIS) for WBN Units 1 & 2 (Table C-21) and the NRC's 2008 Final Supplemental Environmental Impact Statement (FSEIS) for WBN Unit 2 (Table 3.3.1) shows that the Chickamauga Reservoir experienced a 24% decline of freshwater fish species between 1970-73 and 1991-1996. Further, Vital Signs and Biological Monitoring reports from 1994 list 36 fish species that were captured

30

in Upper Chickamauga Reservoir, and reports from 1999-2009 show the number of species declined to between 24 and 31 for a given year, another 14% decline.

19.   Evidence that the fish community near WBN is greatly unbalanced may be found by analyzing TVA electrofishing data in <u>Aquatic Ecological Health Determinations for TVA Reservoirs –1994</u>, Table 8, Page 352, and within <u>Biological Monitoring of the Tennessee River Near Watts Bar Nuclear Plant Discharge, 2008</u>, Table 3, Page 18. These data show that in 1994, bluegill -- a species that thrives in man-made habitats and are thus popular for stocking in small ponds across the United States -- comprised only 27% of all fish in TVA's sampling in Upper Chickamauga Reservoir. However, during 2008 sampling, bluegill comprised 63% of all fish captured in Upper Chickamauga Reservoir at areas downstream of Watts Bar Nuclear Plant Discharge. Upon further examination, Centrarchids in general (the family of fishes that is comprised of bluegill, sunfishes, and black-basses) make up 78% of all fish near WBN. A fish community that is made up of 78% bluegill, sunfishes, and black-basses is more indicative of a farm pond than the most biologically diverse freshwater ecosystem in North America. Further, by adding gizzard shad, another species that may thrive in reservoirs, the percent increases to 91%. This results in a very low abundance, whether stated in terms of percent composition and actual numbers, of other native riverine fish species that should be found in the Tennessee River near WBN. When this is compared to 1994 when Centrarchids comprised only 58% and gizzard shad 10%, there is evidence that the fish community is extremely unbalanced, and the percent of indigenous riverine species has continued to decline since WBN1 became operational.

20.   Thus, these data show that the fish community has undergone significant negative changes since WBN1 became operational and the current health of the fish community is poor. The data certainly do not support the existence of a Balanced Indigenous Population or "BIP."

31

F.   Failure to Discuss Cumulative Impacts

1.   TVA has not addressed the cumulative impacts on the Tennessee River Basin from combined operation of WBN Units 1 and 2. The combined operation will increase cooling water needs and increase thermal and chemical discharge. These consequences of adding yet another energy production facility will have adverse impacts on the whole system with large impacts to the upper-basin tributaries that also support highly diverse and unique fish and mussel species. TVA manages the Tennessee River as one hydrosystem; thus, changes in water consumption or changes in flow to accommodate energy and industrial facilities in one area will affect the rest of the system. Further, the quantity of water available at Watts Bar Dam and then released into Chickamauga Reservoir determines the management of the rest of the hydrosystem, especially water releases from the upper basin. Therefore, if WBN Plant requires flow in order to operate at maximum efficiency and to remain within NPDES permit limits, the entire upper basin or at least the aquatic ecology of 10 different tributaries with a high number of fish and mussels will be affected. This is supported by the following excerpts from TVA's discussion of water management policy on its website (http://www.tva.gov/river/lakeinfo/systemwide.htm):

- "In May 2004, the TVA Board of Directors approved a new policy for operating the Tennessee River and reservoir system. This policy shifts the focus of TVA reservoir operations from achieving specific summer pool elevations on TVA-managed reservoirs to managing the flow of water through the river system. The new policy specifies flow requirements for individual reservoirs and for the system as a whole."

- "System-wide flow requirements ensure that enough water flows through the river system to meet downstream needs."

32

- "When water must be released to meet downstream flow requirements, a fair share of water is drawn from each reservoir. System-wide flows are measured at Chickamauga Dam, located near Chattanooga, Tenn., because this location provides the best indication of the flow for the upper half of the Tennessee River system."

- "If the total volume of water flowing into Chickamauga Reservoir is less than needed to meet system-wide flow requirements, additional water must be released from upstream reservoirs, resulting in some drawdown of these projects. How much water is released depends on the time period and the total volume of water in storage in 10 tributary reservoirs: Blue Ridge, Chatuge, Cherokee, Douglas, Fontana, Nottely, Hiwassee, Norris, South Holston and Watauga."

2.   For all the reasons discussed above, TVA has not resolved the concerns raised by Contention 7. Therefore the contention should not be dismissed.

Under penalty of perjury, I declare that the foregoing facts are true and correct to the best of my knowledge, and that the expressions of opinion are based on my best professional judgment.

Shawn P. Young

_____

Shawn Paul Young, Ph.D.
P.O. Box 507
Bonners Ferry, ID 83805

Dated: December 20, 2011

33

Shawn P. Young, *PhD*

P.O. Box 507
Bonners Ferry, ID 83805
(765) 427 - 3997
syfishhead@msn.com

EDUCATION

PhD   Fisheries Sciences   May 2005   Clemson University, Clemson, SC
MS   Fisheries Sciences   Aug 2001   Clemson University, Clemson, SC
BS   Environmental Studies   May 1996   Northland College, Ashland, WI

PROFESSIONAL EXPERIENCE

Environmental Consultant   Private practice   Jan 2005 – Present
Fish Biologist   Kootenai Tribe of Idaho   July 2011 - Present
Fisheries Researcher   GADNR/Clemson Univ.   Feb 2010 – Nov 2011
Lecturer/Scientist   University of Idaho   Aug 2008 - Sep 2009
Visiting Scientist   University of Iceland   July 2008 - Aug 2008
Visiting Assistant Professor   Purdue University   Aug 2007- May 2008
Postdoctoral Researcher   Clemson University   Oct 2006 - Aug 2007
Fish Biologist/Facility Manager   Clemson University   Jun 1999 - May 2006
Fisheries Technician   Idaho Fish and Game   Apr 1997 - June 1999

Environmental Consultant - Aquatic Ecology / Fisheries Expert
Private Practice, Owner – Shawn Paul Young LLC Environmental Consulting

- *Savannah Harbor, GA*:   Impacts of dredging on Shortnose sturgeon, Atlantic sturgeon, and other native fish
- *Savannah River, GA*:   Flow regulation effects on fish in the Savannah River
- *Wateree River, SC*:   River flows and fish habitat - Wateree Dam FERC Re-licensing
- *Pee Dee River, NC*:   River flows and fish habitat - Tillery Dam FERC Re-licensing
- *Watts Bar Nuclear, TN*: Nuclear reactor impacts to fish and mussel populations - Tennessee River – Chickamauga Reservoir

- *Bellefonte Nuclear, AL*: Nuclear reactor impacts to fish and mussel populations -Tennessee River – Guntersville Reservoir
- *Vogtle Nuclear, GA*:   Nuclear reactor impacts to fish and mussel populations – Savannah River
- *North Anna Nuclear, VA*: Nuclear reactor impacts to fish populations – Pamunkey River - North Anna Reservoir
- *Watts Bar NPDES, TN*:   Permit comments concerning pollution discharged from operation of nuclear reactors
- *Tennessee Water Quality Standards*: Comments to strengthen water quality standards and protections during triennial review

Fish Biologist
Kootenai Tribe of Idaho; Bonners Ferry, ID (July 2011 – Present)

I assist the Kootenai Tribe in all aspects of its Native Fish Program including the aquaculture and restoration of the federally endangered Kootenai River White Sturgeon population. I am one of the technical leads on the design of a new hatchery facility and on the development of a restoration and monitoring strategy for burbot and kokanee. I also represent the Kootenai Tribe on interagency matters including the Kootenai River White Sturgeon Recovery Team and the Kootenai River / Libby Dam Flow Technical Committee. I will also be involved in a large-scale physical habitat restoration effort to restore ecosystem function to a highly altered segment of the Kootenai River in order to rebuild native fish populations.

Fisheries Researcher
Georgia Department of Natural Resources; Albany, GA and Department of Forestry and Natural Resources; Clemson University, Clemson, SC (Feb 2010 – November 2011)

I led a field investigation of spawning Alabama shad in the Apalachicola River, FL. My primary objectives were to estimate spawning population size, to evaluate fish passage at Jim Woodruff Lock and Dam, to determine use of Flint and Chattahoochee Rivers as spawning habitat and juvenile rearing habitat by Alabama shad that passed through the navigation lock, and to determine age, growth, and population structure. I also led the investigation of otolith microchemistry to determine ontogenetic shifts in habitat/anadromy and natal origin, and to determine the role of environmental factors play in recruitment success. My study's ultimate objectives were to halt decline of the Alabama shad and ensure a continued self-sustaining population, with hopes to restore historical abundance.

Researcher – Fisheries Biology and Ecophysiology
University of Idaho; Dept of Fish and Wildlife Resources, Moscow, ID (Dec 2008 – September 2009)

As a member of a a research team, I investigated the physiology of wild and hatchery-raised adult Snake River steelhead kelts through life stages from pre-spawn to outmigration to the Pacific Ocean, and the potential to restore wild Snake River steelhead by captive reconditioning of kelts and transport around the Snake/Columbia River hydrosystem.

Lecturer – Fisheries Management
University of Idaho; Department of Fish and Wildlife Resources, Moscow, ID (Fall 2008)

FISH 418 – Fisheries Management w/ Lab

Visiting Scientist
University of Iceland; Reykjavik, Iceland (July 2008)

I was invited by a colleague to investigate physiological differences between genetically distinct components of the Icelandic Atlantic Cod stocks.

Visiting Assistant Professor - Fisheries and Aquatic Sciences
Purdue University; Department of Forestry and Natural Resources; West Lafayette, IN (Aug 2007 – May 2008)

FNR 546 - Fish Ecology
FNR 545 - Fisheries Management
FNR 501 – Limnology
FNR 371 – Watershed Hydrology Practicum
FNR 103 - Introduction to Environmental Conservation

Post-Doctoral Researcher - Adjunct Professor
Dept. of Forestry and Natural Resources: Clemson University, Clemson, SC (Oct 2006 – August 2007)

My research focused on fish ecology and behavior in altered rivers. I conducted research on anadromous and resident fish species in the Apalachicola River. Research objectives were to estimate Alabama shad spawning population size, monitor behavior/movement during spawning migration, and determine passage efficiency at lock-and-dam facilities. I also studied the age, growth, and reproductive ecology of three catostomids and skipjack herring. As another aspect of studying altered river systems, I conducted studies of freshwater mussels to evaluate tagging methods, movement after relocation, and behavior in

Appendix E

fluctuating flow regimes. *(please refer to Publications).*

Committees:
Age, growth, and fecundity of Alabama shad in the Apalachicola River. Thesis. T. Ingram. 2006.
Population estimate of spawning Alabama shad in the Apalachicola River. Thesis. P. Ely. 2007.
Genotype-specific spawning behavior of striped bass in the Apalachicola River. Thesis. M. Noad. 2007.
Paleochannel delineation of the Neuse River, North Carolina. Thesis. B. Wrege. 2007.

WFB 840 Fish Ecology (Team-taught course)
ENR 302 Natural Resource Measurements (Team-taught course)
WFB 300 Wildlife and Fisheries Biology (Team-taught course)

Research Biologist / Fish – Aquatic Organism Research Facility Manager
Aquatic Animal Research Laboratory; Clemson University, Clemson, SC (June 1999 – May 2006)

I conducted research and managed facilities at a leading fisheries/aquaculture research laboratory. Our research specialized in identifying factors that affect fish and aquatic invertebrate physiology, behavior, and population dynamics. I conducted research on habitat requirements of marine, estuarine, anadromous, and freshwater species at the larval, juvenile, and adult life-history stages. *(Please refer to Publications and Presentations).* I also assisted with the research and preparation of the following:

- Using mixed-ion supplementation in Pacific white shrimp culture. 2007. Thesis. K. Parmenter.
- Multi-scale habitat associations of selected primary burrowing crayfish. 2006. Dissertation. S. M. Welch.
- Low-salinity resistance of juvenile cobia (Rachycentron canadum). 2006. Thesis. K. L. Burkey.
- Responses of Pacific white shrimp (Litopenaeus vannamei) to water containing low concentrations of total dissolved solids. 2005. Thesis. A. D. Sowers.
- Responses of hybrid striped bass exposed to waterborne and dietary copper in fresh- and saltwater. 2003. Dissertation. G. K. Bielmyer.
- Ecology and culture of Procambarus acutus acutus. 2003. Dissertation. Y. Mazlum.
- Effects of environmental and dietary factors on tolerance of Nile tilapia Oreochromis niloticus to low temperature. 2002. Dissertation. H. L. Atwood.
- Low-temperature tolerance of southern flounder Paralichthys lethostigma: effect of salinity. 2000. Thesis. W. E. Taylor.

Through the South Carolina Cooperative Fish and Wildlife Research Unit, I also completed a dissertation and thesis that utilized several telemetry field studies to identify seasonal migration patterns, daily movement patterns, and seasonal habitat selection in relation to reservoir limnology/ hydroelectric generation; sources and magnitude of mortality; temporal and spatial patterns of mortality; and, potential to successfully live-release striped bass angled during fishing tournaments. *(Please refer to Publications and Presentations).* Through graduate coursework, I also acquired extensive knowledge of fisheries science and management; physiology, ecology and conservation of aquatic organisms; limnology and hydrology; and experimental statistics. *(Please see transcripts).*

Through collaboration with the SC Cooperative Fish and Wildlife Research Unit, I also assisted with the following:

- Reproductive ecology and seasonal migrations of robust redhorse (Moxostoma robustum) in the Savannah River, Georgia and South Carolina. 2006. Dissertation. T. B. Grabowski.
- A behavioral comparison of hatchery-reared and wild shortnose sturgeon in the Savannah River, South Carolina-Georgia. 2003. Thesis. D. Trested.

- Diel movement of hatchery-reared and wild shortnose sturgeon in the Savannah River, South Carolina-Georgia. 2003. Thesis. T. E. Griggs.
- Movement of migrating American shad in response to flow near a low head lock and dam. 2003. Thesis. S. T. Finney.
- Population size and movement of American shad at New Savannah Bluff Lock and Dam. 2002. Thesis. M. M. Bailey.
- Seasonal and diel movement of largemouth bass in a South Carolina stream. 2001. Thesis. T. A. Jones.
- Habitat utilization by striped bass in Lake Murray, South Carolina. 2001. Thesis. J. J. Schaffler.

Fisheries Technician
Idaho Dept of Fish & Game; Lewiston & Bonners Ferry, ID (April 1997 - May 1999)

My first appointment was in the Lewiston office where I conducted snorkeling surveys to determine abundance and distribution of anadromous and potadromous salmonids in the Clearwater River Basin.

My second position was in the Bonners Ferry Kootenai River Field station where I assisted research on the effects of hydroelectric operations on behavior and survival of salmonids (rainbow trout and bull trout), burbot, and white sturgeon in the Kootenai River, ID-MT. Major responsibility was to conduct fieldwork for large-scale telemetry and capture studies to acquire knowledge of seasonal movements, migratory behavior, and   recruitment.

PUBLICATIONS:

Fish Ecology and Management:

1.   **Young, S.P.,** T I Ingram, and J Tannehill *(in review)* Passage of spawning Alabama shad at Jim Woodruff Lock and Dam, Apalachicola River, Florida Submittal: Transactions of the American Fisheries Society

2.   **Young, S.P.,** T I Ingram, and J Tannehill *(in review)* Survival and behavior of transported shoal bass *Micropterus cataractae* in the Flint River, Georgia Submitted: North American Journal of Fisheries Management

3.   Ingram, T I, **S. P. Young,** and J Tannehill *(in revision)* Age, growth, and fecundity of spawning Alabama shad at Jim Woodruff Lock and Dam, Apalachicola River, Florida Submittal: Transactions of the American Fisheries Society

4.   **Young, S. P.,** P Ely, T Grabowski, and J J Isely *(in review)* Effects of river flow on age, growth, fecundity, and reproductive strategy of catostomids in the Apalachicola River, Florida Submittal: Environmental Biology of Fishes

5.   **Young, S. P.,** P Ely, M Noad, and J J Isely *(in revision)* Age, growth, and relative abundance of skipjack herring in the Apalachicola River, Florida

6.   **Young, S.P.** 2011 Annual Report – Population size, passage, and spawning behavior of Alabama shad, *Alosa alabamae*, in the Apalachicola River Basin, Florida-Georgia Prepared for Georgia Department of Natural Resources and National Marine Fisheries Service

7.   **Young, S.P.** 2010 Annual Report – Population size, passage, and spawning behavior of Alabama shad, *Alosa alabamae*, in the Apalachicola River Basin, Florida-Georgia Prepared for Georgia Department of Natural Resources and National Marine Fisheries Service

8.   **Young, S.P.,** P Ely, T Grabowski, and J J Isely 2010 First Record of *Carpiodes velifer* (highfin carpsucker) in the Apalachicola River, Florida Southeastern Naturalist 9(1):165-170.

9.   Grabowski, T B, **Young S. P.,** Libungan, L A, Steinarsson, A, and G Marteinsdottir (2009) Evidence of

Appendix E

phenotypic plasticity and local adaption in metabolic rates between components of the Icelandic cod (*Gadus morhua* L.) stock. Environmental Biology of Fishes 86:361-370.

10 Barczak, S., and **S. P. Young**. 2009. Water use impacts from increased energy production on Georgia's aquatic resources. 2009 Georgia Water Resources Conference.

11 Ely, P. and **Young, S. P.**, and J J Isely. 2008. Population size and relative abundance of Alabama shad reaching Jim Woodruff Lock and Dam, Apalachicola River, Florida. North American Journal of Fisheries Management 28:827-831.

12 **Young, S. P.** and J J Isely. 2007. Summer diel behavior of striped bass using tailwater habitat as summer refuge. Transactions of the American Fisheries Society 136: 1104-1112.

13 **Young, S. P.**, and J J Isely. 2006. Post-tournament live-release survival, dispersal, and behavior of adult striped bass. North American Journal of Fisheries Management 26: 1030-1033.

14 **Young, S. P.**, and J J Isely. 2004. Temporal and spatial estimates of adult striped bass mortality from telemetry and transmitter return data. North American Journal of Fisheries Management 24: 1112-1119.

15 **Young, S. P.** and J J Isely. 2002. Striped bass annual site fidelity and habitat utilization in J. Strom Thurmond Reservoir, South Carolina-Georgia. Transactions of the American Fisheries Society 131: 828-837.

16 Isely, J J., **S. P. Young**, T A Jones, and J J Schaffler. 2002. Effects of antenna placement and antibiotic treatment on loss of simulated transmitters and mortality in hybrid striped bass. North American Journal of Fisheries Management 22: 204-207.

Fish physiology and aquaculture:

17 Burkey, K B., **S. P. Young** J R Tomasso, and T I J Smith. 2007. Low-salinity resistance of juvenile cobia. North American Journal of Aquaculture 69: 271-274.

18 **Young, S. P.**, J R Tomasso, and T I J Smith. 2007. Survival and water balance of black sea bass held in a range of salinities and calcium-enhanced environments after abrupt salinity change. Aquaculture 258: 646-649.

19 Atwood, H L., **S. P. Young** J R Tomasso, and T I J Smith. 2004. Resistance of cobia, *Ranchycentron canadum*, juveniles to low salinity, low temperature, and high environmental nitrite concentrations. Journal of Applied Aquaculture 15: 191-195.

20 Atwood, H L.; **S. P. Young** J R Tomasso, and T I J Smith. 2004. Information on selected water quality characteristics for the production of black sea bass, *Centropristis striata*, juveniles. Journal of Applied Aquaculture 15: 183-190.

21 Atwood, H L.; **S. P. Young** J R Tomasso, and T I J Smith. 2003. Effect of temperature and salinity on survival, growth, and condition of juvenile black sea bass. Journal of the World Aquaculture Society 34: 398-402.

22 Atwood, H L.; **S. P. Young** J R Tomasso, and T I J Smith. 2001. Salinity and temperature tolerances of black sea bass juveniles. North American Journal of Aquaculture 63: 285-288.

Aquatic invertebrate conservation:

23 **Young, S. P.** and J J Isely. (2008). Tag retention, relocation probability, and mortality of passive integrated transponder and dummy transmitter tagged *Elliptio complanata* in a South Carolina Piedmont stream. Molluscan Research.

24 **Young, S. P.** and J J Isely. (*in revision*). Behavioral response of the freshwater mussel *Elliptio complanata* to fluctuating water levels. Submittal: Journal of North American Benthological Society.

25 **Young, S. P.** and J J Isely. (*in progress*). Behavior of translocated freshwater mussels *Elliptio complanata* in a South Carolina piedmont stream.

Aquatic invertebrate physiology and aquaculture:

26 Parmenter, K and Bisesi, J, **S.P. Young** S J Klaine, H L Atwood, J R Tomasso, and C L Browdy 2009 Culture of pacific white shrimp *Litopenaeus vannamei* in a variety of mixed-ion solution North American Journal of Aquaculture 71:134-137

27 Sowers, A D and **Young, S.P.**, M Grosell, C L Browdy, and J R Tomasso 2006 Hemolymph osmolality and cation concentrations in *Litopenaeus vannamei* during exposure to low concentrations of dissolved solids: Relationship to potassium flux Comparative Biochemistry and Physiology 145(2): 176-180

28 Sowers, A D, D M Gatlin, **S.P. Young** J J Isely, C L Browdy, and J R Tomasso 2005 Responses of *Litopenaeus vannamei* (Boone) in water containing low concentrations of total dissolved solids Aquaculture Research 36: 819-823

29 Sowers, A D and **Young, S.P.**, J J Isely, C L Browdy, and J R Tomasso 2004 Nitrite toxicity to *Litopenaeus vannamei* in water containing low concentrations of sea salt or mixed salts Journal of the World Aquaculture Society 35: 445-451

30 Atwood, H L; **S.P. Young** J R Tomasso, and C L Browdy 2003 Survival and growth of pacific white shrimp, *Litopenaeus vannamei*, postlarvae in low salinity and mixed-salt environments Journal of the World Aquaculture Society 24: 518-523

SELECTED PRESENTATIONS:

Young, S.P. 2008. Ecophysiology of Iceland's Atlantic cod stocks. University of Idaho. Moscow, ID.

Young, S.P. 2007. Thermal biology of fish. Penn State University. State College, PA.

Young, S.P. 2007. Population estimates and passage of Alabama shad at Jim Woodruff Lock and Dam, Apalachicola River - Florida. Purdue University. West Lafayette, IN.

Young, S.P. 2006. Behavioral thermoregulation and metabolic scope of striped bass in various aquatic environments. Austin Peay University. Clarksville, TN.

Young, S.P. 2006. Behavioral thermoregulation and metabolic scope – Lecture for comparative anatomy and physiology. Clemson University. Clemson, SC.

Young, S.P. and J.J. Isely. 2005. Post-tournament live-release survival, dispersal, and behavior of adult striped bass. American Fisheries Society annual meeting. Anchorage, AK.

Young, S.P. 2005. Behavioral thermoregulation in fish. Lake Superior State University. Sault-sainte Marie, MI.

Young, S.P. and J.J. Isely. 2005. Striped bass ecology and management. Clarks Hill Striped Bass Anglers Association. Augusta, GA.

Young, S.P. and J.J. Isely. 2005. Post-tournament live-release survival, dispersal, and behavior of adult striped bass. Trout Unlimited. Upstate South Carolina Chapter.

Young, S.P. and J.J. Isely. 2004. Temporal and spatial estimates of adult striped bass mortality from telemetry and transmitter return data. Annual meeting of the American Fisheries Society. Madison, WI.

Atwood, H.L.; **S.P. Young**, J.R. Tomasso, and T.I.J. Smith. 2004. Effect of temperature and salinity on survival, growth, and condition of juvenile black sea bass. 28th Annual Larval Fish Conference, Early Life History Section, American Fisheries Society. Clemson, SC.

Atwood, H.L.; **S.P. Young**, J.R. Tomasso, and T.I.J. Smith. 2004. Resistance of cobia juveniles to low salinity and low temperature. 28th Annual Larval Fish Conference, Early Life History Section, American Fisheries Society. Clemson, SC.

Young, S.P. 2004. Learning in Fishes: from three-second memory to culture. Department of Biological Sciences. Clemson University.

Young, S.P. 2003. Life skills training for hatchery fish: Social Learning and Survival. Department of Biological Sciences. Clemson University.

Young, S.P. 2003. Mechanisms for learning during early life stages of fish: Imprinting, Homing, and Con-specific Learning. Dept of Biological Sciences. Clemson University.

Young, S.P. 2002. Strain-specific characteristics to manage sub-populations of fish species. Department of Biological Sciences. Clemson University.

AWARDS:

- Animal Research Committee Excellence Award. 2004. Clemson University.
- Animal Research Committee Excellence Award. 2003. Clemson University.
- Outstanding Classified Employee Award. 2003. Clemson University.
- Employee Performance Award. 2003. Clemson University.

**UNITED STATES OF AMERICA
NUCLEAR REGULATORY COMMISSION
BEFORE THE ATOMIC SAFETY AND LICENSING BOARD**

_____)	
In the Matter of)	
)	
Tennessee Valley Authority)	Docket No. 50-391
)	
(Watts Bar Unit 2))	
_____)	

**MOTION TO ADMIT NEW CONTENTION REGARDING
THE SAFETY AND ENVIRONMENTAL IMPLICATIONS OF
THE NUCLEAR REGULATORY COMMISSION TASK FORCE REPORT ON
THE FUKUSHIMA DAI-ICHI ACCIDENT**

I.   INTRODUCTION

Pursuant to 10 C.F.R. § 2.309, Southern Alliance for Clean Energy ("SACE") hereby

move to admit a new contention challenging the adequacy of the Final Supplemental

Environmental Impact Statement ("FSEIS") for the proposed Watts Bar Unit 2 Nuclear Power

Plant on the basis that it fails to address the extraordinary environmental and safety

implications of the findings and recommendations raised by the Nuclear Regulatory

Commission's Fukushima Task Force (the "Task Force") in its report, "Recommendations for

Enhancing Reactor Safety in the 21st Century: The Near-Term Task Force Review of Insights

From the Fukushima Dai-ichi Accident" (July 12, 2011) ("Task Force Report"). SACE

respectfully submit that admitting the new contention is necessary to ensure that the Nuclear

Regulatory Commission ("NRC" or the "Commission") fulfills its non-discretionary duty under

the National Environmental Policy Act ("NEPA") to consider the new and significant

information set forth in the Task Force Report before it makes a decision regarding the

Tennessee Valley Authority's ("TVA's") application for an operating license.

This motion is supported by a Certificate Required by 10 C.F.R. § 2.323(b).

II.   DISCUSSION

To be admitted for hearing, a new contention must satisfy the six general requirements set forth in 10 C.F.R. § 2.309(f)(1), and the timeliness requirements set forth in either 10 C.F.R. § 2.309(f)(2) (governing timely contentions) or 10 C.F.R. § 2.309(c) (governing non-timely contentions). As provided in the accompanying contention, each of the requirements set forth in 10 C.F.R. § 2.309(f)(1) is satisfied. Furthermore, SACE maintains that this Motion and accompanying contention are timely, and the requirements of 10 C.F.R. § 2.309(f)(2) are also satisfied. In the event this Board determines that this Motion and the accompanying contention are not timely, however, SACE also maintains that the requirements of 10 C.F.R. § 2.309(c) are satisfied.

A.   This Motion and the Accompanying Contention Satisfy the Requirements for Admission of a Timely Contention Set Forth in 10 C.F.R. § 2.309(f)(2).

The NRC has adopted a three-part standard for assessing timeliness. *See* 10 C.F.R. § 2.309(f)(2). The Motion and accompanying contention are timely.

1.   The Information Upon Which the Motion and Accompanying Contention are Based was not Previously Available.

The availability of material information "is a significant factor in a Board's determination of whether a motion based on such information is timely filed." Houston Lighting & Power Co. (South Texas Project, Units 1 & 2), LBP-85-19, 21 NRC 1707, 1723 (1985) (internal citations omitted). This Motion and the accompanying contention are based upon information contained within the Task Force Report, which was not released until July 12, 2011. Before issuance of the Task Force Report, the information material to the contention was simply unavailable.

2. The Information Upon Which the Motion and Accompanying Contention are Based is Materially Different than Information Previously Available.

Only five months ago, a nuclear accident occurred at the Fukushima Dai-ichi Nuclear Power Plant. In the wake of the accident, the Task Force was established and instructed by the NRC to provide:

> A systematic and methodical review of [NRC] processes and regulations to determine whether the agency should make additional improvements to its regulatory system and to make recommendations to the Commission for its policy direction, in light of the accident at the Fukushima Dai-ichi Nuclear Power Plant.

Task Force Report at vii. In response to that directive, the Task Force made twelve "overarching" recommendations to "strengthen the regulatory framework for protection against natural disasters, mitigation and emergency preparedness, and to improve the effectiveness of NRC's programs." *Id.* at viii. In these recommendations the Task Force, for the first time since the Three Mile Island accident occurred in 1979, fundamentally questioned the adequacy of the current level of safety provided by the NRC's program for nuclear reactor regulation.

TVA assumes that compliance with existing NRC safety regulations is sufficient to ensure that the environmental impacts of accidents are acceptable. The information in the Task Force Report refutes this assumption and is materially different from the information upon which the ER is based. *See* attached contention and Declaration of Dr. Arjun Makhijani.

3. The Motion and Accompanying Contention are Timely Based on the Availability of the New Information.

SACE has submitted this Motion and accompanying contention in a timely fashion. The NRC customarily recognizes as timely contentions that are submitted within thirty (30) days of the occurrence of the triggering event. Shaw Areva MOX Services, Inc. (Mixed Oxide Fuel Fabrication Facility), LBP-08-10, 67 NRC 460, 493 (2008). The Task Force Report, upon which the contention is based, was published on July 12, 2001. Because they were filed within thirty (30) days of publication of the Task Force Report, this Motion and accompanying contention are timely.

B.   The New Contention Satisfies the Standards For Non-Timely Contentions Set Forth in 10 C.F.R. § 2.309(c).

Pursuant to § 2.309(c), determination on any "nontimely" filing of a contention must be based on a balancing of eight factors, the most important of which is "good cause, if any, for the failure to file on time." Crow Butte Res., Inc. (North Trend Expansion Project), LBP-08-6, 67 NRC 241 (2008). As set forth below, each of the factors favors admission of the accompanying contention.

1.   Good Cause.

Good cause for the late filing is the first, and most important element of 10 C.F.R. § 2.309(c)(1). Private Fuel Storage, L.L.C. (Independent Spent Fuel Storage Installation), CLI-00-02, 51 NRC 77, 79 (2000). Newly arising information has long been recognized as providing the requisite "good cause." See Consumers Power Co. (Midland Plant, Units 1 & 2), LBP-82-63, 16 NRC 571, 577 (1982), citing Indiana & Michigan Elec. Co. (Donald C. Cook Nuclear Plant, Units 1 & 2), CLI-72-75, 5 AEC 13, 14 (1972). Thus, the NRC has previously found good cause where (1) a contention is based on new information and, therefore, could not have been presented earlier, and (2) the intervenor acted promptly after learning of the new information. Texas Utils. Elec. Co. (Comanche Peak Steam Electric Station, Units 1 & 2), CLI-92-12, 36 NRC 62, 69-73 (1992).

As noted above, the information on which this Motion and accompanying contention are based is taken from the Task Force Report, which was issued on July 12, 2011 and analyzes NRC processes and regulations in light of the Fukushima accident, an event that occurred a mere five months ago. This Motion and accompanying contention are being submitted less than thirty (30) days after issuance of the Task Force Report.

Accordingly, SACE has good cause to submit this Motion and the accompanying contention now.

2. Nature of SACE's Right to be a Party to the Proceeding.

SACE's right to be a party to this proceeding has been recognized by the Licensing Board in admitting SACE as an intervenor.

3. Nature of SACE's Interest in the Proceeding.

Through submission of this contention, SACE seeks to protect its members' health and safety and the health of the environment in which they live, by ensuring that the NRC fulfills its non-discretionary duty under NEPA to consider the new and significant information set forth in the Task Force Report regarding the potential environmental effects of the operation of Watts Bar Unit 2, before it makes a decision regarding the proposed re-licensing of the plant.

4. Possible Effect of an Order on SACE's Interest in the Proceeding.

SACE's interest in a safe, clean, and healthful environment would be served by the issuance of an order requiring the NRC to fulfill its non-discretionary duty under NEPA to consider new and significant information before making a licensing decision. *See* Silva v. Romney, 473 F.2d 287, 292 1st Cir. 1973). Compliance with NEPA ensures that environmental issues are given full consideration in "the ongoing programs and actions of the Federal Government." Marsh v. Oregon Natural Res. Council, 490 U.S. 360, 371 n.14 (1989).

5. Availability of Other Means to Protect SACE' Interests.

With regard to this factor, the question is not whether other parties may protect SACE's interests, but rather whether there are other means by which SACE may protect their own interests. <u>Long Island Lighting Co.</u> (Jamesport Nuclear Power Station, Units 1 & 2), ALAB-292, 2 NRC 631 (1975). Quite simply, no other means exist. Only through this hearing does SACE have have a right that is judicially enforceable to seek compliance by NRC with NEPA before the NRC makes a decision regarding the proposed issuance of the Watts Bar Unit 2 license.

6. Extent to which SACE's Interests are Represented by Other Parties.

There is no other citizen or environmental organization that has been admitted to the Watts Bar Unit 2 licensing proceeding and therefore no other party can represent its interests.

7. Extent That Participation Will Broaden the Issues.

While SACE's participation may broaden or delay the proceeding, this factor may not be relied upon to deny this Motion or exclude the contention because the NRC has a non-discretionary duty under NEPA to consider new and significant information that arises before it makes its licensing decision. <u>Marsh</u>, 490 U.S. at 373-4.

8. Extent to which SACE Will Assist in the Development of a Sound Record.

SACE will assist in the development of a sound record, as their contention is supported by the expert opinion of a highly qualified expert, Dr. Arjun Makhijani. *See* attached Makhijani Declaration. *See also* Pacific Gas & Elec. Co. (Watts Bar Unit 2 Power Plant Independent Spent Fuel Storage Installation), CLI-08-01, 67 NRC 1, 6 (2008) (finding that, when assisted by experienced counsel and experts, participation of a petitioner may be reasonably expected to contribute to the development of a sound record). Furthermore, as a matter of law, NEPA requires consideration of the new and significant information set forth in the Task Force Report. *See* 10 C.F.R. § 51.92(a)(2). A sound record cannot be developed without such consideration.

C. The New Contention Satisfies the Standards For Admission of Contentions Set Forth in 10 C.F.R. § 2.309(f)(1).

As discussed in the accompanying contention, the standards for admission of a contention set forth in 10 C.F.R. § 2.309(f)(1) are satisfied.

III.   CONCLUSION

For the foregoing reasons, this Motion should be granted and the accompanying contention admitted.

Respectfully submitted,

Signed (electronically) by:
Diane Curran
Harmon, Curran, Spielberg & Eisenberg, L.L.P.
1726 M Street N.W. Suite 600
Washington, D.C. 20036
202-328-3500
Fax: 202-328-6918
E-mail: dcurran@harmoncurran.com

August 11, 2011

CERTIFICATE REQUIRED BY 10 C.F.R. § 2.323(b)

I certify that on August 9, 2011, I contacted counsel for TVA and the NRC Staff in an attempt to obtain their consent to this motion. Counsel for TVA stated that TVA objected to the motion and would respond to it. Counsel for the Staff said that the Staff did not object to the filing of the motion but would respond to it with respect to the timeliness and admissibility of the contention.

Electronically signed by
Diane Curran

UNITED STATES OF AMERICA
NUCLEAR REGULATORY COMMISSION
BEFORE THE ATOMIC SAFETY AND LICENSING BOARD

In the Matter of)	
)	
Tennessee Valley Authority)	Docket No. 50-391
)	
(Watts Bar Unit 2))	

**CONTENTION REGARDING NEPA REQUIREMENT TO ADDRESS
SAFETY AND ENVIRONMENTAL IMPLICATIONS OF
THE FUKUSHIMA TASK FORCE REPORT**

I.   INTRODUCTION AND SUMMARY

Pursuant to 10 C.F.R. § 2.309(f)(1), Southern Alliance for Clean Energy ("SACE") asserts a new contention seeking consideration of new and significant information relevant to the environmental analysis for the proposed licensing of Watts Bar Unit 2. In the contention set forth in Section II below, SACE requests a hearing on the significant – indeed extraordinary – safety and environmental implications for the Watts Bar Unit 2 licensing decision of the conclusions and recommendations of the U.S. Nuclear Regulatory Commission's Near-Term Task Force (the "Task Force"). The contention is supported by the expert declaration of Dr. Arjun Makhijani of the Institute for Energy and Environmental Research. The contention is also supported by a Motion to Admit a New Contention.

The Task Force, a group of highly qualified and experienced Nuclear Regulatory Commission ("NRC" or the "Commission") staff members selected by the Commission to evaluate the regulatory implications of the Fukushima Dai-ichi accident, has issued a report recommending the NRC strengthen its regulatory scheme for protecting public health and safety by increasing the scope of accidents that fall within the "design basis" and are therefore subject to mandatory safety regulation. <u>Recommendations for Enhancing Reactor Safety in the 21st Century: The Near-Term Task Force Review of Insights from the Fukushima Dai-ichi Accident</u> at 20-21 (July 12, 2011) ("Task Force Report"). The Task Force's recommendation to establish mandatory safety regulations for severe accidents has extremely grave environmental and safety implications because it would not be logical or necessary to recommend an upgrade to the basic level of protection currently afforded by NRC regulations unless those existing regulations were insufficient to ensure adequate protection of public health, safety, and the environment throughout the licensed life of nuclear reactors. The recommendation is all the more grave because it constitutes the second warning that the Commission has received regarding the need to expand the scope of design basis accidents. The first warning, issued by the Rogovin Report over thirty years ago, following the Three Mile Island accident and explained in more detail in Section II below, essentially went unheeded. *Id.* at 16-17. As the Task Force urges, "the time has come" to make fundamental changes to the NRC's program for establishing minimum safety requirements for nuclear reactors. *Id.* at 18.

Moreover, the Task Force's recommendation that the scope of mandatory safety regulations be expanded to include severe accidents raises significant environmental concerns in this proceeding, including that (1) the risks of operating Watts Bar Unit 2 are higher than estimated in the FSEIS and (2) TVA's previous environmental analysis of the relative costs and benefits of severe accident mitigation alternatives ("SAMAs") is

2

fundamentally inadequate because those measures are, in fact, necessary to assure adequate protection of the public health and safety and, therefore, should be imposed without regard to their cost.

Pursuant to the National Environmental Policy Act ("NEPA"), the analysis demanded by this contention may not be deferred until after Watts Bar Unit 2 is licensed. Given that the NRC Commissioners have postponed taking action on the Task Force's recommendations, admission of this contention constitutes the only way of ensuring that the environmental implications of the Task Force recommendations are taken into account in the licensing decision for Watts Bar Unit 2.

SACE wishes to point out that this contention is substantially similar to contentions and comments that are being filed this week in other pending reactor licensing and re-licensing cases and standardized design certification proceedings. In addition, SACE has joined with other individuals and organizations in a rulemaking petition seeking to suspend any regulations that would preclude full consideration of the environmental implications of the Task Force Report. A copy of the rulemaking petition is attached. Finally, in an Emergency Petition, now pending before the Commission for nearly four months, many of the same organizations and individuals previously asked the Commission to suspend its licensing decisions while it evaluated the environmental implications of the Fukushima accident and to establish procedures for the fair and meaningful consideration of those issues in licensing hearings. Emergency Petition to Suspend All Pending Reactor Licensing Decisions and Related Rulemaking Decisions Pending Investigation of Lessons learned From Fukushima Daiichi Nuclear Power Station Accident (April 14-18, 2011) (the "Emergency Petition").

3

In the aggregate, these contentions, rulemaking comments, and the rulemaking petition follow up on the Emergency Petition's demand that the NRC comply with NEPA by addressing the lessons of the Fukushima accident in its environmental analyses for licensing decisions. Having received no response to their Emergency Petition, the signatories to the Emergency Petition now seek consideration of the Task Force's far-reaching conclusions and recommendations in each individual licensing proceeding, including the instant case.

SACE recognizes that given the sweeping scope of the Task Force conclusions and recommendations, it may be more appropriate for the NRC to consider them in generic rather than site-specific environmental proceedings. That is for the NRC to decide. *Baltimore Gas & Electric Co. v. Natural Resources Defense Council,* 462 U.S. 87, 100 (1983). It is the NRC, and not the public, which is responsible for compliance with NEPA. *Duke Power Co. et al.* (Catawba Nuclear Station, Units 1 and 2), CLI-83-19, 17 NRC 1041, 1049 (1983).

II. SACE'S NEW CONTENTION SATISFIES THE REQUIREMENTS OF 10 C.F.R. § 2.309 (f)(1).

1. Statement of Contention.

The FSEIS for Watts Bar Unit 2 fails to satisfy the requirements of NEPA because it does not address the new and significant environmental implications of the findings and recommendations raised by the NRC's Fukushima Task Force Report. As required by NEPA and the NRC regulations, these implications must be addressed in the ER.

2. Brief Explanation of the Basis for the Contention.

The Task Force Report.

4

This contention is based on the Task Force Report, in which the Commission instructed the Task Force to provide:

> A systematic and methodical review of [NRC] processes and regulations to determine whether the agency should make additional improvements to its regulatory system and to make recommendations to the Commission for its policy direction, in light of the accident at the Fukushima Dai-ichi Nuclear Power Plant.

Task Force Report at vii. In response to that directive, the Task Force prepared a detailed history of the NRC's program for regulation of safety and public health and evaluated that program in light of the experience of the Fukushima accident.

The Task Force then assessed the risk posed by "continued operation and continued licensing activities" for U.S. nuclear plants. Applying the NRC's standard for whether nuclear plants pose an "imminent risk" such that they should be shut down immediately, *see, e.g., Yankee Atomic Electric Co.* (Yankee Nuclear Power Station), CLI-96-6, 43 NRC 123, 128 (1996) (finding no "imminent hazard" that would warrant shutdown of a reactor), the Task Force found that no imminent risk was posed by operation or licensing. *Id.* at 18. In addition, the Task Force concluded that U.S. reactors meet the statutory standard for security, *i.e.,* they are "not inimical to the common defense and security." *Id.* at 18; *see also* 42 U.S.C. § 2133(d) (forbidding the NRC from licensing reactors if their operation would be "inimical to the common defense and security"). Notably, however, the Task Force did not report a conclusion that licensing of reactors would not be "inimical to public health and safety," as the AEA requires for licensing of reactors. 42 U.S.C. § 2133.

Instead, the Task Force concluded that the regulatory system on which the NRC relies to make the safety findings that the AEA requires for licensing of reactors must be

5

strengthened by raising the level of safety that is minimally required for the protection of public health and safety:

> In response to the Fukushima accident and the insights it brings to light, the Task Force is recommending actions, some general, some specific, that it believes would be a reasonable, well-formulated set of actions *to increase the level of safety associated with adequate protection of the public health and safety.*

Id. at 18 (emphasis added). In particular, the Task Force found that "the NRC's safety approach is incomplete without a strong program for dealing with the unexpected, including severe accidents." *Id.* at 20. Therefore, the Task Force recommended that the NRC incorporate severe accidents into the "design basis" and subject it to mandatory safety regulations. In order to upgrade the design basis, the Task Force also recommended that the NRC undertake new safety investigations and impose design changes, equipment upgrades, and improvements to emergency planning and operating procedures. *See, e.g.,* Task Force Report at 73-75.[1]

The Task Force also found that the Fukushima accident was not the first warning the NRC had received that it needed to strengthen its safety program in order to provide an adequate level of protection to public health and safety. After the Three Mile Island accident in 1979, an independent body appointed to investigate the accident's implications, headed by Mitchell Rogovin of the NRC's Special Inquiry Group, recommended that the NRC "[e]xpand the spectrum of design basis accidents." *Id.* at 16. But the NRC did little to follow the recommendations of the Rogovin Report. While it "encouraged licensees to search for vulnerabilities" in their plant designs through Individual Plant Examination ("IPE") and Individual Plant Examination for External

---

[1] The Task Force Report contains twelve "overarching" recommendations, which are summarized on pages 69-70.

6

Events ("IPEEE") programs and encouraged the development of severe accident

mitigation guidelines ("SAMGs"), "the Commission did not take action to require the

IPEs, IPEEEs, or SAMGs." *Id.* Thus, the Task Force concluded that:

> While the Commission has been partially responsive to recommendations calling
> for requirements to address beyond-design-basis accidents, the NRC has not made
> fundamental changes to the regulatory approach for beyond-design-basis events
> and severe accidents for operating reactors.

Id. at 17. Looking back on the Commission's failure to heed the Rogovin Report's

recommendations, the Task Force urged that "the time has come" when NRC safety

regulations must be "reviewed, evaluated and changed, as necessary, to insure (sic) that

they continue to address the NRC's requirements to provide reasonable assurance of

adequate protection of public health and safety." *Id.* at 18.

To finally fulfill the Rogovin Report's recommendation, a need now re-confirmed

by the Fukushima Task Force, would require a major re-evaluation and overhaul of the

NRC's regulatory program. As the Task Force recognized, the great majority of the

NRC's current regulations do not impose mandatory safety requirements on severe

accidents, and severe accident measures are adopted only on a "voluntary" basis or

through a "patchwork" of requirements. *Id.*

The lack of an NRC program for mandatory regulation of severe accidents is

clearly evident from the regulations themselves. The Part 50 regulations, which establish

fundamental safety requirements for all reactors (including the current generation and the

proposed new generation), are based on a "design basis" that does not include severe

accidents. Task Force Report at 16. While NRC NEPA regulations require consideration

of severe accident mitigation measures, they need not be adopted unless they are found to

be cost-beneficial. *See, e.g., Entergy Nuclear Operations, Inc.* (Indian Point Nuclear

7

Generating Station, Units 2 and 3), LBP-11-17, slip op. at 17 (July 14, 2011). Because

the imposition of severe accident mitigation measures is based on cost considerations,

they are not part of the design basis for adequate protection of public health and safety.

Union of Concerned Scientists v. NRC, 824 F.2d 108, 120 (D.C. Cir. 1987).[2]

Therefore, the NRC's current regulatory scheme requires significant re-evaluation

and revision in order to expand or upgrade the design basis for reactor safety as

recommended by the Task Force Report. The fact that this effort has been postponed for

thirty years makes the scope of the required undertaking all the more massive and urgent.

The National Environmental Policy Act.

The contention is also based on NEPA, "our basic national charter for protection

of the environment." 40 C.F.R § 1500.1(a). NEPA requires a federal agency to prepare

an Environmental Impact Statement for any "major Federal action significantly affecting

the quality of the human environment." 42 U.S.C. § 4332(2)(C)(i). This duty to

carefully consider information regarding a project's environmental impacts is non-

discretionary. *Silva v. Romney,* 473 F.2d 287, 292 (1st Cir. 1973). Federal agencies are

---

[2] Even the NRC's Part 52 regulations for new reactors do not contain mandatory requirements for severe accident mitigation features. While the Part 52 regulations require combined license applicants to submit analyses of measures to mitigate severe accidents, Part 52 contains no standards for the adequacy of such analyses. In addition, the Commission has also stated that Part 52 severe accident mitigation measures, which must be described under the NRC's safety regulations in 10 C.F.R. §§ 52.47(a)(23) and 52.79(a)(38), are subject to cost-benefit analysis. *See, e.g.,* Statement of Considerations ("SOC") for AP1000 design certification rule, 10 C.F.R. Part 52 Appendix B, 71 Fed. Reg. 4,464, 4,469 (January 27, 2006): As stated in that notice:

> Westinghouse's evaluation of various design alternatives to prevent and mitigate severe accidents does not constitute design requirements. The Commission's assessment of this information is discussed in Section VII (sic) of this SOC on environmental impacts.

8

held to a "strict standard of compliance" with the Act's requirements. *Calvert Cliff's Coordinating Commission v. AEC*, 449 F.2d 1109, 1112 (D.C. Cir. 1971).

NEPA and the Council on Environmental Quality ("CEQ") regulations implementing NEPA are intended to ensure that environmental considerations are "infused into the ongoing programs and actions of the Federal Government." *Marsh v. Oregon Natural Res. Council*, 490 U.S. 360, 371 n.14 (1989). Thus, NEPA imposes on agencies a continuing obligation to gather and evaluate new information relevant to the environmental impact of its actions. *Warm Springs Dam Task Force v. Gribble*, 621 F.2d 1017, 1023-24 (9th Cir. 1980) (citing 42 U.S.C. 4332(2)(A), (B); *Essex County Preservation Ass'n v. Campbell*, 536 F.2d 956, 960-61 (1st Cir. 1976); *Society for Animal Rights, Inc. v. Schlesinger*, 512 F.2d 915, 917-18 (D.C. Cir. 1975)). "An agency that has prepared an EIS cannot simply rest on the original document. The agency must be alert to new information that may alter the results of its original environmental analysis, and continue to take a "hard look" at the environmental effects of [its] planned action, even after a proposal has received initial approval." *Friends of the Clearwater v. Dombeck*, 222 F.3d 552, 557-58 (9th Cir. 2000) (quoting *Marsh*, 490 U.S. at 373-74).

In order to aid the Commission in complying with NEPA, each applicant shall submit to the Commission an environmental report ("ER"). *See* 10 C.F.R. §§ 51.14; 51.45. In this case, the TVA's FSEIS serves the same purpose of an ER, *i.e.,* to provide the applicant's initial analysis of the environmental impacts of the proposed nuclear power plant operation. The ER must contain a description of the proposed action, a statement of its purposes, and a description of the environment affected. *Id.* § 51.45 (b). Further, the ER must discuss the impact of the proposed action on the environment, any

9

adverse environmental effects which cannot be avoided should the proposal be implemented, alternatives to the proposed action, the relationship between local short-term uses of man's environment and the maintenance and enhancement of long-term productivity, and any reversible and irretrievable commitments of resources which would be involved in the proposed action should it be implemented. *Id.* § 51.45(b)(5). The ER must also contain an analysis that considers and balances the environmental effects of the proposed action, the environmental impacts of alternatives to the proposed action, and alternatives available for reducing or avoiding adverse environmental effects. *Id.* § 51.45(c). An environmental report for the licensing action contemplated in this instance must also include consideration of the economic, technical, and other benefits and costs of the proposed action and its alternatives. *Id.* The environmental report must to the fullest extent practicable, quantify the various factors considered and contain sufficient data to aid the Commission in its development of an independent analysis. *Id.*

Within this regulatory framework, "[t]he Commission recognizes *a continuing obligation* to conduct its domestic licensing and related regulatory functions in a manner which is both receptive to environmental concerns and consistent with the Commission's responsibility as an independent regulatory agency for protecting the radiological health and safety of the public." *Id.* § 51.10(b) (emphasis added).

The Environmental Report Does Not Consider the Significant New Information Contained in the Task Force Report and the FSEIS Must Be Supplemented to Comply with NEPA.

NEPA requires federal agencies to supplement their NEPA documentation when "there are significant new circumstances or information relevant to environmental concerns and bearing on the proposed action or its impacts." 40 C.F.R. § 1509(c)(1)(ii).

10

A federal agency's *continuing duty* to take a "hard look" at the environmental effects of their actions requires they consider, evaluate, and make a reasoned determination about the significance of this new information and prepare supplemental NEPA documentation accordingly. *Warm Springs Task Force v. Gribble*, 621 F.2d at 1023-24; *Stop H-3 Association v. Dole,* 740 F.2d 1442, 1463-64 (9th Cir. 1984). The need to supplement under NEPA when there is new and significant information is also found throughout the NRC regulations. *See* 10 C.F.R. §§ 51.92 (a)(2), 51.50(c)(iii), 51.53(b), 51.53(c)(3)(iv).

The conclusions and recommendations presented in the Task Force Report constitute "new and significant information" whose environmental implications must be considered before the NRC may make a decision that approves the licensing of Watts Bar Unit 2. First, the information is "new" because it stems directly from the Fukushima accident, which occurred only five months ago and for which the special study commissioned by the Commission has only just been issued.

Second, the information is "significant" because it raises an extraordinary level of concern regarding the manner in which the proposed operation of Watts Bar Unit 2 "impacts public health and safety." *See* 40 C.F.R. § 1508.27(b)(2). For the first time since the Three Mile Island accident occurred in 1979, a highly respected group of scientists and engineers within the NRC Staff has fundamentally questioned the adequacy of the current level of safety provided by the NRC's program for nuclear reactor regulation. NEPA demands that federal agencies "insure the professional integrity, including the scientific integrity, of the discussions and analyses" included in an EIS[3] and disclose "all major points of view on the environmental impacts" including any

---

[3] 40 C.F.R. § 1502.24.

11

"responsible opposing view."[4] Courts have found that an EIS that fails to disclose and respond to expert opinions concerning the hazards of a proposed action, particularly those opinions of the agency's own experts, are "fatally deficient" and run contrary to NEPA's "hard look" requirement.[5] As a result, the NRC must revisit any conclusions in the FSEIS for Watts Bar Unit 2 which is based on the assumption that compliance with NRC safety regulations is sufficient to ensure that environmental impacts of accidents are acceptable.

The Task Force Report Reveals that the Full Spectrum of All Design-Basis Accidents Has Not Been Assessed and the FSEIS Must Be Supplemented to Consider Additional Design-Basis Accidents that Have the Potential for Releases to the Environment.

In Section 3.12.1 of the FSEIS, TVA asserts that:

The term "accident" refers to any unintentional event (i.e., outside the normal or expected plant operation envelope) that results in a release or a potential for a release of radioactive material to the environment. The NRC categorizes accidents as either design basis or severe. Design basis accidents are those for which the risk is great enough that NRC requires the plant design and construction to prevent unacceptable accident consequences. Severe accidents are those that NRC considers too unlikely to warrant normal design controls.

---

[4] 40 C.F.R. §§ 1502.9(a), (b)

[5] *Center for Biological Diversity v. United States Forest Service*, 349 F.3d 1157 (9th Cir. 2003) (finding an EIS's failure to disclose and discuss responsible opposing scientific viewpoints violated NEPA and the implementing regulations); *Seattle Audubon Society v. Moseley,* 798 F.Supp. 1473, 1479 (W.D. Wa. 1992) aff'd sub nom *Seattle Audubon Society v. Espy,* 998 F.2d 699 (9th Cir. 1993) (quoting *Friends of the Earth v. Hall,* 693 F.Supp. 904, 934 (W.D. Wa. 1988) ("[a]n EIS that fails to disclose and respond to 'the opinions held by well respected scientists concerning the hazards of the proposed action...is fatally deficient.")); *Western Watersheds Project v. Kraayenbrink,* 632 F.3d 472, 487 (9th Cir. 2010) (finding that agency failed to take a "hard look" under NEPA when it ignored concerns raised *by its own experts*). See also *Blue Mtns. Biodiversity Project v. Blackwood,* 161 F.3d 1208, 1213 (9th Cir. 1998) (noting that an agency's failure to discuss and consider an independent scientific report's recommendations "lends weight to [plaintiff's] claim that the [agency] did not take the requisite 'hard look' at the environmental consequences" of the project).

12

FSEIS at 73. Thus, TVA bases its environmental analysis on the assumption that compliance with the NRC's regulatory program for protection against design basis accidents is sufficient to maintain environmental impacts from design basis accidents at an acceptable or insignificant level, and that severe accidents are too unlikely to merit inclusion in the design basis. The findings of the Task Force Report call these assumptions into serious question. *See* Makhijani Declaration, pars. 7-10. If, as suggested by the Task Force Report, the design basis for the reactor does not incorporate accidents that should be considered in order to satisfy the adequate protection standard, then it is not possible to reach a conclusion that the design of the reactor adequately protects against accident risks.

The FEIS Must Be Supplemented in Light of the Task Force Findings that Certain Accidents Formerly Classified as Severe Should Be Incorporated into the Design Basis.

By recommending the incorporation of accidents formerly classified as "severe" or "beyond design basis" into the design basis, the Task Force effectively recommends a complete overhaul of the NRC's system for mitigating severe accidents through consideration of SAMAs. *See* 10 C.F.R. § 51.45(c). As the Task Force recognizes, currently the NRC does not impose measures for the mitigation of severe accidents unless they are shown to be cost-beneficial or unless they are adopted voluntarily. Task Force Report at 15. *See also* 10 C.F.R. §§ 51.71(d); 51.75(c)(2) (allowing EISs for combined license applications ("COLAs") that rely on certified standardized designs to reference the severe accident mitigation analyses for those designs).[6] But the Task Force

_____

[6] *See also* Memorandum from NRC Staff to AP1000 and ESBWR design-Centered Working Groups re: Summary of the March 22 and 23, 2007, Meeting to Discuss pre-Combined License Application Issues (April 23, 2007) (suggesting that some SAMAs for

13

recommends that severe accident mitigation measures should be adopted into the design basis, *i.e.,* the set of regulations adopted *without regard to their cost* as fundamentally required for all NRC standards that set requirements for adequate protection of health and safety. *Union of Concerned Scientists v. NRC,* 824 F.2d 108, 120 (D.C. Cir. 1987). Thus, the values assigned to the cost-benefit analysis for Watts Bar Unit 2, as described in TVA's SAMA analysis (*see* Memorandum to File from Robert Lutz, Westinghouse, re: Watts Bar Unit 2 Severe Accident Mitigation Analysis (January 29, 2009)), must be re-evaluated in light of the Task Force's conclusion that the value of SAMAs is so high that they should be elected as a matter of course.

Were SAMAs imposed as mandatory measures, the outcome of the FSEIS and subsequently the NRC's EIS for Watts Bar Unit 2 could be affected significantly in two major respects. First, severe accident mitigative measures now rejected as too costly may be required, thus substantially improving the safety of the Watts Bar Unit 2 operation if it is licensed. Second, consideration of the costs of mandatory mitigative measures could affect the overall cost-benefit analysis for the reactor.[7] As discussed in Dr. Makhijani's declaration, these costs may be significant, showing that other alternatives such as the no-action alternative and other alternative electricity production sources may be more attractive.[8] As the fundamental purposes of NEPA are: (1) to guarantee that the

---

proposed reactors with standardized designs should be included in the design application and some should be included in COLAs).

[7] *See* 10 C.F.R. § 51.45 (c) (explaining that environmental reports should also include consideration of the economic, technical, and other benefits and costs of the proposed action and its alternatives).

[8] NEPA requires the NRC to include in its EIS a "detailed statement . . . on . . . alternatives to the proposed action." 42 U.S.C. § 4332(C)(iii). The alternatives analysis should address "the environmental impacts of the proposal and the alternatives in comparative form, thus sharply defining the issues and providing a clear basis for the

14

government takes a "hard look" at all of the environmental consequences of proposed federal actions before the actions occur, *Robertson v. Methow Valley Citizens Council*, 490 U.S. 332, 350 (1989); and (2) to "guarantee[] that the relevant information will be made available to the larger audience that may also play a role in both the decisionmaking process and the implementation of that decision," *id.* at 349, the NRC cannot meet the fundamental purposes of NEPA if it does not include all of the costs associated with required mitigative measures. *See Sierra Club v. Sigler*, 695 F.2d 957, 979 (5th Cir. 1983) ("There can be no 'hard look' at the costs and benefits unless all costs are disclosed.").

The FSEIS Must Be Supplemented to Include a Discussion of the Task Force Report's Recommended Measures to Ensure the Plant's Protection From Seismic and Flooding Events.

Following the devastating events in Japan, the Task Force Report explained the importance of protecting structures, systems and components (SSCs) of nuclear reactors from natural phenomena, including seismic and flooding hazards:

> Protection from natural phenomena such seismic and flooding is critical for safe operation of nuclear power plants due to potential common-cause failures and significant contribution to core damage frequency from external events. Failure to adequately protect SSC's important to safety from appropriate design-basis natural phenomena with appropriate safety margins has the potential for common-

---

choice among options by the decisionmaker and the public." 40 C.F.R. § 1502.14. This analysis must "rigorously explore and objectively evaluate all reasonable alternatives." 40 C.F.R. § 1502.14(a). Agencies must consider three types of alternatives, which include a no action alternative, other reasonable courses of actions, and mitigation measures not in the proposed action. 40 C.F.R. § 1508.25. The purpose of this section is "to insist that no major federal project should be undertaken without intense consideration of other more ecologically sound courses of action, including shelving the entire project, or of accomplishing the same result by entirely different means." *Environmental Defense Fund v. Corps of Engineers*, 492 F.2d 1123, 1135 (5th Cir. 1974). "The existence of a viable but unexamined alternative renders an [EIS] inadequate." *Natural Resources Defense Council v. U.S. Forest Service*, 421 F.3d 797, 813 (9th Cir. 2005) (quoting *Citizens for a Better Henderson v. Hodel*, 768 F.2d 1051, 1057 (9th Cir. 1985)).

15

cause failures and significant consequences as demonstrated at Fukushima. Task Force Report at 30.

Yet, the Task Force found that significant differences may exist between plants in the way they protect against design-basis natural phenomena (including seismic and flooding hazards) and the safety margin provided. Task Force Report at 29. For instance, while tsunami hazards have been considered in the design basis for operating plants sited on the Pacific Ocean, the same cannot be said for those sited on the Atlantic Ocean and Gulf of Mexico. *Id.* Accordingly, the Task Force recommended that licensees reevaluate the seismic and flooding hazards at their sites and if necessary update the design basis and SSCs important to safety to protect against the updated hazards. Task Force Report at 30.

The FSEIS must be supplemented in light of this new and significant information. The Task Force's findings and recommendations are directly relevant to environmental concerns and have a bearing on the proposed action and its impacts as they point to the need for a reevaluation of the seismic and flooding hazards at the Watts Bar Unit 2 site, a "hard look" at the environmental consequences such hazards could pose, and an examination of what, if any, design measures could be implemented (i.e. through NEPA's requisite "alternatives" analysis) to ensure that the public is adequately protected from these risks.

The FSEIS Must Be Supplemented to Include a Discussion of the Additional Mitigation Measures Recommended by the Task Force Report.

"The discussion of steps that can be taken to mitigate adverse environmental consequences plays an important role in the environmental analysis under NEPA." *Robertson v. Methow Valley Citizens Council*, 490 U.S. 332, 351 (1989); *see also*

16

1502.16(h) (stating that an EIS must contain "means to mitigate adverse environmental impacts"). There must be a "reasonably complete discussion of possible mitigation measures." *Robertson*, 490 U.S. at 352. Mitigation measures may be found insufficient when the agency fails to study the efficacy of the proposed mitigation, fails to take certain steps to ensure the efficacy of the proposed mitigation (such as including mandatory conditions in permits), or fails to consider alternatives in the event that the mitigation measures fail. *Id.*

The Task Force Report makes several significant findings when it comes to increasing and improving mitigation measures at new reactors and recommends a number of specific steps licensees could take in this regard. These recommendations include strengthening SBO mitigation capability at all operating and new reactors for design-basis and beyond-design-basis external events, (Section 4.2.1), requiring reliable hardened vent designs in BWR facilities with Mark I and Mark II containments (Section 4.2.2), enhancing spent fuel pool makeup capability and instrumentation for the spent fuel pool (Section 4.2.4), strengthening and integrating onsite emergency response capabilities such as EOPs, SAMGs, and EDMGs (Section 4.2.5) and addressing multi-unit accidents. *See also* Makhijani Declaration, pars. 18-24. Accordingly, the FSEIS must be supplemented to consider the use of these additional mitigation measures to reduce the project's environmental impacts. *See* 40 C.F.R. §§ 1502.14(f), 1502.16), 1508.25 (b)(3)).

Requirement for Prior Consideration of Environmental Impacts.

The Task Force urges that some of its recommendations be considered before certain licensing decisions are made. For instance, the Task Force recommends that the

17

operating license review for Watts Bar Unit 2 should include "all of the near-term actions and any of the recommended rule changes that have been completed at the time of licensing." Task Force Report at 72. Similarly, the Task Force recommends that Recommendation 4 (proposing new requirements for prolonged station blackout ("SBO") mitigation) and Recommendation 7 (proposing measures for spent fuel pool makeup capability and instrumentation) should apply to all design certifications or to COL applicants if the recommended requirements are not addressed in the referenced certified design. Task Force Report at 71. The Task Force recommends that design certifications and COLs under active staff review address this recommendation "before licensing." *Id.* at 72.

SACE respectfully submits that NEPA does not give the NRC the discretion to postpone consideration of any of the Task Force recommendations until after the licensing of Watts Bar or any other reactor for which a licensing decision is before the agency. NEPA requires the NRC to address the environmental implications of the Task Force's analysis *before* making a re-licensing decision for Watts Bar Unit 2, in order to ensure that "important effects [of the licensing decision] will not be overlooked or underestimated only to be discovered after resources have been committed or the die otherwise cast." *Robertson,* 490 U.S. at 349. *See also* 40 C.F.R. §§ 1500.1(c), 1502.1, 1502.14. The NRC's obligation to comply with NEPA in this respect is independent of and in addition to the NRC's responsibilities under the AEA, and must be enforced to the "fullest extent possible." *Calvert Cliffs Coordinating Committee,* 449 F.2d at 1115. *See also Limerick Ecology Action v. NRC,* 869 F.2d 719, 729 (3rd Cir. 1989) (citing *Public Service Co. of New Hampshire v. NRC,* 582 F.2d 77, 86 (1st Cir. 1978)). Under NEPA,

18

therefore, the Commission is required to address the Task Force's findings and recommendations as they pertain to Watts Bar Unit 2 before making a licensing decision, regardless of whether it does or does not choose to do so in the context of its AEA-based regulations.

Of course the Commission could moot the contention by adopting all of the Task Force's recommendations. *See Citizens for Safe Power v. NRC*, 524 F.2d 1291, 1299 (D.C. Cir. 1975). However, a majority of the Commissioners has voted not to do so immediately. *See* Notation Vote Response Sheets re: SECY-11-0093, Near-Term Report and Recommendations for Agency Actions Following the Events in Japan, posted on the NRC's website at http://www.nrc.gov/reading-rm/doc-collections/commission/cvr/2011/. Thus, while the NRC may eventually address the Task Force's recommendations in the context of its AEA-based regulatory scheme, the Commission has given no indication that it intends to address any of the Task Force's conclusions in its prospective licensing decisions. In the absence of any AEA-based review of the Task Force's conclusions, the FSEIS for Watts Bar Unit 2 must be supplemented in order to meet NEPA's goal that the NRC's licensing decision for Watts Bar Unit 2 will be "based on an accurate understanding of the environmental consequences of [its] actions." *Indian Point*, LBP-11-17, slip op. at 17.

3. **Demonstration that the Contention is Within the Scope of the Proceeding.**

The contention is within the scope of the proceeding because it seeks compliance with NEPA and NRC-implementing regulations, which must be complied with before Watts Bar Unit 2 may be licensed.

19

4.   Demonstration that the Contention is Material to the Findings NRC Must Make to License Watts Bar Unit 2.

As demonstrated above in Section B, this contention challenges TVA's failure to fully comply with NEPA and federal regulations for the implementation of NEPA in its FSEIS for the proposed licensing of Watts Bar Unit 2. TVA's FSEIS is akin to an environmental report and therefore it must comply with NEPA supplementation requirements in the same way that ERs prepared by applicants or environmental impact statements prepared by the NRC must comply. Unless TVA and the NRC comply with the procedural requirements of NEPA that are discussed in the contention, the NRC cannot make a valid finding that Watts Bar Unit 2 should be licensed. Therefore the contention is material to the findings the NRC must make in order to license this facility.

SACE recognizes that some issues raised by the Task Force Report may be appropriate for generic rather than case-specific resolution. The determination of whether it is appropriate to address the issues raised in this contention generically or on a case-specific basis is a discretionary matter for the NRC to decide. *Baltimore Gas & Electric Co. v. Natural Resources Defense Council,* 462 U.S. at 100. Nevertheless, any generic resolution of the issues must be reached *before* the licensing decision in this case is made, and must be applied to this licensing decision. *Robertson,* 490 U.S. at 350.

5.   Concise Statement of the Facts or Expert Opinion Supporting the Contention, Along With Appropriate Citations to Supporting Scientific or Factual Materials.

SACE relies on the facts and opinions of the Task Force members as set forth in their Task Force Report and as summarized above in Section B. The high level of technical qualifications of the Task Force members has been recognized by the

20

Commission. *See* Transcript of May 12, 2011, briefing at 5, in which Commissioner Magwood refers to the Task force as the NRC's "A-team."

Additional technical support is provided by the attached Declaration of Dr. Arjun Makhijani, which confirms the environmental significance of the Task Force's findings and recommendations with respect to the environmental analyses for all pending nuclear reactor licensing cases and design certification applications including the instant case.

6. Sufficient Information to Show the Existence of a Genuine Dispute With the Applicant and the NRC.

Based on the complete failure of the NRC to address the environmental implications of the Task Force Report for the proposed licensing of Watts Bar Unit 2, it appears that the parties have a dispute as to whether the FSEIS for the facility must be revised to address those implications. As demonstrated above in Section B, the Task Force Report and Dr. Makhijani's Declaration provide sufficient information to show the genuineness and materiality of the dispute.

III. CONCLUSION

For the foregoing reasons, the contention is admissible and should be admitted for a hearing.

Respectfully submitted,

Signed (electronically) by:
Diane Curran
Harmon, Curran, Spielberg & Eisenberg, L.L.P.
1726 M Street N.W. Suite 600
Washington, D.C. 20036
202-328-3500
Fax: 202-328-6918
E-mail: dcurran@harmoncurran.com

August 11, 2011

21

UNITED STATES OF AMERICA
NUCLEAR REGULATORY COMMISSION
BEFORE THE ATOMIC SAFETY AND LICENSING BOARD
AND THE SECRETARY

	)	
In the Matter of	)	Rulemaking Docket No.__
	)	
Tennessee Valley Authority	)	Docket No. 50-391
	)	
(Watts Bar Unit 2)	)	
	)	

**RULEMAKING PETITION TO RESCIND PROHIBITION
AGAINST CONSIDERATION OF ENVIRONMENTAL IMPACTS
OF SEVERE REACTOR AND SPENT FUEL POOL ACCIDENTS
AND REQUEST TO SUSPEND LICENSING DECISION**

I.   INTRODUCTION

Pursuant to 10 C.F.R. § 2.802, Southern Alliance for Clean Energy ("SACE")

petitions the U.S. Nuclear Regulatory Commission ("NRC") to rescind regulations in 10

C.F.R. Part 51 that make generic conclusions about the environmental impacts of severe

reactor and spent fuel pool accidents and that preclude consideration of those issues in

individual licensing proceedings. This petition also requests the NRC to suspend the

above-captioned licensing proceeding while the NRC considers this petition and the

environmental issues raised in the attached Contention Regarding NEPA Requirement to

Address Safety and Environmental Implications of the Fukushima Task Force Report

("Contention").

This petition is captioned in both the rulemaking docket and the docket for the

Watts Bar Unit 2 licensing proceeding because it seeks relief that is both generic and

applicable to the individual proceeding. The rulemaking petition is also being filed by

other organizations and individuals who have submitted contentions regarding the safety and environmental implications of the NRC's report entitled <u>Recommendations for Enhancing Reactor Safety in the 21st Century: The Near-Term Task Force Review of Insights from the Fukushima Dai-ichi Accident at 20-21</u> (July 12, 2011) ("Task Force Report").

II.   DISCUSSION

   A. General Solution

The general solution sought by SACE is to rescind all regulations in 10 C.F.R. Part 51 to the extent that they reach generic conclusions about the environmental impacts of severe reactor and/or spent fuel pool accidents and therefore prohibit consideration of those impacts in reactor licensing proceedings. These regulations include 10 C.F.R. Part 51, Appendix B; 10 C.F.R. §§ 51.45, 51.53, and 51.95.

   B. SACE's Grounds for and Interest in the Action Requested.

SACE seeks rescission of any NRC regulations that would prevent the NRC from complying with its obligation under the National Environmental Policy Act ("NEPA") and NRC implementing regulations to consider, in the operating license proceeding for Watts Bar Unit 2, the environmental implications of new and significant information discussed in the Task Force Report regarding the regulatory implications of the Fukushima Dai-ichi nuclear accident. Our legal and technical grounds for seeking consideration of new and significant information in the Task Force Report are discussed at length in the attached Contention, which is attached and incorporated herein by reference.

   C. Support for Petition

2

This petition for rulemaking is supported by the Task Force Report and also by the attached Declaration of Dr. Arjun Makhijani (August 8, 2011). As demonstrated in both of those documents, the Fukushima accident has significant regulatory implications with respect to both severe reactor accidents and spent fuel pool accidents, because the Task Force Report recommends that mitigative measures for both of these types of accidents, which are not currently included in the design basis for nuclear reactors, should be added to the design basis and subject to mandatory safety regulation.

D. Request for Suspension of Licensing Proceeding

As discussed in the attached Contention, NEPA requires that agencies consider the environmental impacts of their actions *before* they are taken, in order to ensure that "important effects [of the licensing decision] will not be overlooked or underestimated only to be discovered after resources have been committed or the die otherwise cast." *Robertson,* 490 U.S. 332, 349 (1989). *See also* 40 C.F.R. §§ 1500.1(c), 1502.1, 1502.14. The NRC's obligation to comply with NEPA in this respect is independent of and in addition to the NRC's responsibilities under the Atomic Energy Act, and must be enforced to the "fullest extent possible." *Calvert Cliffs Coordinating Committee,* 449 F.2d at 1115. *See also Limerick Ecology Action v. NRC,* 869 F.2d 719, 729 (3rd Cir. 1989) (citing *Public Service Co. of New Hampshire v. NRC,* 582 F.2d 77, 86 (1st Cir. 1978)). The NRC's obligation to delay licensing decisions until after it has considered the environmental impacts of those decisions is also nondiscretionary. *Silva v. Romney,* 473 F.2d 287, 292 (1st Cir. 1973). Therefore the NRC has a non-discretionary duty to suspend the Watts Bar Unit 2 operating license proceeding while it considers the

3

environmental impacts of that decision, including the environmental implications of the

Task Force Report with respect to severe reactor and spent fuel pool accidents.

III.   CONCLUSION

For the foregoing reasons, the Commission should grant this rulemaking petition.

Respectfully submitted,

Signed (electronically) by:
Diane Curran
Harmon, Curran, Spielberg & Eisenberg, L.L.P.
1726 M Street N.W. Suite 600
Washington, D.C. 20036
202-328-3500
Fax: 202-328-6918
E-mail: dcurran@harmoncurran.com

August 11, 2011

4

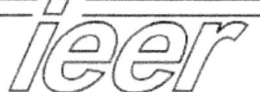

**INSTITUTE FOR ENERGY AND
ENVIRONMENTAL RESEARCH**

6935 Laurel Avenue, Suite 201
Takoma Park, MD 20912

Phone: (301) 270-5500
FAX: (301) 270-3029
e-mail: ieer@ieer.org
http://www.ieer.org

DECLARATION OF DR. ARJUN MAKHIJANI
REGARDING SAFETY AND ENVIRONMENTAL SIGNIFICANCE OF
NRC TASK FORCE REPORT REGARDING LESSONS LEARNED FROM
FUKUSHIMA DAIICHI NUCLEAR POWER STATION ACCIDENT[1]

I, Arjun Makhijani, declare as follows:

Introduction and Statement of Qualifications

1.   I am President of the Institute for Energy and Environmental Research ("IEER") in
Takoma Park, Maryland. Under my direction, IEER produces technical studies on a wide range
of energy and environmental issues to provide advocacy groups and policy makers with sound
scientific information and analyses as applied to environmental and health protection and for the
purpose of promoting the understanding and democratization of science. A copy of my
curriculum vita is attached.

2.   I am qualified by training and experience as an expert in the fields of plasma physics,
electrical engineering, nuclear engineering, the health effects of radiation, radioactive waste
management and disposal (including spent fuel), estimation of source terms from nuclear
facilities, risk assessment, energy-related technology and policy issues, and the relative costs and
benefits of nuclear energy and other energy sources. I am the principal author of a report on the
1959 accident at the Sodium Reactor Experiment facility near Simi Valley in California,
prepared as an expert report for litigation involving radioactivity emissions from that site. I am
also the principal author of a book, *The Nuclear Power Deception: U.S. Nuclear Mythology from
Electricity "Too Cheap to Meter" to "Inherently Safe' Reactors"* (Apex Press, New York, 1999,
co-author, Scott Saleska), which examines, among other things, the safety of various designs of
nuclear reactors.

3.   I have written or co-written a number of other books, reports, and publications analyzing
the safety, economics, and efficiency of various energy sources, including nuclear power. I am
also the author of *Securing the Energy Future of the United States: Oil, Nuclear and Electricity*

---

[1] Task Force Review (*Recommendations for Enhancing Reactor Safety in the 21st Century: The Near Term Task
Force Review of Insights from the Fukushima Dai ichi Accident*, Nuclear Regulatory Commission, July 12, 2011, at
http://pbadupws.nrc.gov/docs/ML1118/ML111861807.pdf)

Vulnerabilities and a Post September 11, 2001 Roadmap for Action (Institute for Energy and Environmental Research, Takoma Park, Maryland, December 2001). In 2004, I wrote "Atomic Myths, Radioactive Realities: Why nuclear power is a poor way to meet energy needs," *Journal of Land, Resources, & Environmental Law,* v. 24, no. 1 at 61-72 (2004). The article was adapted from an oral presentation given on April 18, 2003, at the Eighth Annual Wallace Stegner Center Symposium entitled, "Nuclear West: Legacy and Future," held at the University of Utah S.J. Quinney College of Law. In 2008, I prepared a report for the Sustainable Energy & Economic Development (SEED) Coalition entitled *Assessing Nuclear Plant Capital Costs for the Two Proposed NRG Reactors at the South Texas Project Site.*

4.   I am generally familiar with the basic design and operation of U.S. nuclear reactors and with the safety and environmental risks they pose. I am also generally familiar with materials from the press, the Japanese government, the Tokyo Electric Power Company, the French government safety authorities, and the U.S. Nuclear Regulatory Commission ("NRC") regarding the Fukushima Daiichi (hereafter Fukushima) accident and its potential implications for the safety and environmental protection of U.S. reactors. I have also read *Recommendations for Enhancing Reactor Safety in the 21st Century: The Near Term Task Force Review of Insights from the Fukushima Dai ichi Accident,* July 12, 2011 (hereafter the "Task Force Review"), published by the NRC.

5.   On April 19, 2011, I prepared a declaration stating my opinion that although the causes, evolution, and consequences of the Fukushima accident were not yet fully clear a month after the accident began, it was already presenting new and significant information regarding the risks to public health and safety and the environment posed by the operation of nuclear reactors. My declaration was submitted to the NRC by numerous individuals and environmental organizations in support of a legal petition to suspend licensing decisions while the NRC investigated the regulatory implications of the Fukushima accident. Emergency Petition to Suspend All Pending Reactor Licensing Decisions and Related Rulemaking Decisions Pending Investigation of Lessons learned From Fukushima Daiichi Nuclear Power Station Accident (April 14-18, 2011). In my declaration I also stated my belief that the integration of new information from the Fukushima accident into the NRC's licensing process could affect the outcome of safety and environmental analyses for reactor licensing and relicensing decisions by resulting in the denial of licenses or license extensions or the imposition of new conditions and/or new regulatory requirements. I also expressed the opinion that the new information could also affect the NRC's evaluation of the fitness of new reactor designs for certification. *Id.,* par. 5.

Purpose

6.   The purpose of my declaration is to explain why the Task Force Review provides further support for my opinions that the Fukushima accident presents new and significant information regarding the risks to public health and safety and the environment posed by the operation of nuclear reactors and that the integration of this new information into the NRC's licensing process could affect the outcome of safety and environmental analyses for reactor licensing and relicensing decisions and the NRC's evaluation of the fitness of new reactor designs for certification.

2

Agreement With Task Force Review's Conclusions Regarding Need to Expand Design Basis

7.   In my opinion, the Task Force reasonably concludes that substantial revisions to the very framework of NRC regulations are needed to adequately protect public health and the environment. I also agree that a major overarching step that needs to be taken is to integrate into the design basis for NRC safety requirements an expanded list of severe accidents and events, based on current scientific understanding and evaluations. This would ensure that potential mitigation measures are evaluated on the basis of whether they are needed for safety and not whether they are merely desirable. Should the NRC fail to incorporate an expanded list of severe accident requirements in the design basis of reactors, then a conclusion that the design provides for adequate protection to the public against severe accident risks could not be justified. The necessity for an expanded list of design basis requirements should be viewed in light of the Fukushima experience and the nuclear accident experience which preceded Fukushima, including Three Mile Island and Chernobyl accidents. Specifically, adequate protection of the public is incompatible with the NRC's continued reliance on voluntary evaluation of severe external and internal events, voluntary adoption of mitigation measures, or the use of cost-benefit analysis to evaluate their desirability.

8.   I believe my opinion is consistent with the Task Force's statement that:

> Adequate protection has been, and should continue to be, an evolving safety standard supported by new scientific information, technologies, methods, and operating experience. This was the case when new information about the security environment was revealed through the events of September 11, 2001. Licensing or operating a nuclear power plant with no emergency core cooling system or without robust security protections, while done in the past, would not occur under the current regulations. As new information and new analytical techniques are developed, safety standards need to be *reviewed, evaluated, and changed, as necessary, to insure that they continue to address the NRC's requirements to provide reasonable assurance of adequate protection of public health and safety. The Task Force believes, based on its review of the information currently available from Japan and the current regulations, that the time has come for such change.* [p. 18, italics added]

9.   I am concerned that over the past three decades or more, the NRC has not conducted the type of review of the adequacy of its safety regulations that is necessary to update its requirements so as to ensure that NRC safety requirements will provide the minimum level of protection required by the Atomic Energy Act. For instance, the Task Force Review points out that, over 30 years ago, the Rogovin Commission recommended that the scope of the design basis should be expanded to include a greater range of severe accidents. The Rogovin Commission explicitly stated that "[m]odification is definitely needed in the current philosophy that there are some accidents ("Class Nine accidents") [2] so unlikely that reactor designs need not

---

[2] Class Nine accidents are now called "severe accidents." (Task Force Review p. 16)

3

provide for mitigating their consequences."[3]. This recommendation was effectively disregarded by the NRC. Instead of imposing and enforcing mandatory requirements for prevention and mitigation of severe accidents, the NRC accepted voluntary measures and the use of cost-benefit assessments by licensees to exclude requirements for a range of preventive or mitigative measures. As a result the Task Force Review concluded that despite including some requirements for beyond-design-basis accidents, "the NRC *has not made fundamental changes to the regulatory approach for beyond design basis events and severe accidents* for operating reactors." (p. 17, italics added). Even the installation of hardened vents on Mark I and Mark II BWRs was left to the voluntary discretion of the licensees. Given the NRC's failure to make the needed changes in its basic regulatory requirements for safety since the Rogovin Commission report was issued over thirty years ago, and in light of the disastrous consequences of the Fukushima accident, which continues nearly five months after it started, I consider the current inadequacies in the NRC's program for regulation of basic reactor safety to be extraordinarily grave problems.

Potential Effects of Task Force Review on Environmental Analyses for New Reactors, Existing Reactor License Renewal, and Standardized Design Certification

10.   If the Task Force's recommendation to incorporate severe accidents into the design basis for NRC safety requirements is considered in environmental analyses for reactor licensing decisions or standardized design certifications, I think it would have very significant effects on the outcome of those analyses, in three key respects. First, the environmental analysis would have to consider the implication of the Task Force Review that compliance with current NRC safety requirements does not adequately protect public health and safety from severe accidents and their environmental effects. Second, for reactors that are unable to comply with new mandatory requirements, it could result in the denial of licenses. Third, the cost of adopting mandatory measures necessary to significantly improve the safety of currently operating reactors and proposed new reactors is likely to be significant.

Change to Estimate of Environmental Risk

11.   An analysis of the environmental implications of the Task Force Review would have to consider the ramifications of the Task Force's implicit conclusion that compliance with current NRC safety standards does not adequately protect public health and safety from severe accidents and their environmental effects. For instance, the Task Force Review indicates that seismic and flooding risks as well as risks of seismically-induced fires and floods may be greater than previously understood by the NRC in some cases. Therefore in its environmental analyses, the NRC would have to revise its analysis to reflect the new understanding that the risks and radiological impacts of accidents are greater than previously thought.

Potential Denial of License Applications Based on Environmental Risk Analyses

12.   The Task Force Review implicitly raises the potential that some reactors will be unable to

---

[3] Rogovin Commission report (*Three Mile Island. A Report to the Commissioners and to the Public*, by Mitchell Rogovin and George T. Frampton, et al. NUREG/CR 1250 1980. (Rogovin, Stern & Huge, Washington, DC, January 1980), v. 1, p. 151

4

comply with new mandatory requirements, thus resulting in the denial of licenses. For instance, this would be the case if a reactor cannot be adequately backfitted to comply with present-day assessment of ground shaking induced by earthquakes. Similarly, multi-unit siting may not be allowed in certain cases due to the impracticality of meeting upgraded emergency management requirements.

Significant Changes to Cost-Benefit Analyses

13.   The cost of adopting mandatory measures necessary to significantly improve the safety of currently operating reactors and proposed new reactors is likely to be significant. Adoption of a coherent regulatory framework as recommended by the Task Force, including periodic reassessments of whether the design basis is up to date with scientific assessments of flooding and seismic threats, is likely to result in significantly increased costs for nuclear reactors.

14.   The Task Force Review contains numerous recommendations for consideration of new mandatory requirements for increasing the capability of the reactors, equipment, and personnel to handle and to respond to a range of severe accidents. Adoption of such measures could have high costs. This, in turn, will affect the overall cost-benefit analysis for reactors, especially the comparisons of nuclear power with alternative sources of electricity. Examples of potentially significant costs if severe accident mitigation measures are adopted follow in paragraphs 15 through 24 below:

15.   If the Task Force recommendations are adopted, all existing reactors will be required to make changes to extend their capacity to handle station blackouts. This design upgrade is likely to have significant costs.

16.   Similar considerations apply to new reactor combined construction and operating license applications. For instance, the Task Force recommends adding station blackout requirements to the Advanced Boiling Water Reactor, which would also likely result in increased costs. (p. 72).

17.   Even where the Task Force deems some narrow issues to be already resolved by COL (combined license) applications and/or design certification applications, the interplay of other Task Force recommendations may raise environmental issues and cost concerns. For instance, while the Task Force found that the AP1000 and ESBWR designs already have a 72-hour provision for passive emergency core cooling, thereby satisfying the design requirement recommendations for station blackouts (pp. 71-72), other statements in the Task Force Review indicate the existence of environmental concerns that should be addressed in an EIS. For instance, the Task Force recommendations relating to the provision of backup power during the time beyond 72 hours relate mainly to prepositioning equipment offsite (Recommendation 4.1, p. 38) and therefore were regarded as not relevant to AP1000 and ESBWR design certifications but only to the COL process (p. 72). However, in the context of emergency preparedness, the Task Force Review notes that "[i]n the case of large natural disasters such as earthquakes, hurricanes, and floods, the phenomena challenging the plant will also have affected the local community. In these cases, *prearranged resources may not be available because of their inability to reach the plant site….*" (p. 60, italics added). Therefore the designs of the AP1000 and the ESBWR need to be reviewed in the context of their ability to mitigate the environmental impacts of station

5

blackout lasting more than 72 hours. The potential for destruction of infrastructure that would prevent prestaged offsite equipment from reaching the site would also needs to be taken into account in environmental analyses for COLs and license extension applications.

18.   Similarly, while the Task Force concludes that COL and Early Site Permit (ESP) applications already satisfy Recommendation 2.1 with respect to analysis of seismic and flooding risks (p. 71), it does not appear that all of the seismic and flooding-related implications of the Review have been addressed. Specifically, the flooding and fires that may be induced by earthquakes was closed by the NRC without imposing new requirements; the Task Force Review recommends reopening this issue (p. 32). These are issues that combine site characteristics and reactor design. For instance, the passive cooling features of AP1000s and ESBWRs involve pools of water located above the reactors. In addition, the ESBWR design has a buffer spent fuel pool in roughly the same position relative to the reactor as the Mark I design reactors (i.e., above the reactor vessel). Hence it is important to revisit this issue for these two reactor designs since they may be built at seismically active sites, including in the central and eastern United States (see paragraph 22 below), where there are active COL applications pending.

19. In the context of existing reactors, the Task Force Review recommends incorporating the latest understanding of seismic impacts and flooding (Recommendation 2, p. 30), and reopening the issue seismically induced flooding and fires (Recommendation 3, p. 32). This reassessment may also involve increased costs due to required backfits.

20.   Taken as a whole, the Task Force Review's recommendations implicitly call for a review of all new reactor design certifications regarding station blackout (SBO) arrangements, including mitigation measures for SBO events that extend beyond 72 hours and spent fuel pool instrumentation and make up water supply capability. The effects of seismically induced flooding and fires on spent fuel pool arrangements should also be reviewed. All of these reviews could result in the imposition of costly prevention or mitigation measures, affecting comparisons with the alternatives.

21..   In view of the events leading to the hydrogen explosions in Units 1, 3, and 4 at Fukushima, the reliability of the existing hardened vent system in Mark I and Mark II reactors has been thrown into question. The Task Force Review recommends installation of *reliable* hardened vents in all Mark I and Mark II BWRs (Recommendation 5, p. 41). Because such vents have not yet been designed and tested, their costs are unknown. However, they are likely to be substantial. These costs must be determined and evaluated for NEPA purposes for all 23 Mark I reactors and all eight Mark II reactors.

22.   The recommended mandatory review of the flooding and seismic design basis of existing reactors to evaluate whether they meet the design basis safety requirements could result in greatly increased costs in some or many cases. The establishment of the Shoreline Fault just offshore the Diablo Canyon Power Plant and the Oceanside thrust in the area of the San Onofre Nuclear Generating Station provides examples of recent developments that could lead to large expenditures for restoring the design basis safety margins for these reactors. As a reflection of the uncertainty, Pacific Gas & Electric (PG&E), which owns Diablo Canyon has itself requested and obtained a delay of 52 months in its license extension application so that the necessary

6

seismic studies can be completed. Another example relates to seismic hazard assessments in the central and eastern United States. In that case, the NRC has concluded that "[u]pdates to seismic data and models indicate that estimates of the seismic hazard, at some operating nuclear power plant sites in the Central and Eastern United States, have increased."[4] The NRC does not have enough data at present to determine what, if any, backfits may be called for, but intends to use a cost-benefit approach in deciding whether they should be implemented. It specifically states that "[i]n order to progress with the Regulatory Analysis Stage, a comprehensive list of candidate plant backfits must be identified for subsequent value-impact analysis."[5] "Value-impact analysis" is the NRC's terminology for a cost-benefit analysis.[6] However, if backfitting for more severe earthquakes than were incorporated into the original design were *required* for safety rather than left to a cost-benefit analysis, the implications for comparison with the alternatives could be considerable for existing reactors in the Central and Eastern United States.

23.   The Task Force noted that the same concern applies to flooding hazards, where "the assumptions and factors that were considered in flood protection at operating plants vary. In some cases, the design basis does not consider the probable maximum flood (PMF)." (p. 29) Again, protection of reactors against updated flood hazards could involve significant costs, depending on the outcome of the updated evaluations.

24.   Finally, the Task Force Review points out the importance of considering mitigation measures associated with multi-unit events. Such events had not been considered before and therefore were assigned zero probability for all intents and purposes. The Task Force review recommends a revision of regulations to cover multi-unit events, for instance, to ensure adequate emergency core and spent fuel cooling for more than one unit at a time:

> As part of the revision to 10 CFR 50.63, the NRC should require that the *equipment* and personnel necessary to implement the minimum and extended coping strategies shall include *sufficient capacity to provide core and spent fuel pool cooling, and reactor cooling system and primary containment integrity for all units at a multiunit facility.* The staff should also make the appropriate revisions to the definitions of "station blackout" and "alternate ac source" in 10 CFR 50.2. [p. 39, italics added]

Because most new applicants for COLs, such as Vogtle 3 and 4, propose to locate the new units at sites that already have reactors, the entire basis of emergency response adequacy, station-blackout related requirements, and emergency core and spent fuel pool cooling needs to be

---

[4] *Implications of Updated Probabilistic Seismic Hazard Estimates in Central and Eastern United States on Existing Plants Safety/Risk Assessments*, Generic Issue 199 (GI 199), Nuclear Regulatory Commission, August 2010, at http://pbadupws.nrc.gov/docs/ML1002/ML100270639.pdf, p. 30

[5] GI 199 p. 30

[6] NRC guidelines require "that the value impact of an alternative be quantified as the "net value" (or "net benefit"). To the extent possible, all attributes, whether values or impacts, are quantified in monetary terms and added together (with the appropriate algebraic signs) to obtain the net value in dollars. The net value calculation is generally favored over other measures, such as a value impact ratio or internal rate of return (RWG 1996, Section III.A.2)." (*Regulatory Analysis Technical Evaluation Handbook: Final Report*, NUREG/BR 0184, Nuclear Regulatory Commission, Office of Nuclear Regulatory Research, January 1997, p. 5.2. Link at http://www.osti.gov/energycitations/product.biblio.jsp?osti_id_446391.

7

reconsidered for the total number of units proposed at the site. The design and cost implications could be significant and must be reconsidered and reevaluated.

Conclusions

25.   I agree with the conclusions of the Task Force that significant changes to the NRC's regulatory system are needed in order to ensure that the operation of new reactors and re-licensed existing reactors does not pose unacceptable safety and environmental risks to the public. In light of the disastrous and ongoing events at Fukushima since March 11, 2011, it is clear that the issues of public safety raised by the Task Force are exceptionally grave. I also believe that it is highly likely that consideration of the Task Force's conclusions and recommendations in environmental analyses for new reactor licensing, existing reactor re-licensing, and design certification rulemakings, would materially affect the outcome of many and possibly all those studies.

The facts presented above are true and correct to the best of my knowledge, and the opinions expressed therein are based on my best professional judgment.

Date: 8 August 2011

Dr. Arjun Makhijani

8

**INSTITUTE FOR ENERGY AND
ENVIRONMENTAL RESEARCH**

6935 Laurel Avenue, Suite 201
Takoma Park, MD 20912

Phone: (301) 270-5500
FAX: (301) 270-3029
e-mail: ieer@ieer.org
http://www.ieer.org

Curriculum Vita of Arjun Makhijani

Address and Phone:
Institute for Energy and Environmental Research
6935 Laurel Ave., Suite 201
Takoma Park, MD 20912
Phone: 301-270-5500
e-mail: arjun@ieer.org
Website: www.ieer.org

A recognized authority on energy issues, Dr. Makhijani is the author and co-author of numerous reports and books on energy and environment related issues, including two published by MIT Press. He was the principal author of the first study of the energy efficiency potential of the US economy published in 1971. He is the author of *Carbon Free and Nuclear Free: A Roadmap for U.S. Energy Policy* (2007).

In 2007, he was elected Fellow of the American Physical Society. He was named a Ploughshares Hero, by the Ploughshares Fund (2006); was awarded the Jane Bagley Lehman Award of the Tides Foundation in 2008 and the Josephine Butler Nuclear Free Future Award in 2001; and in 1989 he received The John Bartlow Martin Award for Public Interest Magazine Journalism of the Medill School of Journalism, Northwestern University, with Robert Alvarez. He has many published articles in journals and magazines as varied as *The Bulletin of the Atomic Scientists, Environment, The Physics of Fluids, The Journal of the American Medical Association*, and *The Progressive*, as well as in newspapers, including the *Washington Post*.

Dr. Makhijani has testified before Congress, and has appeared on ABC World News Tonight, the CBS Evening News, CBS 60 Minutes, NPR, CNN, and BBC, among others. He has served as a consultant on energy issues to utilities, including the Tennessee Valley Authority, the Edison Electric Institute, the Lawrence Berkeley Laboratory, and several agencies of the United Nations.

Education:

- Ph.D. University of California, Berkeley, 1972, from the Department of Electrical Engineering. Area of specialization: plasma physics as applied to controlled nuclear fusion. Dissertation topic: multiple mirror confinement of plasmas. Minor fields of doctoral study: statistics and physics.
- M.S. (Electrical Engineering) Washington State University, Pullman, Washington, 1967. Thesis topic: electromagnetic wave propagation in the ionosphere.
- Bachelor of Engineering (Electrical), University of Bombay, Bombay, India, 1965.

Current Employment:

- 1987-present: President and Senior Engineer, Institute for Energy and Environmental Research, Takoma Park, Maryland. (part-time in 1987).
- February 3, 2004-present, Associate, SC&A, Inc., one of the principal investigators in the audit of the reconstruction of worker radiation doses under the Energy Employees Occupational Illness Compensation Program Act under contract to the Centers for Disease Control and Prevention, U.S. Department of Health and Human Services.

Other Long-term Employment

- 1984-88: Associate Professor, Capitol College, Laurel, Maryland (part-time in 1988).
- 1983-84: Assistant Professor, Capitol College, Laurel, Maryland.
- 1977-79: Visiting Professor, National Institute of Bank Management, Bombay, India. Principal responsibility: evaluation of the Institute's extensive pilot rural development program.
- 1975-87: Independent consultant (see page 2 for details)
- 1972-74: Project Specialist, Ford Foundation Energy Policy Project. Responsibilities included research and writing on the technical and economic aspects of energy conservation and supply in the U.S.; analysis of Third World rural energy problems; preparation of requests for proposals; evaluation of proposals; and the management of grants made by the Project to other institutions.
- 1969-70: Assistant Electrical Engineer, Kaiser Engineers, Oakland California. Responsibilities included the design and checking of the electrical aspects of mineral industries such as cement plants, and plants for processing mineral ores such as lead and uranium ores. Pioneered the use of the desk-top computer at Kaiser Engineers for performing electrical design calculations.

Professional Societies:

- Institute of Electrical and Electronics Engineers and its Power Engineering Society
- American Physical Society (Fellow)
- Health Physics Society
- American Association for the Advancement of Science

Awards and Honors:

- The John Bartlow Martin Award for Public Interest Magazine Journalism of the Medill School of Journalism, Northwestern University, 1989, with Robert Alvarez
- The Josephine Butler Nuclear Free Future Award, 2001
- Ploughshares Hero, Ploughshares Fund, 2006
- Elected a Fellow of the American Physical Society, 2007, *"For his tireless efforts to provide the public with accurate and understandable information on energy and environmental issues"*
- Jane Bagley Lehman Award of the Tides Foundation, 2007/2008

2

Invited Faculty Member, Center for Health and the Global Environment, Harvard Medical School: Annual Congressional Course, *Environmental Change: The Science and Human Health Impacts*, April 18-19, 2006, Lecture Topic: An Update on Nuclear Power - Is it Safe?

Consulting Experience, 1975-1987
Consultant on a wide variety of issues relating to technical and economic analyses of alternative energy sources; electric utility rates and investment planning; energy conservation; analysis of energy use in agriculture; US energy policy; energy policy for the Third World; evaluations of portions of the nuclear fuel cycle.

Partial list of institutions to which I was a consultant in the 1975-87 period:

- Tennessee Valley Authority
- Lower Colorado River Authority
- Federation of Rocky Mountain States
- Environmental Policy Institute
- Lawrence Berkeley Laboratory
- Food and Agriculture Organization of the United Nations
- International Labour Office of the United Nations
- United Nations Environment Programme
- United Nations Center on Transnational Corporations
- The Ford Foundation
- Economic and Social Commission for Asia and the Pacific
- United Nations Development Programme

Languages: English, French, Hindi, Sindhi, and Marathi.

Reports, Books, and Articles (Partial list)

(Newsletter, newspaper articles, excerpts from publications reprinted in books and magazines or adapted therein, and other similar publications are not listed below)

Hower, G.L., and A. Makhijani, "Further Comparison of Spread-F and Backscatter Sounder Measurements," *Journal of Geophysical Research*, 74, p. 3723, 1969.

Makhijani, A., and A.J. Lichtenberg, *An Assessment of Energy and Materials Utilization in the U.S.A.*, University of California Electronics Research Laboratory, Berkeley, 1971.

Logan, B. G., A.J. Lichtenberg, M. Lieberman, and A. Makhijani, "Multiple-Mirror Confinement of Plasmas," *Physical Review Letters*, 28, 144, 1972.

Makhijani, A., and A.J. Lichtenberg, "Energy and Well-Being," *Environment*, 14, 10, June 1972.

Makhijani, A., A.J. Lichtenberg, M. Lieberman, and B. Logan, "Plasma Confinement in Multiple Mirror Systems. I. Theory," *Physics of Fluids*, 17, 1291, 1974.

3

A Time to Choose: America's Energy Future, final report of the Ford Foundation Energy Policy Project, Ballinger, Cambridge, 1974. One of many co-authors.

Makhijani, A., and A. Poole, *Energy and Agriculture in the Third World*, Ballinger, Cambridge, 1975.

Makhijani, A., *Energy Policy for the Rural Third World*, International Institute for Environment and Development, London, 1976.

Kahn, E., M. Davidson, A. Makhijani, P. Caeser, and S. Berman, *Investment Planning in the Energy Sector*, Lawrence Berkeley Laboratory, Berkeley, 1976.

Makhijani, A., "Solar Energy for the Rural Third World," *Bulletin of the Atomic Scientists*, May 1977.

Makhijani, A., "Energy Policy for Rural India," *Economic and Political Weekly*, 12, Bombay, 1977.

Makhijani, A., *Some Questions of Method in the Tennessee Valley Authority Rate Study*, Report to the Tennessee Valley Authority, Chattanooga, 1978.

Makhijani, A., *The Economics and Sociology of Alternative Energy Sources*, Economic and Social Commission for Asia and the Pacific, 1979.

Makhijani, A., *Energy Use in the Post Harvest Component of the Food Systems in Ivory Coast and Nicaragua*, Food and Agriculture Organization of the United Nations, Rome, 1982.

Makhijani, A., *Oil Prices and the Crises of Debt and Unemployment: Methodological and Structural Aspects*, International Labour Office of the United Nations, Final Draft Report, Geneva, April 1983.

Makhijani, A., and D. Albright, *The Irradiation of Personnel at Operation Crossroads*, International Radiation Research and Training Institute, Washington, D.C., 1983.

Makhijani, A., K.M. Tucker, with Appendix by D. White, *Heat, High Water, and Rock Instability at Hanford*, Health and Energy Institute, Washington, D.C., 1985.

Makhijani, A., and J. Kelly, *Target: Japan   The Decision to Bomb Hiroshima and Nagasaki*, July 1985, a report published as a book in Japanese under the title, Why Japan?, Kyoikusha, Tokyo, 1985.

Makhijani, A., *Experimental Irradiation of Air Force Personnel During Operation Redwing 1956*, Environmental Policy Institute, Washington, D.C., 1985.

Makhijani, A., and R.S. Browne, "Restructuring the International Monetary System," *World Policy Journal*, New York, Winter, 1985-86.

4

Makhijani, A., R. Alvarez, and B. Blackwelder, *Deadly Crop in the Tank Farm: An Assessment of Management of High Level Radioactive Wastes in the Savannah River Plant Tank Farm*, Environmental Policy Institute, Washington, D.C., 1986.

Makhijani, A., "Relative Wages and Productivity in International Competition," *College Industry Conference Proceedings*, American Society for Engineering Education, Washington, D.C., 1987.

Makhijani, A., *An Assessment of the Energy Recovery Aspect of the Proposed Mass Burn Facility at Preston, Connecticut*, Institute for Energy and Environmental Research, Takoma Park, 1987.

Makhijani, A., R. Alvarez, and B. Blackwelder, *Evading the Deadly Issues: Corporate Mismanagement of America's Nuclear Weapons Production*, Environmental Policy Institute, Washington, D.C., 1987.

Franke, B. and A. Makhijani, *Avoidable Death: A Review of the Selection and Characterization of a Radioactive Waste Repository in West Germany*, Health & Energy Institute, Washington, DC; Institute for Energy and Environmental Research, Takoma Park, November 1987.

Makhijani, A., *Release Estimates of Radioactive and Non Radioactive Materials to the Environment by the Feed Materials Production Center, 1951 85*, Institute for Energy and Environmental Research, Takoma Park, 1988.

Alvarez, R., and A. Makhijani, "The Hidden Nuclear Legacy," *Technology Review*, 91, 42, 1988.

Makhijani, A., Annie Makhijani, and A. Bickel, *Saving Our Skins: Technical Potential and Policies for the Elimination of Ozone Depleting Chlorine Compounds*, Environmental Policy Institute and Institute for Energy and Environmental Research, Takoma Park, 1988.

Makhijani, A., Annie Makhijani, and A. Bickel, *Reducing Ozone Depleting Chlorine and Bromine Accumulations in the Stratosphere: A Critique of the U.S. Environmental Protection Agency's Analysis and Recommendations*, Institute for Energy and Environmental Research and Environmental Policy Institute/Friends of the Earth, Takoma Park, 1989.

Makhijani, A., and B. Franke, *Addendum to Release Estimates of Radioactive and Non Radioactive Materials to the Environment by the Feed Materials Production Center, 1951 85*, Institute for Energy and Environmental Research, Takoma Park, 1989.

Makhijani, A., *Global Warming and Ozone Depletion: An Action Program for States*, Institute for Energy and Environmental Research, Takoma Park, 1989.

Makhijani, A., *Managing Municipal Solid Wastes in Montgomery County*, Prepared for the Sugarloaf Citizens Association, Institute for Energy and Environmental Research, Takoma Park, 1990.

Saleska, S., and A. Makhijani, *To Reprocess or Not to Reprocess: The Purex Question   A Preliminary Assessment of Alternatives for the Management of N Reactor Irradiated Fuel at the*

5

U.S. Department of Energy's Hanford Nuclear Weapons Production Facility, Institute for Energy and Environmental Research, Takoma Park, 1990.

Makhijani, A., "Common Security is Far Off," *Bulletin of the Atomic Scientists*, May 1990.

Makhijani, A., *Draft Power in South Asian Agriculture: Analysis of the Problem and Suggestions for Policy*, prepared for the Office of Technology Assessment, Institute for Energy and Environmental Research, Takoma Park, 1990.

Mehta, P.S., S.J. Mehta, A.S. Mehta, and A. Makhijani, "Bhopal Tragedy's Health Effects: A Review of Methyl Isocyanate Toxicity," *JAMA* 264, 2781, December 1990.

Special Commission of International Physicians for the Prevention of Nuclear War and the Institute for Energy and Environmental Research, *Radioactive Heaven and Earth: The Health and Environmental Effects of Nuclear Weapons Testing In, On, and Above the Earth*, Apex Press, New York, 1991. One of many co-authors.

Makhijani, A., and S. Saleska, *High Level Dollars Low Level Sense: A Critique of Present Policy for the Management of Long Lived Radioactive Waste and Discussion of an Alternative Approach*, Apex Press, New York, 1992.

Makhijani, A., *From Global Capitalism to Economic Justice: An Inquiry into the Elimination of Systemic Poverty, Violence and Environmental Destruction in the World Economy*, Apex Press, New York, 1992.

Special Commission of International Physicians for the Prevention of Nuclear War and the Institute for Energy and Environmental Research, *Plutonium: Deadly Gold of the Nuclear Age*, International Physicians Press, Cambridge, MA, 1992. One of several co-authors.

Makhijani, A., "Energy Enters Guilty Plea," *Bulletin of the Atomic Scientists*, March/April 1994.

Makhijani, A., "Open the Files," *Bulletin of the Atomic Scientists*, Jan./Feb. 1995.

Makhijani, A., " 'Always' the Target?" *Bulletin of the Atomic Scientists*, May/June 1995.

Makhijani, A., and Annie Makhijani, *Fissile Materials in a Glass, Darkly: Technical and Policy Aspects of the Disposition of Plutonium and Highly Enriched Uranium*, IEER Press, Takoma Park, 1995.

Makhijani, A., and K. Gurney, *Mending the Ozone Hole: Science, Technology, and Policy*, MIT Press, Cambridge, MA, 1995.

Makhijani, A., H. Hu, K. Yih, eds., *Nuclear Wastelands: A Global Guide to Nuclear Weapons Production and the Health and Environmental Effects*, MIT Press, Cambridge, MA, 1995.

6

Zerriffi, H., and A. Makhijani, *The Nuclear Safety Smokescreen: Warhead Safety and Reliability and the Science Based Stockpile Stewardship Program*, Institute for Energy and Environmental Research, Takoma Park, May 1996.

Zerriffi, H., and A. Makhijani, "The Stewardship Smokescreen," *Bulletin of the Atomic Scientists*, September/October 1996.

Makhijani, A., *Energy Efficiency Investments as a Source of Foreign Exchange*, prepared for the International Energy Agency Conference in Chelyabinsk, Russia, 24-26 September 1996.

Makhijani, A., "India's Options," *Bulletin of the Atomic Scientists*, March/April 1997.

Ortmeyer, P. and A. Makhijani, "Worse than We Knew," *Bulletin of the Atomic Scientists*, November/December 1997.

Fioravanti, M., and A. Makhijani, *Containing the Cold War Mess: Restructuring the Environmental Management of the U.S. Nuclear Weapons Complex*, Institute for Energy and Environmental Research, Takoma Park, October 1997.

Principal author of three chapters in Schwartz, S., ed., *Atomic Audit: The Costs and Consequences of U.S. Nuclear Weapons Since 1940*, Brookings Institution, Washington, D.C., 1998.

Franke, B., and A. Makhijani, *Radiation Exposures in the Vicinity of the Uranium Facility in Apollo, Pennsylvania*, Institute for Energy and Environmental Research, Takoma Park, February 2, 1998.

Fioravanti, M., and A. Makhijani, *Supplement to Containing the Cold War Mess IEER's Response to the Department of Energy's Review*, Institute for Energy and Environmental Research, Takoma Park, March 1998.

Makhijani, A., "A Legacy Lost," *Bulletin of the Atomic Scientists*, July/August 1998.

Makhijani, A., and Hisham Zerriffi, *Dangerous Thermonuclear Quest: The Potential of Explosive Fusion Research for the Development of Pure Fusion Weapons*, Institute for Energy and Environmental Research, Takoma Park, July 1998.

Makhijani, A., and Scott Saleska, *The Nuclear Power Deception U.S. Nuclear Mythology from Electricity "Too Cheap to Meter" to "Inherently Safe" Reactors*, Apex Press, New York, 1999.

Makhijani, A., "Stepping Back from the Nuclear Cliff," *The Progressive*, vol. 63, no. 8, August 1999.

Makhijani, A., Bernd Franke, and Hisham Zerriffi, *Preliminary Partial Dose Estimates from the Processing of Nuclear Materials at Three Plants during the 1940s and 1950s*, Institute for Energy and Environmental Research, Takoma Park, September 2000. (Prepared under contract to the newspaper USA Today.)

7

Makhijani, A., and Bernd Franke, *Final Report of the Institute for Energy and Environmental Research on the Second Clean Air Act Audit of Los Alamos National Laboratory by the Independent Technical Audit Team*, Institute for Energy and Environmental Research, Takoma Park, December 13, 2000.

Makhijani, A., *Plutonium End Game: Managing Global Stocks of Separated Weapons Usable Commercial and Surplus Nuclear Weapons Plutonium*, Institute for Energy and Environmental Research, Takoma Park, January 2001.

Makhijani, A., Hisham Zerriffi, and Annie Makhijani, "Magical Thinking: Another Go at Transmutation," *Bulletin of the Atomic Scientists*, March/April 2001.

Makhijani, A., *Ecology and Genetics: An Essay on the Nature of Life and the Problem of Genetic Engineering.* New York: Apex Press, 2001.

Makhijani, A., "Burden of Proof," *Bulletin of the Atomic Scientists*, July/August 2001.

Makhijani, A., "Reflections on September 11, 2001," in Kamla Bhasin, Smitu Kothari, and Bindia Thapar, eds., *Voices of Sanity: Reaching Out for Peace*, Lokayan, New Delhi, 2001, pp. 59-64.

Makhijani, A., and Michele Boyd, *Poison in the Vadose Zone: An examination of the threats to the Snake River Plain aquifer from the Idaho National Engineering and Environmental Laboratory*, Institute for Energy and Environmental Research, Takoma Park, October 2001.

Makhijani, A., *Securing the Energy Future of the United States: Securing the Energy Future of the United States: Oil, Nuclear, and Electricity Vulnerabilities and a post September 11, 2001 Roadmap for Action*, Institute for Energy and Environmental Research, Takoma Park, November 2001.

Makhijani, A., and Sriram Gopal, *Setting Cleanup Standards to Protect Future Generations: The Scientific Basis of Subsistence Farmer Scenario and Its Application to the Estimation of Radionuclide Soil Action Levels (RSALs) for Rocky Flats*, Institute for Energy and Environmental Research, Takoma Park, December 2001.

Makhijani, A., "Some Factors in Assessing the Response to September 11, 2001," *Medicine and Global Survival*, International Physicians for the Prevention of Nuclear War, Cambridge, Mass., February 2002.

Makhijani, Annie, Linda Gunter, and A. Makhijani, *Cogema: Above the Law?: Concerns about the French Parent Company of a U.S. Corporation Set to Process Plutonium in South Carolina.* A report prepared by Institute for Energy and Environmental Research and Safe Energy Communication Council. Takoma Park, MD, May 7, 2002.

Deller, N., A. Makhijani, and J. Burroughs, eds., *Rule of Power or Rule of Law? An Assessment of U.S. Policies and Actions Regarding Security Related Treaties*, Apex Press, New York, 2003.

8

Makhijani, A., "Nuclear targeting: The first 60 years," *Bulletin of the Atomic Scientists*, May/June 2003.

Makhijani, A., "Strontium," *Chemical & Engineering News*, September 8, 2003.

Makhijani, A., and Nicole Deller, *NATO and Nuclear Disarmament: An Analysis of the Obligations of the NATO Allies of the United States under the Nuclear Non Proliferation Treaty and the Comprehensive Test Ban Treaty*, Institute for Energy and Environmental Research, Takoma Park, Maryland, October 2003.

Makhijani, A., *Manifesto for Global Democracy: Two Essays on Imperialism and the Struggle for Freedom*, Apex Press, New York, 2004.

Makhijani, A., "Atomic Myths, Radioactive Realities: Why nuclear power is a poor way to meet energy needs," *Journal of Land, Resources, & Environmental Law*, v. 24, no. 1, 2004, pp. 61-72. Adapted from an oral presentation given on April 18, 2003, at the Eighth Annual Wallace Stegner Center Symposium titled "Nuclear West: Legacy and Future," held at the University of Utah S.J. Quinney College of Law."

Makhijani, A., and Michele Boyd, *Nuclear Dumps by the Riverside: Threats to the Savannah River from Radioactive Contamination at the Savannah River Site*, Institute for Energy and Environmental Research, Takoma Park, Maryland, March 2004.

Makhijani, A., and Brice Smith, *The Role of E.I. du Pont de Nemours and Company (Du Pont) and the General Electric Company in Plutonium Production and the Associated I 131 Emissions from the Hanford Works*, Institute for Energy and Environmental Research, Takoma Park. Maryland, March 30, 2004.

Makhijani, A., Peter Bickel, Aiyou Chen, and Brice Smith, *Cash Crop on the Wind Farm: A New Mexico Case Study of the Cost, Price, and Value of Wind Generated Electricity*, Institute for Energy and Environmental Research, Takoma Park, Maryland, April 2004.

Makhijani, A., Lois Chalmers, and Brice Smith, *Uranium Enrichment: Just Plain Facts to Fuel an Informed Debate on Nuclear Proliferation and Nuclear Power*, Institute for Energy and Environmental Research, Takoma Park, Maryland, October 15, 2004.

Makhijani, A., and Brice Smith, *Costs and Risks of Management and Disposal of Depleted Uranium from the National Enrichment Facility Proposed to be Built in Lea County New Mexico by LES*, Institute for Energy and Environmental Research, Takoma Park, Maryland, November 24, 2004.

Makhijani, A., project director, *Examen critique du programme de recherche de l'ANDRA pour déterminer l'aptitude du site de Bure au confinement géologique des déchets à haute activité et à vie longue: Rapport final*, prepared for le Comité ocal d'Information et de Suivi; coordinator: Annie Makhijani; authors: Detlef Appel, Jaak Daemen, George Danko, Yuri Dublyansky, Rod Ewing, Gerhard Jentzsch, Horst Letz, Arjun Makhijani, Institute for Energy and Environmental Research, Takoma Park, Maryland, December 2004

9

Institute for Energy and Environmental Research, *Lower Bound for Cesium 137 Releases from the Sodium Burn Pit at the Santa Susana Field Laboratory*, IEER, Takoma Park, Maryland, January 13, 2005. (Authored by A. Makhijani and Brice Smith.)

Institute for Energy and Environmental Research, *Iodine 131 Releases from the July 1959 Accident at the Atomics International Sodium Reactor Experiment*, IEER, Takoma Park, Maryland, January 13, 2005. (Authored by A. Makhijani and Brice Smith.)

Makhijani, A., and Brice Smith. *Update to Costs and Risks of Management and Disposal of Depleted Uranium from the National Enrichment Facility Proposed to be Built in Lea County New Mexico by LES*. Institute for Energy and Environmental Research, Takoma Park, Maryland, July 5, 2005.

Makhijani, A., "A Readiness to Harm: The Health Effects of Nuclear Weapons Complexes," *Arms Control Today*, **35**, July/August 2005.

Makhijani, A., *Bad to the Bone: Analysis of the Federal Maximum Contaminant Levels for Plutonium 239 and Other Alpha Emitting Transuranic Radionuclides in Drinking Water*, Institute for Energy and Environmental Research, Takoma Park, Maryland, August 2005.

Makhijani, A., and Brice Smith, *Dangerous Discrepancies: Missing Weapons Plutonium in Los Alamos National Laboratory Waste Accounts*, Institute for Energy and Environmental Research, Takoma Park, Maryland, April 21, 2006.

Makhijani, Annie, and A. Makhijani, *Low Carbon Diet without Nukes in France: An Energy Technology and Policy Case Study on Simultaneous Reduction of Climate Change and Proliferation Risks*, Institute for Energy and Environmental Research, Takoma Park, Maryland, May 4, 2006.

Makhijani, Annie, and A. Makhijani. *Shifting Radioactivity Risks: A Case Study of the K 65 Silos and Silo 3 Remediation and Waste Management at the Fernald Nuclear Weapons Site*, Institute for Energy and Environmental Research, Takoma Park, Maryland, August 2006.

Smith, Brice, and A. Makhijani, "Nuclear is Not the Way," *Wilson Quarterly*, v.30, p. 64, Autumn 2006.

Makhijani, A., Brice Smith, and Michael C. Thorne, *Science for the Vulnerable: Setting Radiation and Multiple Exposure Environmental Health Standards to Protect Those Most at Risk*, Institute for Energy and Environmental Research, Takoma Park, Maryland, October 19, 2006.

Makhijani, A., *Carbon Free and Nuclear Free: A Roadmap for U.S. Energy Policy*, IEER Press, Takoma Park, Maryland; RDR Books, Muskegon, Michigan, 2007.

Makhijani, A., *Assessing Nuclear Plant Capital Costs for the Two Proposed NRG Reactors at the South Texas Project Site*, Institute for Energy and Environmental Research, Takoma Park, Maryland, March 24, 2008.

10

Makhijani, A., *Energy Efficiency Potential: San Antonio's Bright Energy Future,* Institute for Energy and Environmental Research, Takoma Park, Maryland, October 9, 2008.

Makhijani, A., *The Use of Reference Man in Radiation Protection Standards and Guidance with Recommendations for Change,* Institute for Energy and Environmental Research, Takoma Park, Maryland, December 2008.

Makhijani, A., *Comments of the Institute for Energy and Environmental Research on the U.S. Nuclear Regulatory Commission's Proposed Waste Confidence Rule Update and Proposed Rule Regarding Environmental Impacts of Temporary Spent Fuel Storage,* Institute for Energy and Environmental Research, Takoma Park, Maryland, February 6, 2009.

Makhijani, A., *Technical and Economic Feasibility of a Carbon Free and Nuclear Free Energy System in the United States,* Institute for Energy and Environmental Research, Takoma Park, Maryland, March 4, 2009.

Fundación Ideas para el Progreso, *A New Energy Model For Spain: Recommendations for a Sustainable Future* (originally: *Un nuevo modelo energético para España: Recomendaciones para un futuro sostenible*), by the Working Group of Foundation Ideas for Progress on Energy and Climate Change, Fundación Ideas , Madrid, May 20, 2009. Arjun Makhijani contributed Section 2.2. The cost of nuclear energy and the problem of waste.

Makhijani, A., *IEER Comments on the Nuclear Regulatory Commission's Rulemaking Regarding the "Safe Disposal of Unique Waste Streams Including Significant Quantities of Depleted Uranium,"* Institute for Energy and Environmental Research, Takoma Park, Maryland, October 30, 2009.

Makhijani, A., *The Mythology and Messy Reality of Nuclear Fuel Reprocessing,* Institute for Energy and Environmental Research, Takoma Park, Maryland, April 8, 2010.

CV updated October 11, 2010

11

Tennessee Valley Authority
1101 Market Street
Chattanooga, Tennessee 37402

Brenda E. Brickhouse
Vice President
Environmental Permitting & Compliance

January 27, 2012

Ms. Cindy Bladey
Chief, Rules Announcements and Directives Branch
Office of Administration
Mail Stop: TWB-05-B01M
U.S. Nuclear Regulatory Commission
Washington, DC 20555-0001

Dear Ms. Bladey:

DOCKET ID NRC-2008-0369 - COMMENTS ON DRAFT SUPPLEMENT 2 TO FINAL
ENVIRONMENTAL STATEMENT (SFES) RELATED TO THE OPERATION OF WATTS BAR
NUCLEAR PLANT, UNIT 2, TENNESSEE VALLEY AUTHORITY (TVA)

In response to the Notice of Availability published in the Federal Register Volume 76, No. 218
dated November 10, 2011, TVA has reviewed NUREG-0498, Final Environmental Statement 2,
Related to the Operation of Watts Bar Nuclear Plant (WBN), Unit 2 - Draft Report for Comment
(Draft SFES), dated October 2011.

TVA finds that the NRC's comprehensive assessment of the operation of WBN 2 contained in
the Draft SFES meets standards for an adequate environmental impact statement under
regulations of the Council on Environmental Quality. Our reviewers noted that, with the
incorporation of the enclosed comments, the information in the Draft SFES accurately
corresponds to the Environmental Report and supporting information provided in TVA's facility
operating license application. TVA appreciates the opportunity to review and provide the
enclosed comments on the Draft SFES.

Sincerely,

Brenda E. Brickhouse

ABH:DLS

Enclosure
cc: Gordon Arent, LP 5A-C
      Aaron B. Nix, BR 4A-C
      Diedre B. Nida, BR 4A-C
      Edward J. Vigluicci, WT 6A-K
      EDMC, BR 4A-C

Watts Bar Unit 2 Operation
Draft Final Environmental Statement (DFES) issued November 2011

TVA Comments

No.	Page No./ Line No.	Comment
1	xviii / 3	Insert space between "used" and "information"
2	2-101/10-12 and 18-21	As the DFES states, in 2008, TVA submitted its Final Supplemental Environmental Impact Statement (FSEIS) dated 2007, as an environmental report (ER) in the application. In the list of references for Chapter 2, "TVA 2007a" and "TVA 2008a" are both citations for the same FSEIS. In the DFES, TVA's document is referred to as TVA's FSEIS and TVA's ER. TVA suggests that a single reference for TVA's 2007 FSEIS would improve clarity for the reader.
3	1-5 / 8	Text indicates there is a 45-day comment period on the DFES. On page xix, line 8, text indicates a 75-day comment period.
4	2-17 / 18	Consider replacing or supplementing the statement, "During summer, gray bats are known to roost in two caves within 8 km (5 mi) from the WBN site," with the following more specific information from the FSEIS, Section 3.4.3, page 60, "Small numbers (less than 500) of gray bats continue to roost in a cave approximately 3.3 miles from the project."
5	2-69 / 13	In Section 2.4.2.7, Tax Revenues, consider adding information about TVA's Mitigation payments (see SEIS pg. 68, Section 3.8.7, second paragraph) under which TVA makes additional payments to local governments impacted by TVA activities during the period of constructing WBN. TVA notes that these mitigation payments are mentioned on pages 4-42 and 4-43 of the DFES. Adding a statement about mitigation payments in Section 2.4.2.7 would clarify that areas around WBN receive funds in addition to the annual in-lieu of tax payments.
6	2-88 / 7	The following statement is no longer accurate: "The [Watts Bar] fossil plant currently is not operating, but could be reactivated in the future." TVA demolished the Watts Bar Fossil plant in December 2011. Accordingly, TVA recommends this section be updated with information available in TVA's environmental assessment of Watts Bar Fossil Plant Deconstruction http://www.tva.com/environment/reports/wbf_deconstruction/index.htm

Watts Bar Unit 2 Operation
Draft Final Environmental Statement (DFES) issued November 2011

7	4-15 / 39-40	TVA recommends the following revision "While TVA does not conduct studies of avian mortality, no noticeable events of avian mortality associated with the existing transmission system have been recorded by TVA."
8	2-80/ 27	In the statement "... NPDES temperature limits for WBN outfalls to the Tennessee River are at or below 95°C, which...." the unit of measure should be changed to Fahrenheit.
9	3-13/ 21	In the statement "The WBN site is located on a 2.7-m (9-ft)- wide navigable channel...." "wide" should be replaced with "deep."

2

Appendix F

Key Consultation Correspondence Regarding the Watts Bar Nuclear Unit 2 Operating License

Appendix F

Key Consultation Correspondence Regarding the Watts Bar Nuclear Unit 2 Operating License

The Endangered Species Act of 1973, as amended, the Magnuson-Stevens Fisheries Conservation and Management Act of 1996, as amended; and the National Historic Preservation Act require that Federal agencies consult with applicable State and Federal agencies and groups prior to taking action that may affect threatened and endangered species, essential fish habitat, or historic and archaeological resources, respectively. This appendix contains consultation documentation.

Table F-1 provides a list of the consultation documents sent between the U.S. Nuclear Regulatory Commission (NRC) and other agencies. The NRC staff is required to consult with these agencies based on the National Environmental Policy Act of 1969 requirements.

Table F-1. Consultation Correspondences

Author	Recipient	Date of Letter/Email
U.S. Nuclear Regulatory Commission (J. Wiebe)	U.S. Fish and Wildlife Service (M. Jennings)	September 2, 2009 (ML092100088)
U.S. Nuclear Regulatory Commission (J. Wiebe)	Advisory Council on Historic Preservation (D. Klima)	September 10, 2009 (ML092120105)
U.S. Nuclear Regulatory Commission (J. Wiebe)	Cherokee Nation (R. Allen)	September 10, 2009 (ML092110475)
U.S. Nuclear Regulatory Commission (J. Wiebe)	Eastern Band of the Cherokee Indians (T. Howe)	September 10, 2009 (ML092110475)
U.S. Nuclear Regulatory Commission (J. Wiebe)	Eastern Band of the Cherokee Indians (R. Townsend)	September 10, 2009 (ML092110475)
U.S. Nuclear Regulatory Commission (J. Wiebe)	United Keetoowah Band Headquarters (L. Larue-Stopp)	September 10, 2009 (ML092110475)
U.S. Nuclear Regulatory Commission (J. Wiebe)	The Chickasaw Nation (V. (Gingy) Nail)	September 10, 2009 (ML092110475)

Table F-1. (contd)

Author	Recipient	Date of Letter/Email
U.S. Nuclear Regulatory Commission (J. Wiebe)	Choctaw Nation of Oklahoma (T. Cole)	September 10, 2009 (ML092110475)
U.S. Nuclear Regulatory Commission (J. Wiebe)	Choctaw Nation of Oklahoma (G. Pyle)	September 10, 2009 (ML092110475)
U.S. Nuclear Regulatory Commission (J. Wiebe)	Jena Band of Choctaw Indians (L. Strange)	September 10, 2009 (ML092110475)
U.S. Nuclear Regulatory Commission (J. Wiebe)	Muscogee (Creek) Nation of Oklahoma (J. Bear)	September 10, 2009 (ML092110475)
U.S. Nuclear Regulatory Commission (J. Wiebe)	Alabama-Coushatta Tribe of Texas (B. Battise)	September 10, 2009 (ML092110475)
U.S. Nuclear Regulatory Commission (J. Wiebe)	Alabama-Quassarte Tribal Town (A. Asbury)	September 10, 2009 (ML092110475)
U.S. Nuclear Regulatory Commission (J. Wiebe)	Kialegee Tribal Town (E. Bucktrot and G. Bucktrot)	September 10, 2009 (ML092110475)
U.S. Nuclear Regulatory Commission (J. Wiebe)	Thlopthlocco Tribal Town (C. Coleman)	September 10, 2009 (ML092110475)
U.S. Nuclear Regulatory Commission (J. Wiebe)	Absentee Shawnee Tribe of Oklahoma (K. Kaniatobe)	September 10, 2009 (ML092110475)
U.S. Nuclear Regulatory Commission (J. Wiebe)	Eastern Shawnee Tribe of Oklahoma (R. DuShane)	September 10, 2009 (ML092110475)
U.S. Nuclear Regulatory Commission (J. Wiebe)	Eastern Shawnee Tribe of Oklahoma (G.J. Wallace)	September 10, 2009 (ML092110475)
U.S. Nuclear Regulatory Commission (J. Wiebe)	Shawnee Tribe (R. Sparkman)	September 10, 2009 (ML092110475)
U.S. Nuclear Regulatory Commission (J. Wiebe)	Shawnee Tribe (B. Pryor)	September 10, 2009 (ML092110475)
U.S. Nuclear Regulatory Commission (J. Wiebe)	Tennessee Historical Commission (J.Y. Garrison)	September 10, 2009 (ML092120097)
U.S. Nuclear Regulatory Commission (J. Wiebe)	U.S. Army Corps of Engineers (R. Gatlin)	September 10, 2009 (ML092110147)
U.S. Nuclear Regulatory Commission (J. Wiebe)	Office of Environment Policy and Compliance, Department of Interior (G.L. Hogue)	September 10, 2009 (ML092110147)
U.S. Nuclear Regulatory Commission (J. Wiebe)	Sam Nunn Atlanta Federal Center (A.S. Meiburg and S. Gordon)	September 10, 2009 (ML092110147)

Table F-1. (contd)

Author	Recipient	Date of Letter/Email
U.S. Nuclear Regulatory Commission (J. Wiebe)	Tennessee Department of Environment and Conservation (M. Apple)	September 10, 2009 (ML092110147)
U.S. Nuclear Regulatory Commission (J. Wiebe)	Tennessee Department of Environment and Conservation (S. Baxter)	September 10, 2009 (ML092110147)
U.S. Nuclear Regulatory Commission (J. Wiebe)	Tennessee Department of Environment and Conservation (B. Bowen)	September 10, 2009 (ML092110147)
U.S. Nuclear Regulatory Commission (J. Wiebe)	Tennessee Department of Economic and Community Development (M. Atchinson)	September 10, 2009 (ML092110147)
U.S. Nuclear Regulatory Commission (J. Wiebe)	Commissioners Office, Tennessee Department of Transportation	September 10, 2009 (ML092110147)
U.S. Nuclear Regulatory Commission (J. Wiebe)	Environment and Planning Environmental Division (E. Cole)	September 10, 2009 (ML092110147)
U.S. Nuclear Regulatory Commission (J. Wiebe)	Tennessee Department of Agriculture (K. Givens)	September 10, 2009 (ML092110147)
U.S. Nuclear Regulatory Commission (J. Wiebe)	Tennessee Department of Environment and Conservation (P. Davis)	September 10, 2009 (ML092110147)
U.S. Nuclear Regulatory Commission (J. Wiebe)	Water Supply (R. Foster)	September 10, 2009 (ML092110147)
U.S. Nuclear Regulatory Commission (J. Wiebe)	Tennessee Department of Environment and Conservation (J. Fyke)	September 10, 2009 (ML092110147)
U.S. Nuclear Regulatory Commission (J. Wiebe)	Division of Radiological Health (L.E. Nanney)	September 10, 2009 (ML092110147)
U.S. Nuclear Regulatory Commission (J. Wiebe)	Tennessee Department of Environment and Conservation (B. Stephens)	September 10, 2009 (ML092110147)
U.S. Nuclear Regulatory Commission (J. Wiebe)	Tennessee Department of Environment and Conservation (M. Tummons)	September 10, 2009 (ML092110147)
U.S. Nuclear Regulatory Commission (J. Wiebe)	Groundwater (A. Schwendimann)	September 10, 2009 (ML092110147)
U.S. Nuclear Regulatory Commission (J. Wiebe)	Tennessee Wildlife Resource Agency (E. Carter)	September 10, 2009 (ML092110147)

Table F-1. (contd)

Author	Recipient	Date of Letter/Email
U.S. Nuclear Regulatory Commission (J. Wiebe)	Resource Management Division (A. Marshall)	September 10, 2009 (ML092110147)
U.S. Nuclear Regulatory Commission (J. Wiebe)	Commissioners Office Tennessee Department of Transportation	September 10, 2009 (ML092110147)
Tennessee Historical Commission (E.P. McIntyre)	U.S. Nuclear Regulatory Commission (J. Wiebe)	September 22, 2009 (ML093510985)
Eastern Band of Cherokee Indians (T. Howe)	U.S. Nuclear Regulatory Commission	September 29, 2009 (ML092860591)
U.S. Fish and Wildlife Services (M. Jennings)	U.S. Nuclear Regulatory Commission (J. Wiebe)	October 9, 2009 (ML092930182)
Tennessee Historic Commission (E. Patrick McIntyre, Jr.)	U.S. Nuclear Regulatory Commission (J. Wiebe)	March 5, 2010 (ML100770290)
U.S. Nuclear Regulatory Commission (S. Campbell)	Absentee Shawnee Tribe of Oklahoma (K. Kaniatobe)	November 1, 2011 (ML11301A320)
U.S. Nuclear Regulatory Commission (S. Campbell)	Tennessee Historical Commission (E.P. McIntyre, Jr)	November 2, 2011 (ML11304A040)
U.S. Nuclear Regulatory Commission (S. Campbell)	Advisory Council on Historic Preservation (R. Nelson)	November 2, 2011 (ML11305A245)
U.S. Nuclear Regulatory Commission (S. Campbell)	U.S. Fish and Wildlife Service (M. Jennings)	November 2, 2011 (ML11304A083)
U.S. Nuclear Regulatory Commission (S. Campbell)	U.S. Fish and Wildlife Service (M. Jennings)	November 2, 2011 (ML11304A083)
Tennessee Valley Authority (E. Freeman)	U.S. Nuclear Regulatory Commission	November 21, 2011 (ML11329A001)
Choctaw Nation of Oklahoma (I. Thompson)	U.S. Nuclear Regulatory Commission (P. Milano)	November 21, 2011 (ML12053A441)
U.S. Fish and Wildlife Service (M. Jennings)	U.S. Nuclear Regulatory Commission (S. Campbell)	December 20, 2011 (ML12004A167)
Chickasaw Nation (J. Keel)	U.S. Nuclear Regulatory Commission (P. Milano)	January 19, 2012 (ML12053A439)
U.S. Department of the Interior (J. Stanley)	U.S. Nuclear Regulatory Commission (J. Poole)	January 23, 2012 (ML12023A185)

Appendix F.1

Biological Assessment

U.S. Fish and Wildlife Service

Watts Bar Unit 2 Nuclear Power Plant

Rhea County, Tennessee

U.S. Nuclear Regulatory Commission Operating License Application
Docket No. 50-391

Gray Bat (*Myotis gresescens*)
Pink mucket (*Lampsilis abrupta*)
Eastern fanshell pearlymussel (*Cyprogenia stegaria*)
Rough pigtoe (*Pleurobema plenum*)
Dromedary pearlymussel (*Dromus dromas*)
Orangefoot pimpleback (*Plethobasus cooperianus*)
Snail darter (*Percina tanasi*)

September 2011

U.S. Nuclear Regulatory Commission
Rockville, Maryland

Acronyms

°C	degree(s) Celsius
°F	degree(s) Fahrenheit
ac	acre(s)
BA	Biological Assessment
Btu	British thermal units
Btu/hr	British thermal unit(s) per hour
CCW	condenser circulating water
cfs	cubic feet per second
cm	centimeter(s)
EIS	Environmental Impact Statement
ER	Environmental Report
ERCW	essential raw cooling water
ESA	Endangered Species Act
FES-CP	Final Environmental Statement related to the construction permit for WBN Units 1 and 2
FES-OL	Final Environmental Statement related to Operation
FSAR	Final Safety Analysis Report
ft	foot (feet)
FWS	U.S. Fish and Wildlife Service
gpm	gallon(s) per minute
ha	hectare(s)
hr	hour
in.	inch(es)
IPS	intake pumping station
km	kilometer(s)
kV	kilovolt(s)
L/s	liter(s) per second
m	meter(s)
m^3/s	cubic meter(s) per second
mi	mile(s)
MW(e)	megawatts electric
NPDES	National Pollutant Discharge Elimination System
NRC	U.S. Nuclear Regulatory Commission
ppm	parts per million
PWR	pressurized water reactor
RAI	request for additional information
RCW	raw cooling water
s	second(s)

SCCW   Supplemental Condenser Cooling Water
TRM   Tennessee River Mile
TVA   Tennessee Valley Authority
WBN   Watts Bar Nuclear

Contents

Figures

Tables

1.0 Introduction and Purpose

Under the Endangered Species Act of 1973, as amended, each Federal agency shall, in consultation with, and with the assistance of the Secretary of the Interior, the Secretary of Commerce, or the Secretary of Agriculture (as appropriate), ensure that any action authorized by such agency is not likely to jeopardize the continued existence of any endangered species or threatened species or result in the destruction or adverse modification of habitat of such species. Each agency shall use the best scientific and commercial data available. Each Federal agency requests of the Secretary information about whether any species that is listed or proposed to be listed may be present in the area of such proposed action. If the Secretary advises, based on the best scientific and commercial data available, that such species may be present, such agency shall conduct a biological assessment (BA) for the purpose of identifying any endangered species or threatened species that is likely to be affected by such action.

The Federal agency uses the the BA to determine whether formal consultation or a conference is required. If the BA indicates that there are no listed species or critical habitat present that are likely to be adversely affected by the action and the Director (U.S. Fish and Wildlife Service [FWS] regional director, or the appropriate authorized representative) concurs, then formal consultation is not required. If the BA indicates that the action is not likely to jeopardize the continued existence of proposed species or result in the destruction or adverse modification of proposed critical habitat, and the Director concurs, then a conference is not required. Note that the Director may use the results of the BA in (1) determining whether to request the Federal agency to initiate formal consultation or a conference, (2) formulating a biological opinion, or (3) formulating a preliminary biological opinion.

The U.S. Nuclear Regulatory Commission (NRC) is currently considering a request by the Tennessee Valley Authority (TVA) for an operating license for Watts Bar Nuclear (WBN) Unit 2, located on the northwest shore of Chickamauga Reservoir (on the Tennessee River) in Rhea County, Tennessee (see Figure 1-1). The site has two Westinghouse-designed pressurized-water reactors (PWRs). In early 1996, the NRC issued an operating license for WBN Unit 1. The TVA operates the WBN site. TVA has not yet completed WBN Unit 2. On August 3, 2007, TVA informed the NRC of its intention to complete construction activities at WBN Unit 2 under the existing construction permit (TVA 2007a). On March 4, 2009, TVA submitted to the NRC a request to reactivate its application for a license to operate a second light-water nuclear reactor at the WBN site (TVA 2008).

The NRC staff requested in a letter dated September 2, 2009 (NRC 2009) that the FWS provide information on Federally-listed endangered or threatened species, proposed or candidate species, and designated critical habitats that may occur in the vicinity of the WBN site. The FWS responded to NRC's request in a letter dated October 9, 2009 (FWS 2009), which provided a list of seven Federally listed threatened and endangered species near the WBN site.

This BA examines the potential impacts of the proposed actions on the seven Federally listed species within FWS's jurisdiction (see Table 1-1). The list included one mammal, the gray bat (*Myotis grisescens*); one fish, the snail darter (*Percina tanasi*); and five species of mussel. The mussels include the pink mucket (*Lampsilis abrupta*), the Eastern fanshell pearly mussel (*Cyrpogenia stegaria*), the rough pigtoe (*Pleurobema plenum*), the dromedary pearlymussel (*Dromus dromas*), and the orangefoot pimpleback (*Plethobasus cooperianus*). No critical habitat areas are designated near the WBN site. FWS indicated that the staff "should assess potential impacts and determine if the proposed project may affect these species."

On January 19, 2011, the sheepnose mussel (*Plethobasus cyphyus*) was proposed for listing (76 FR 3392). The sheepnose mussel occurs in the Southeast and the Midwest, but has been eliminated from two-thirds of the streams where it had been known to occur. The sauger is the only known host for the sheepnose mussel (FWS 2011). The sheepnose mussel is known to occur in the vicinity of the WBN site. In September 2010, TVA found a specimen, judged to be approximately 20 years old, during sampling (TVA 2011a).

Therefore, the NRC prepared this BA to support the draft supplemental final environmental statement related to the operating license for WBN Unit 2.

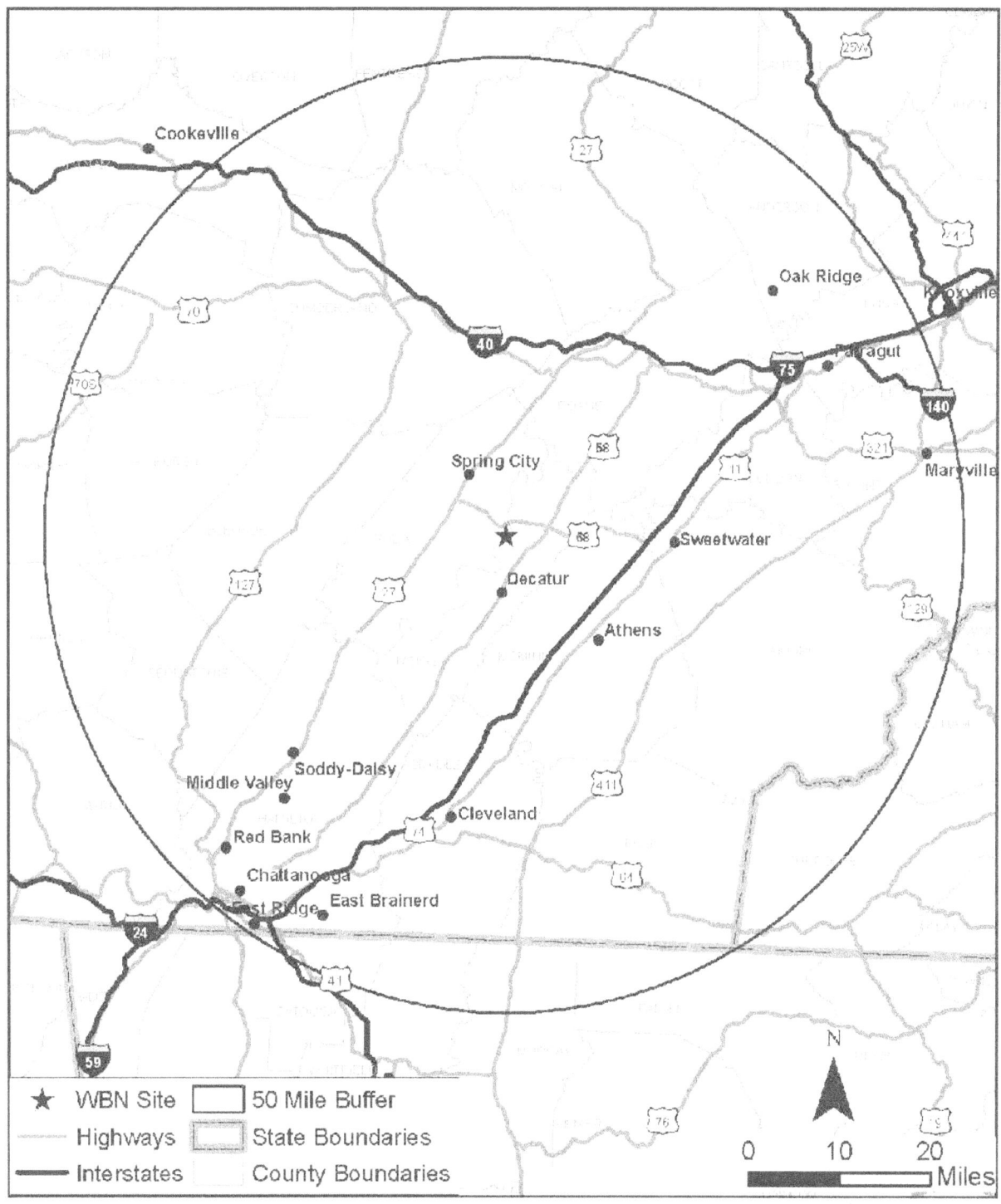

Figure 1-1. The WBN Site and the 80-km (50-mi) Vicinity

Table 1-1. Federally Listed Terrestrial Species Occurring in the Vicinity of the WBN Site

Scientific Name	Common Name	Federal Status
Terrestrial Species		
Mammals		
Myotis grisescens	gray bat	E
Aquatic Species		
Fish		
Percina tanasi	snail darter	T
Freshwater mussels		
Lampsilis abrupta	pink mucket	E
Cyprogenia stegaria	Eastern fanshell pearly mussel	E
Pleurobema plenum	rough pigtoe	E
Dromus dromas	dromedary pearlymussel	E
Plethobasus cooperianus	orange pimpleback	E

2.0 Proposed Action and History

The proposed action is for the NRC to issue an operating license for WBN Unit 2 at the WBN site.

WBN Units 1 and 2 possess a unique licensing history and regulatory framework. On May 14, 1971, TVA submitted a request for issuance of construction permits for WBN Units 1 and 2. TVA issued its Final Environmental Statement related to the construction permit for WBN Units 1 and 2 (FES-CP) in November 1972 (TVA 1972). The FES mentioned the bald eagle (*Haliaeetus leucocephalus*) as a relatively common visitor to the WBN area and addressed potential impacts on freshwater mussel species. On January 23, 1973, the Atomic Energy Commission issued Construction Permits for WBN Units 1 and 2.

In late 1976, TVA submitted an application requesting operating licenses for Units 1 and 2 (TVA 1976). Subsequently, on December 1, 1978, the NRC issued the 1978 Final Environmental Statement related to Operation (FES-OL), which evaluated operation of WBN Units 1 and 2 (NRC 1978). The 1978 FES-OL addressed the bald eagle and two endangered freshwater mussel species (pink mucket and dromedary pearly mussel). NRC concluded that operation of WBN would not affect these species (TVA 1972).

In 1994 following several construction delays, NRC determined that the units were nearing completion. In a letter dated April 1, 1995, NRC issued Supplement No. 1 to the 1978 FES-OL re-examining environmental considerations before issuance of an operating license for WBN Units 1 and 2 (NRC 1995). NRC entered into Section 7 consultation with FWS by submitting a BA, completed by TVA, to FWS on October 28, 1994. The BA included four species of freshwater mussel (i.e., pink mucket, dromedary pearly mussel, Eastern fanshell pearly mussel, and rough pigtoe), the snail darter, the bald eagle, and the gray bat. It also identified three additional aquatic species that FWS had designated as active candidates. TVA concluded that the operation of WBN Units 1 and 2 was not likely to affect individuals or populations of any of the listed species or candidate species or their critical habitats. NRC agreed with the "no effect" determination but requested a formal consultation. On January 25, 1995, NRC indicated that its staff and TVA had become aware of the existence of a fourth candidate species in the vicinity of the WBN site. In a biological opinion, FWS indicated that the action was not likely to jeopardize the continued existence of the listed species. TVA received the full power-operating license for Unit 1 on February 7, 1996.

As indicated in Section 1.0, TVA submitted an updated application on March 4, 2009 for a facility-operating license from NRC to possess, use, and operate WBN Unit 2 (TVA 2009a) and the NRC requested consultation with the FWS in a letter dated September 2, 2009 (NRC 2009).

3.0 WBN Site Description

TVA owns the 427 ha (1,055 ac) WBN site, located in southeastern Tennessee. The WBN site contains structures to support the operation of two nuclear units. WBN Unit 1 is currently operating and WBN Unit 2 is partially constructed. Figure 3-1 shows the layout of the site. A rural road, Morrison Lane, and forested land form the western border of the site, while TN-68 (also known as Watts Bar Highway) makes up the northern border. The WBN site is bounded by Chickamauga Reservoir (an impoundment of the Tennessee River) to the east and south of the site. The WBN site lies entirely within an unincorporated area of Rhea County, Tennessee, approximately 13 km (8 mi) southeast of Spring City.

TVA originally designed the WBN site as a two-unit PWR nuclear plant with a total electrical generating capacity of 2,540 megawatts (MWe). Unit 1 began operating in 1996. In addition to the reactors, the WBN site consists of two reactor containment buildings, a diesel generator building, a training facility, a turbine building, a service building, an intake pumping station, a water treatment plant, two cooling towers, 500-kV and 161-kV switchyards, and associated parking facilities. Figure 3-2 shows the reactor buildings and associated facility layout (NRC 1995). The United States owns the existing facilities at the WBN site, and TVA is the custodian (TVA 2008).

TVA terminated construction of Unit 2 in 1985 when the unit was 80 percent complete (TVA 2008). Since then, TVA has used many Unit 2 components to replace portions of Unit 1 and other TVA facilities. As a result, at the time of the operating license application, Unit 2 was approximately 60 percent complete. Completing Unit 2 may result in some additional ground-disturbing activities, but these activities would be mostly restricted to the existing disturbed portion of the property (TVA 2008). Because the facility (including the intakes and discharge systems used by Unit 1) was essentially completed, the only impacts that will affect aquatic and terrestrial biota include those from operations.

The original cooling system constructed for the WBN units was a closed-cycle system to transfer heat from the main condenser of each unit to the natural-draft cooling tower basin associated with that unit. In its 2008 environmental report (ER) (TVA 2008), TVA identified this system as the condenser circulating water (CCW) system. During normal plant operation, the CCW system for each unit would dissipate up to 7.8×10^9 Btu/hr of waste heat (TVA 1972, TVA 2009b). The Essential Raw Cooling Water (ERCW) system and the Raw Cooling Water (RCW) system remove additional heat from the plant components. Water from both of these systems discharges to the cooling tower basins for the CCW.

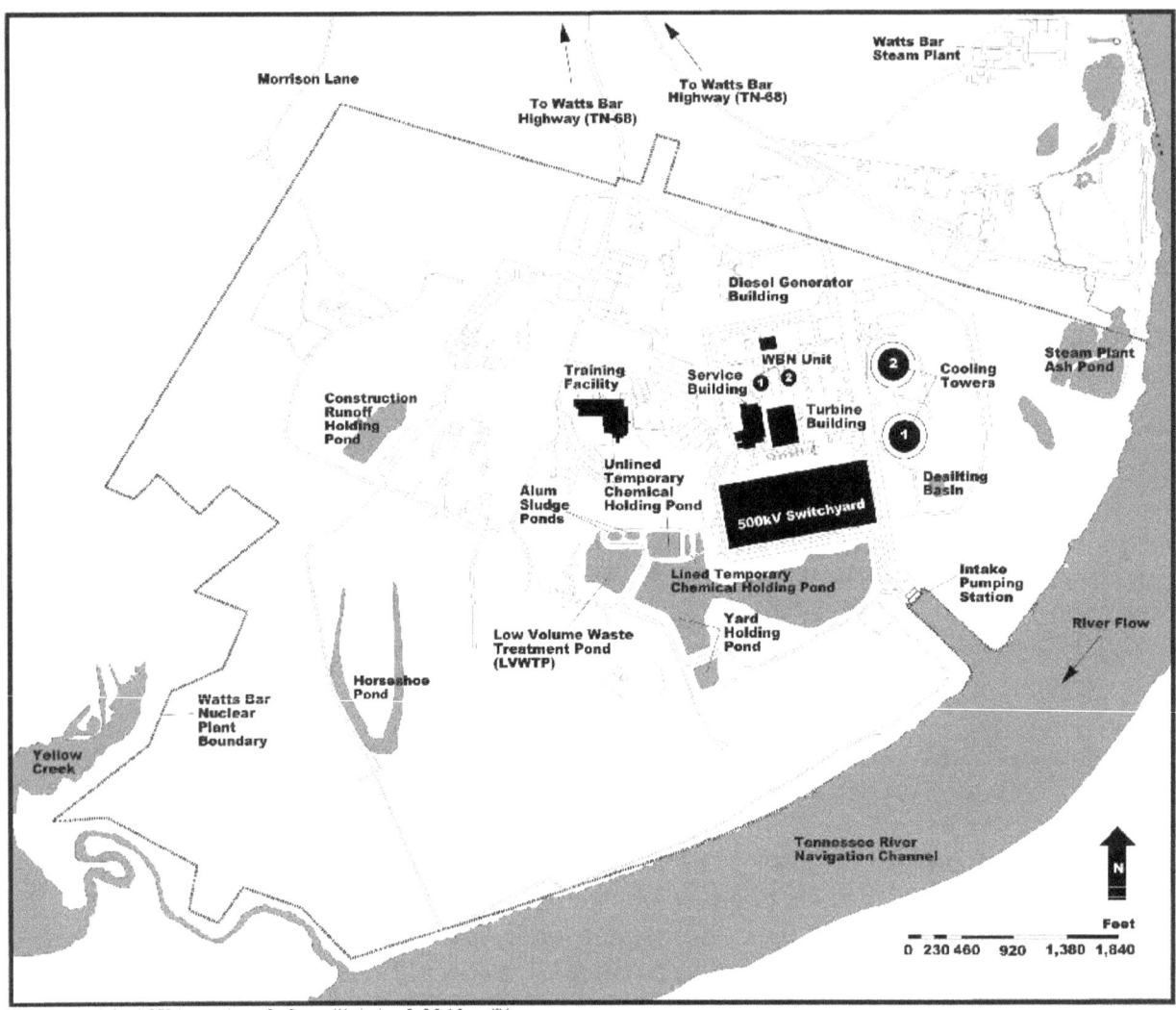

(To convert feet [ft] to meters [m], multiply by 0.3048 m/ft)

Figure 3-1. WBN Site (TVA 2008)

The WBN cooling water system uses natural-draft cooling towers to dissipate waste heat from the plant. Two single cooling towers, one for each unit, would serve the WBN site. Each tower is 108 m (354 ft) in diameter and 146 m (478 ft) high (TVA 1972). Most excess heat in the cooling water transfers to the atmosphere by evaporative and conductive cooling in the cooling tower. In addition to evaporative losses, a small percentage of water is lost in the form of droplets (drift) from the cooling tower. The water that does not evaporate or drift from the tower routes back to the cooling tower basin.

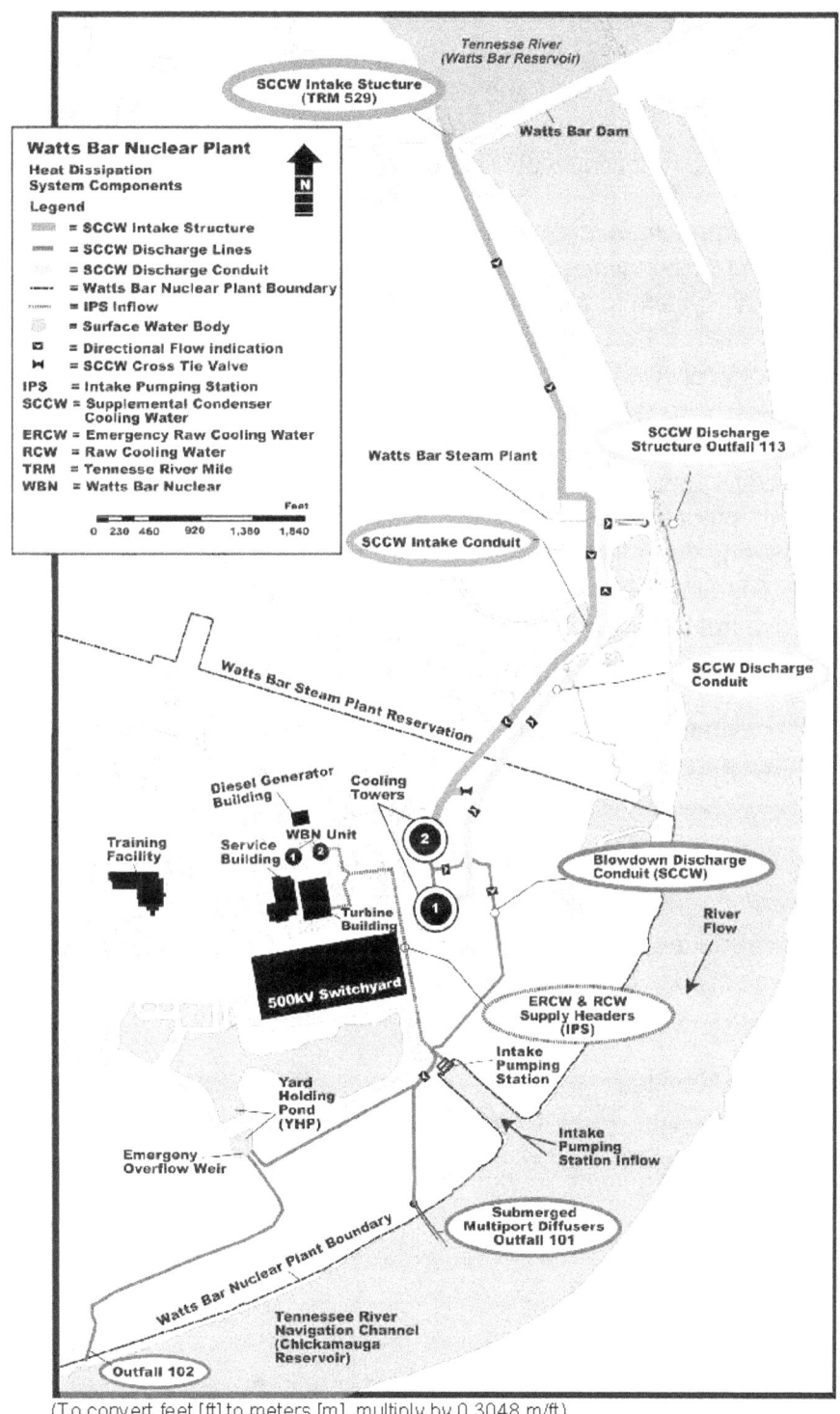

(To convert feet [ft] to meters [m], multiply by 0.3048 m/ft)

Figure 3-2. Major Components of the Cooling System for WBN Units 1 and 2 (TVA 2008)

Evaporation of cooling water system water from the cooling tower increases the concentration of dissolved solids in the cooling water system. In most closed-cycle wet-cooling systems, a portion of the cooling water is removed and replaced with makeup water from the source (for WBN, the Tennessee River) to limit the concentration of dissolved solids in the cooling system and in the discharge to the receiving water body.

Because the WBN cooling towers cannot remove the desired amount of heat from the circulating water during certain times of the year, TVA added the Supplemental Condenser Cooling Water (SCCW) system to the cooling system for the WBN reactors. The SCCW draws water from behind Watts Bar Dam and delivers it by gravity flow to the cooling tower basins to supplement cooling of WBN Unit 1. Unit 1 currently uses the SCCW system. Unit 2 will also use the SCCW system. The temperature of the water from the SCCW intake is usually lower than the temperature of the water in the cooling tower basin and, as a result, lowers the temperature of the water used to cool the steam in the condensers. Approximately the same volume of water that enters the cooling tower basins through the SCCW intake leaves the cooling tower basins and flows through the SCCW discharge structure into Chickamauga Reservoir (TVA 2008). Since the SCCW has been operating, elevated total dissolved solids in blowdown water have not been a concern because a large volume of water enters and leaves the cooling tower basins continually (PNNL 2009).

Table 3-1 lists the anticipated water usage parameters associated with current operation of Unit 1, the anticipated parameters for Unit 2 and the increment from the added operation of Unit 2.

3.1 Intakes

WBN Unit 1 uses two intakes. The first is the SCCW intake, which withdraws water from Watts Bar Reservoir. The second is the intake pumping station (IPS) for the CCW, which withdraws water from Chickamauga Reservoir. Unit 2 would also operate with two intakes.

The intake for the SCCW system, which TVA originally used for its Watts Bar Fossil Plant, is located above Watts Bar Dam. The intake canal for the IPS, which supplies water to the CCW system, is located at Tennessee River Mile (TRM) 528.0, which is approximately 3.1 km (1.9 mi) below the dam.

Table 3-1. Anticipated Water Use

Item	WBN Unit 1 Current Operations	Anticipated WBN Units 1 and 2	WBN Unit 2 Added Increment
Circulating Water System			
Heat discharged	7.8×10^9 Btu/hr[c]	1.5×10^{10} Btu/hr[c]	7.7×10^9 Btu/hr
Waste heat to atmosphere	6.9×10^9 Btu/hr[c]	1.4×10^{10} Btu/hr[c]	7.1×10^9 Btu/hr
Waste heat via liquid discharges to outfall 101	1.5×10^8 Btu/hr[b]	1.7×10^8 Btu/hr[b]	2×10^7 Btu/hr[b]
Intake Pumping Station			
Normal maximum makeup water flow rate	2.5 m³/s (88 cfs)[c]	4.93 m³/s (174 cfs)[c]	2.4 m³/s (86 cfs)
Consumptive use			
Evaporation rate	0.82 m³/s (29 cfs)[c]	1.73 m³/s (61.1 cfs)[c]	0.87 m³/s (31 cfs)
Drift rate	2.8 L/s (45 gpm)[a]	5.7 L/s (90 gpm)[a]	2.8 L/s (45 gpm)
Blowdown Flow Rate			
Normal	1.5 m³/s (53 cfs)[c]	1.8 m³/s (64 cfs)[c]	0.3 m³/s (11 cfs)
Maximum when discharging from yard holding pond and cooling tower basins	3.82 m³/s (135 cfs)[b]	4.81 m³/s (170 cfs)[b]	0.99 m³/s (35 cfs)
Maximum allowable blowdown temperature	35°C (95°F)[b]	35°C (95°F)[b]	No change
SCCW System			
Waste heat via liquid discharges	7.5×10^8 Btu/hr[b]	8.6×10^8 Btu/hr[b]	1.1×10^8 Btu/hr[b]
Intake flow rate	7.31 m³/s (258 cfs)[c]	7.1 m³/s (250 cfs)[c]	Intake flow rate will decline because elevation of water surface in Unit 2 cooling tower will be higher when plant is in operation.
Discharge flow rate	7.48 m³/s (264 cfs)[c]	8.46 m³/s (299 cfs)[c]	A portion of the water entering the system through the IPS will be discharged through the SCCW discharge
Temperature of discharge	35°C (95°F) also 33.5°C (92.3°F) in receiving stream bottom)[b]	35°C (95°F) also 33.5°C (92.3°F) in receiving stream bottom)[b]	No change

(a) 1972 FES-CP (TVA 1972)
(b) TVA (2008)
(c) TVA (2010a)

3.1.1   Water Consumption

The maximum normal makeup water flow rate through the IPS from Chickamauga Reservoir would be 4.93 m³/s (174 cfs) (TVA 2010a), which is 0.6 percent of the mean annual flow of the Tennessee River at Watts Bar Dam (i.e., 778 m³/s [27,500 cfs]). The average monthly intake flow rate through the SCCW intake from above Watts Bar Dam in the Watts Bar Reservoir would be 7.1 m³/s (250 cfs), which is slightly less than that currently withdrawn for WBN Unit 1 and is 0.91 percent of the mean flow of the Tennessee River at the dam (TVA 2010a). Combined, this total withdrawal is 1.3 percent of the mean flow of the Tennessee River at Watts Bar Dam. However, much of this water returns to the river in the discharge. The maximum annual plant consumption rate represents 0.1 percent of the mean annual flow of the Tennessee River at Watts Bar Dam. The NRC staff considers the total withdrawal and the consumptive withdrawal to have a slight, if any, affect on the aquatic biota in Watts Bar Reservoir, Chickamauga Reservoir, and the Tennessee River downstream. Data collected during the preoperational and operational periods for Unit 1 also indicate that the number of species in the reservoir and numbers of individuals per species in the reservoir did not change significantly from the preoperational period to the operational period.

3.1.2   Intake Pumping Station

TVA originally designed the IPS to supply water to both WBN Units 1 and 2; however, since 1996, it has supplied water only to WBN Unit 1. It is located about 3.1 km (1.9 mi) below Watts Bar Dam at TRM 528.0. The IPS is located at the end of an intake channel approximately 240 m (800 ft) from the shoreline of the reservoir (TVA 2009b). The IPS has two sump areas with two intake bays each. Each intake bay is 1.58 m (5.17 ft) wide at the traveling screens and 5.3 m (17.5 ft) high, resulting in an opening of 8.40 m² (90.4 ft²). The open area through the trash racks at each bay opening in the IPS is approximately 8.8 m² (95.1 ft²), for a total of 35.3 m² (380 ft²) open for the passage of water through the trash racks.

Currently, Unit 1 withdraws approximately 2.5 m³/s (88 cfs) of water from Chickamauga Reservoir for normal operations (TVA 2010a). TVA estimates normal maximum operations for WBN Units 1 and 2 would require withdrawal of 4.93 m³/s (174 cfs) of water from the reservoir (TVA 2010a). Under these conditions, while drawing water through all four bays in the IPS, the maximum water velocity through the openings in the traveling screens would be 0.21 m/s (0.67 ft/s) in the winter and 0.19 m/s (0.62 ft/s) in the summer for the portion of the intake structure with four RCW pumps operating (TVA 2011b). The maximum water velocity through the openings in the traveling screens would be 0.24 m/s (0.8 ft/s) (TVA 2010b).

3.1.3   Supplemental Condenser Cooling Water Intake

The intake facility for the SCCW is located above Watts Bar Dam at TRM 529.9. The SCCW has six intake bays and uses three for operation of WBN Unit 1. No additional bays are required for operation of both units. Each intake bay is 2.17 m (7.13 ft) wide at the traveling screens and 9.37 m (30.75 ft) high, resulting in an opening of 20.3 m^2 (219.1 ft^2). The traveling screens and their support structures occupy a portion of the opening leaving 9.16 m^2 (98.6 ft^2) open to the passage of water in each bay for a total of 27.48 m^2 (295.8 ft^2) for the passage of water through the screens into the SCCW intake. The open area through the trash racks at each bay opening in the SCCW intake structure is approximately 11.5 m^2 (124 ft^2), for a total of 34.6 m^2 (372 ft^2) (TVA 2010a). Figure 3-2 shows the locations of the IPS and SCCW water intakes.

The SCCW system operates by gravity flow, so the flow through the intake structure fluctuates as the water-level elevation in Watts Bar Reservoir changes. TVA estimates that the average monthly SCCW intake flow from Watts Bar Reservoir to Unit 1 is approximately 7.31 m^3/s (258 cfs) (TVA 2010a). For the operation of both Units 1 and 2, TVA estimates that the average monthly flow through the SCCW intake would be 7.1 m^3/s (250 cfs) of water from Watts Bar Reservoir (TVA 2010a). The lower flow rate for two units in operation is anticipated because water moves through the system under gravity flow, and the water level in the cooling tower basin for Unit 2 would be 0.6 m (2 ft) higher when the unit is operating (TVA 2010a). This reduces the water level elevation difference between Watts Bar Reservoir and the cooling tower basin, resulting in a reduction of flow rate.

The normal intake flow rates are higher in the summer months when TVA maintains the elevation of Watts Bar Reservoir at 225.7 m (740.5 ft) above mean sea level. Normal flow rates during summer months with both units operating would be approximately 7.6 m^3/s (270 cfs), resulting in a water velocity of 0.22 m/s (0.73 ft/s) through the open areas in the trash racks in the SCCW. The water velocity through the openings in the traveling screens at the SCCW would be 0.28 m/s (0.91 ft/s) under these conditions (TVA 2010a).

3.2 Discharge Systems

WBN Unit 1 uses three discharge systems and three outfalls for discharge from the cooling water systems. TVA holds permits through the National Pollutant Discharge Elimination System (NPDES) permit process for the three outfalls. All three outfalls empty into Chickamauga Reservoir. The outfalls include Outfall 101, which uses discharge diffusers; Outfall 102, which uses a shoreline discharge; and Outfall 113, which also uses an emergency overflow weir that flows into a local stream channel and empties into Chickamauga Reservoir.

3.2.1   Outfall 101 – Discharge Diffusers

TVA plans to discharge cooling water from the main cooling-water system for WBN Units 1 and 2 to Chickamauga Reservoir through a diffuser system located approximately 3.2 km (2 mi) below Watts Bar Dam at TRM 527.9 (TVA 2008). The National Pollutant Discharge Elimination

System (NPDES) permit for the WBN site identifies the diffuser discharge as Outfall 101 (TDEC 2011). TVA (1997) describes this diffuser system as consisting of two pipes branching from a central conduit at the right bank of Chickamauga Reservoir and extending perpendicular to the river flow of the Tennessee River. Each pipe is controlled by a butterfly valve located a short distance from the junction with the central conduit.

The downstream leg of the diffuser consists of 49 m (160 ft) of unpaved 1.37-m (4.5-ft)-diameter corrugated steel diffuser pipe at the end of approximately 91 m (297 ft) of paved corrugated steel approach pipe of the same diameter. The diffuser pipe is half buried in the river bottom and has two 2.54-cm (1-in.)-diameter ports per corrugation. The centroid of the ports is angled up at 45 degrees from horizontal in a downstream direction (TVA 1997).

The upstream leg of the diffuser system consists of 24 m (80 ft) of unpaved 1.07-m (3.5-ft)-diameter corrugated steel diffuser pipe at the end of approximately 136 m (447 ft) of paved corrugated steel approach pipe of the same diameter. The upstream diffuser pipe section is half buried in the river bottom and extends its entire length beyond the dead end of the downstream diffuser pipe section. The port diameter, spacing, and orientation of the upstream leg are the same as those of the downstream leg (TVA 1997). TVA document Figure 3 (1977) illustrates the diffuser configuration. TVA does not plan to make any upgrades or changes to the diffuser design in preparation for operating Unit 2 (TVA 2010c).

TVA maintains operational procedures for this system to ensure adequate dilution of the plant effluent. The 2008 TVA ER explains the process as follows:

> To provide adequate dilution of the plant effluent, discharge from the diffusers is permitted only when the release from Watts Bar Dam is at least 3,500 cubic feet per second (cfs). To ensure this happens, an interlock is provided between the dam and WBN that automatically closes the diffusers when the flow from the hydroturbines at Watts Bar Dam drops below 3,500 cfs. To provide temporary storage of water during these events, the blowdown discharge conduit also is connected to a yard holding pond. When the flow from Watts Bar Dam drops below 3,500 cfs, thereby closing the diffuser valves, the blowdown is automatically routed to the yard holding pond. When hydro operations resume with releases of at least 3,500 cfs, the interlock is 'released' and the diffuser valves can be opened. When this occurs, the discharge from the diffusers would contain blowdown from the cooling towers and blowdown from the yard holding pond. To protect the site from the consequences of exceeding the capacity of the yard holding pond, an emergency overflow weir is provided for the pond, which delivers the water to a local stream channel that empties into the Tennessee River at TRM 527.2. The operation of Watts Bar Dam and the WBN blowdown system are very carefully coordinated to avoid unexpected overflows from the yard holding pond (TVA 2008).

3.2.2   Outfall 113 – SCCW Discharge

The SCCW system discharges water through a discharge structure originally constructed for the Watts Bar Fossil Plant. The NPDES permit for the WBN site identifies the SCCW discharge as Outfall 113 (TVA 2008). Water leaving the cooling tower basins flows through a pipe to the discharge structure approximately 1.8 km (1.1 mi) upstream of the IPS. TVA describes the discharge structure as an "open discharge canal, an overflow weir drop structure, and a below water discharge tunnel" (TVA 1998a). TVA describes the discharge tunnel as a "rectangular culvert 7 feet wide by 10 feet high at the discharge point" (TVA 1998a). The elevation of the culvert outlet is 205.7 m (675 ft) above mean sea level. To reduce the impact of the discharge on the river bottom, TVA installed a concrete incline to direct flow toward the river surface as it leaves the outfall (TVA 1998a; PNNL 2009).

TVA designed and constructed the SCCW system so it could operate the cooling system for WBN Units 1 and 2 with or without the SCCW. If the temperature of the discharge water exceeds allowable release limits, TVA can shut down the SCCW system. TVA also included a crosstie and control valve in the system that allows part of the flow from the SCCW intake to bypass the cooling tower basins and mix with the effluent in the discharge pipeline. When the possibility of exceeding the NPDES river temperature limit exists, TVA opens a bypass valve to allow cooler water in the intake pipeline to mix with water in the discharge line, thus cooling the effluent before it is discharged to the reservoir (TVA 2008). The bypass is necessary during winter months when the water temperature in the Tennessee River is cooler, and a possibility exists of exceeding the instream temperature rate of change limit in the NPDES permit. TVA opens the crosstie around November 1, and it remains open until the end of April (PNNL 2009).

3.2.3   Outfall 102 – Yard Holding Pond Emergency Overflow

TVA uses the unlined yard holding pond (Figure 3-2), which is approximately 8.9 ha (22 ac) in area (TVA 2005a), for temporary storage of cooling tower blowdown when the flow from the hydroturbines at Watts Bar Dam is less than 99 m^3/s (3500 cfs). When dam operations resume with releases of at least 99 m^3/s (3,500 cfs), diffuser valves allow the yard-holding pond to discharge into Chickamauga Reservoir through the diffusers (TVA 2008).

The yard-holding pond has an emergency overflow weir at 215.3 m (706. 5 ft) above mean sea level. This weir design prevents the yard-holding pond from overflowing the capacity of the pond. In the event that water rises above the height of the weir, it flows into a local stream channel that empties into Chickamauga Reservoir at TRM 527.2 (TVA 2008). The NPDES permit for the WBN site identifies this discharge as Outfall 102 (TVA 2008).

3.2.4   Thermal Effects from Discharges

WBN Unit 2 would continue to discharge water via three outfalls. Table 3-2 shows the current NPDES temperature limits for the three outfalls used during operation for Unit 1. The NPDES permit issued by the State of Tennessee for Unit 1 specifies limits on the amount of thermal effluent the plant may discharge into the Tennessee River. The permit also establishes an active mixing zone and defines in-stream monitoring and reporting requirements necessary to comply with effluent limitations. Table 3-1 provided the increment added for waste heat discharged to the river for both Outfall 113 (i.e., the SCCW system shoreline discharge) and Outfall 101 (i.e., the diffuser discharge). The additional increment for flow is approximately 14 percent of the current amount of heat discharged. The mixing zone dimensions for the outfall to the SCCW (i.e., Outfall 113) are based on a physical hydrothermal model test of the discharge. TVA has confirmed the model output with actual measurements (TVA 2005b, 2006, 2007b, 2007c). The model and measurements indicate that the plume rises after hitting the concrete pad located at the end of the discharge. The model results also predict a zone of passage for fish along the bottom of the river especially in the area of the navigation channel (TVA 2004). The location of the plume from the SCCW discharge does not prohibit fish from swimming past the plant, and the plume would likely not reach the river's mussel beds.

Table 3-2. NPDES Temperature Limits for WBN Outfalls to the Tennessee River from TVA

Outfall	Effluent Parameter	Daily Report	Limit
101	Effluent Temperature	Daily Avg	35.0°C (95°F)
102	Effluent Temperature	Grab	35.0°C (95°F)
113	Instream Temperature[a]	Max Hourly Avg	30.5°C (86.9°F)
	Instream Temperature Rise[b]	Max Hourly Avg	3.0°C (5.4°F)
	Instream Temperature Rate-of-Change[a]	Max Hourly Avg	±2°C/hr (±3.6°F/hour)
	Instream Temperature Receiving Stream Bottom[c]	Max Hourly Avg	33.5°C (92.3°F)

Source: TVA 2010d
(a) Downstream edge of mixing zone.
(b) Upstream ambient to downstream edge of mixing zone.
(c) Mussel relocation zone at SCCW outlet.

TVA relocated freshwater mussels from an area 46 m by 46 m (150 ft by 150 ft) at Outfall 113. TVA relocated the mussels to the mussel bed directly across the river in order to prevent adverse impacts during operation of the SCCW. In addition, TVA placed a ramp on the invert of the SCCW outfall to deflect the discharge upward, and away from the bottom of the river (TVA 2004). The analysis of instream data collected by TVA for Outfall 113 showed that heat from the SCCW effluent does not reach the river bottom in significant amounts (TVA 2004).

Discharge from the emergency overflow (i.e., Outfall 102) is infrequent. The current NPDES permit also specifies a discharge temperature limit of 35°C (95°F) for Outfall 102 (TVA 2008).

3.2.5   Physical Effects from Scouring at the Discharges

No impacts are anticipated to benthic organisms in the vicinity of, or immediately downstream of, the outfalls from scouring of the bottom of the reservoir by adding WBN Unit 2. TVA indicates that water flow from the SCCW discharge would not increase, and the concrete structure at the discharge of the SCCW (i.e., Outfall 113) continues to reduce the affect the discharge has on the river bottom and directs the flow of water toward the river surface as it leaves the outfall (TVA 1998a). The use of a diffuser that discharges at an angle of 45 degrees above horizontal in the downstream direction for Outfall 101 minimizes the amount of scouring discharge from this outfall. Use of Outfall 102, which discharges emergency outflow from the yard holding pond, has been infrequent. This outfall discharges into a local stream channel that empties into the Chickamauga Reservoir. The NRC staff determines that physical changes at the outfalls as a result of the additional operation of Unit 2 would not affect the aquatic biota of Watts Bar Reservoir.

3.2.6   Chemical Discharges from Outfalls

Another discharge-related stressor involves chemical treatment of the cooling water. TVA would control water chemistry for various plant water uses by adding biocides, algaecides, corrosion inhibitors, pH buffering, scale inhibitors, and dispersants. The NPDES permit requires that TVA follow the TDEC-approved Biocide/Corrosion Treatment Plan (B/CTP) (TDEC 2011). WBN's current B/CTP was approved in 2009 (TDEC 2011) based on the list of chemicals included in the permit modification request submitted by TVA in April 2009 (TVA 2010e). Table 3-3 lists chemicals and their discharge quantities included in the WBN site's NPDES permit request submitted for the WBN site on April 2009 (TVA 2009c).

TVA discharges water containing chemical and biocidal additives for the condenser cooling system and the SCCW system to the Chickamauga Reservoir through Outfalls 101 and 113, respectively. Chemical and biocidal additives and waste streams from various other water-treatment processes and drains are returned to the Yard Holding Pond (YHP) where they are subjected to dilution, aeration, vaporization, and chemical reactions. The plant then discharges the YHP water to Chickamauga Reservoir through Outfall 101 or 102, subject to the limitations of the WBN site's existing NPDES permit (TDEC 2011).

The NPDES permit (TDEC 2011) provides additional detail about the chemicals that may be in water discharged through the outfalls. In addition to the chemicals added as biocide and for corrosion-treatment, other chemical additives are used in a variety of plant processes. These chemicals may occur in trace quantities at Outfall 101 or Outfall 102. The potential discharge of these chemicals is through the cooling-tower blowdown line to Outfalls 101 and 102 so Outfall 113 would not receive these discharges. The summary of potential chemicals discharged by NPDES outfall number is shown in Table 3-4.

Table 3-3. Raw Water Chemical Additives at WBN

Product	Purpose	Frequency of Discharge	Active Ingredients	Discharge Concentration[a] (ppm active ingredients)
Depositrol PY5200 (replaces Nalco 73200)[b]	Dispersant to facilitate iron corrosion inhibition	Continuous	copolymer	< 0.2
Inhibitor AZ8100 (replaces Nalco 1336)[b]	Copper corrosion Inhibition	Periodic	sodium tolyltriazole	< 0.25
Sprectrus ED 1500 (replaces Nalco 73551)[b]	Surfactant to facilitate oxidizing biocides	Periodic	nonionic surfactant	< 2.0
Towerbrom 60 m (replaces Towerbrom 960)[b]	Oxidizing biocide (chlorination)	Periodic	sodium bromide and sodium dichloroisocyanurate	0.10 chlorine (total residual)
Spectrus OX 1200 (replaces Nalco 901 G)[b]	Oxidizing biocide (chlorination)	Continuous	bromo-chloro, dimethyl hydantoin	0.10 chlorine (total residual)
Spectrus DT 1404 (replaces Nalco CA-3S)[b]	De-chlorination	Periodic[c]	sodium bisulfite	< 10
Spectrus CT1300[d] (replaces H150M)[b] or	Nonoxidizing biocide (mollusk control)	Periodic	Alkyl dimethyl benzyl ammonium chloride	< 0.001 active ingredient in stream after mixing < 0.05 measured in effluent
Spectrus NX1104[4] (replaces Spectrus NX 104)[b]	Nonoxidizing biocide (mollusk control)	Periodic	dimethylbenzylammonium chloride and dodecylguanidine hydrochloride	< 0.001 total active ingredient in stream after mixing < 0.031 quaternary ammonium compound measured in effluent
Bentonite clay[b]	Detoxification of nonoxidizing biocides	Periodic[c]	sodium silicate (bentonite clay)	< 10
Liquid bleach[b]	Oxidizing biocide (chlorination)	Continuous	sodium hypochlorite	0.10 chlorine (total residual)
H150M[e]	Nonoxidizing biocide	Minimum of 4 times per year	25 percent dimethyl benzyl ammonium chloride and 25 percent dimethyl ethylbenzyl ammonium chloride.	< 0.05 ppm

Table 3-3. (contd)

Product	Purpose	Frequency of Discharge	Active Ingredients	Discharge Concentration (ppm active ingredients)
Flogard MS6209 (replaces MSW-109, 2010)[g]	Iron Corrosion Inhibitor	Continuous when river temperature is above 15.6°C (60°F).	zinc chloride, orthophosphate	< 0.2 total zinc < 0.2 total phosphorus

Source: From Table in TVA (2009d)

(a) The maximum discharge concentration is indicated except where noted. Concentrations are achieved through a combination of dilution and dechlorination with sodium bisulfite or detoxification with bentonite clay.

(b) Denotes chemicals previously approved by the division (Tennessee Department of Environment & Conservation, Division of Water Pollution Control).

(c) Dechlorination and detoxification chemicals are applied as needed to ensure the discharge limitations identified in this table are met.

(d) Non-oxidizing biocide treatments are not applied at the same time as oxidizing biocide treatments.

(e) Application information from TVA (2008).

(f) SCCW and river flow conditions have a significant impact on these discharge concentrations.

(g) Active ingredient information from TVA (2008).

Table 3-4. Potential Chemical Discharge to NPDES Outfalls at the WBN Site

No.	Outfall Description	Chemical
101	Diffuser Discharge	ammonium hydroxide, ammonium chloride, alpha cellulose, asbestos after 5 micron filter, boric acid, sodium tetraborate, bromine, chlorine, copolymer dispersant, ethylene oxide, propylene oxide copolymer, ethylene glycol, hydrazine, laboratory chemical wastes, lithium, molybdate, monoethanolamine, molluscicide, oil and grease, phosphates, phosphate cleaning agents, paint compounds, sodium bisulfite, sodium hypochlorite, sodium hydroxide, surfactant, tolyltriazole, x-ray film processing rinse water, zinc chloride orthophosphate, zinc sulfate, phosphino-carboxylic acid copolymer, diethylenetriaminepenta-methylene phosphonic acid, sodium salt, sodium chloride, ethylenediamine tetracetic acid.
102	YHP Overflow Weir	Alternate discharge path for Outfall 101
103	Low-Volume Waste Treatment Pond	ammonium hydroxide, ammonium chloride, boric acid, sodium tetraborate, bromine, chlorine copolymer dispersant, ethylene glycol, hydrazine, laboratory chemical wastes, lithium, molybdate, monoethanolamine, molluscicide, oil and grease, phosphates, phosphate cleaning agents, paint compounds, sodium hydroxide, surfactant, tolyltriazole, x-ray film processing rinse water, zinc sulfate
107	Lined Pond and Unlined Pond	metals – mainly iron and copper, acids and caustics, ammonium hydroxide, ammonium chloride, asbestos after 5 micron filter, boric acid, sodium tetraborate, bromine, chlorine, copolymer dispersant, hydrazine, laboratory chemical wastes, molybdate, molluscicide, oil and grease, phosphates, phosphate cleaning agents, sodium, sodium hydroxide, surfactant, tolyltriazole, zinc sulfate
113	SCCW Discharge	some contact with chemicals listed for outfall 101, alpha cellulose, bromine, chlorine, copolymer, molluscicide, zinc chloride orthophosphate

Source: TDEC 2011

4.0 Assessment of Listed Species

4.1 Gray Bat (Myotis grisescens)

4.1.1 Life History of the Gray Bat

The gray bat, listed as endangered by FWS (41 FR 17736) and the State of Tennessee, is a migrant colonial bat. The distribution of gray bats is centered by limestone karst areas within the southeastern United States (Brady et al. 1982). The gray bat possesses very specific microclimate requirements and use caves during both winter and summer. Colonies may travel over 100 km (60 mi) between winter and summer habitats (NatureServe 2010). Summer colonies occupy traditional home ranges that include a maternal cave and several roost caves usually within 1 km (0.6 mi) of a river or reservoir (NatureServe 2010).

Adult gray bats feed on insects almost exclusively over water bodies (Brady et al. 1982). They have been known to forage more than 19 km (12 mi) from summer roost caves and are known to forage over and along the Tennessee River. FWS has not designated critical habitat for the gray bat.

4.1.2 Status of the Gray Bat in the Vicinity of the WBN Site

Gray bats have not been observed on the WBN Site. In 1982, three caves in the State of Tennessee served as major winter hibernacula for gray bats (Brady et al. 1982). Two caves (see Figure 4-1) within 16 km (10 mi) from the WBN site serve as summer roosts for gray bats (NRC 1995). A cave located approximately 4 km (2.5 mi) from of the WBN site contained 385 gray bats in 2002, while another cave almost 13 km (8 mi) from the WBN site contained 340 gray bats during the same year (Harvey and Britzke 2002). Although no direct observations of gray bats foraging over the Tennessee River immediately adjacent to the WBN site or under transmission lines that service the site have been recorded, the staff concludes gray bats routinely forage at these locations based on habitat preferences and proximity to known active summer roost caves.

4.2 Aquatic Biota

Federally listed aquatic biota that could potentially reside in the vicinity of the WBN site include freshwater mussels (pink mucket mussel [*Lampsilis abrupta*], Eastern fanshell pearlymussel [*Cyprogenia stegaria*], rough pigtoe [*Pleurobema plenum*], dromedary pearlymussel [*Dromus dromas*] and orangefoot pimpleback [*Plethobasus cooperianus*]) and the snail darter (*Percina tanasi*).

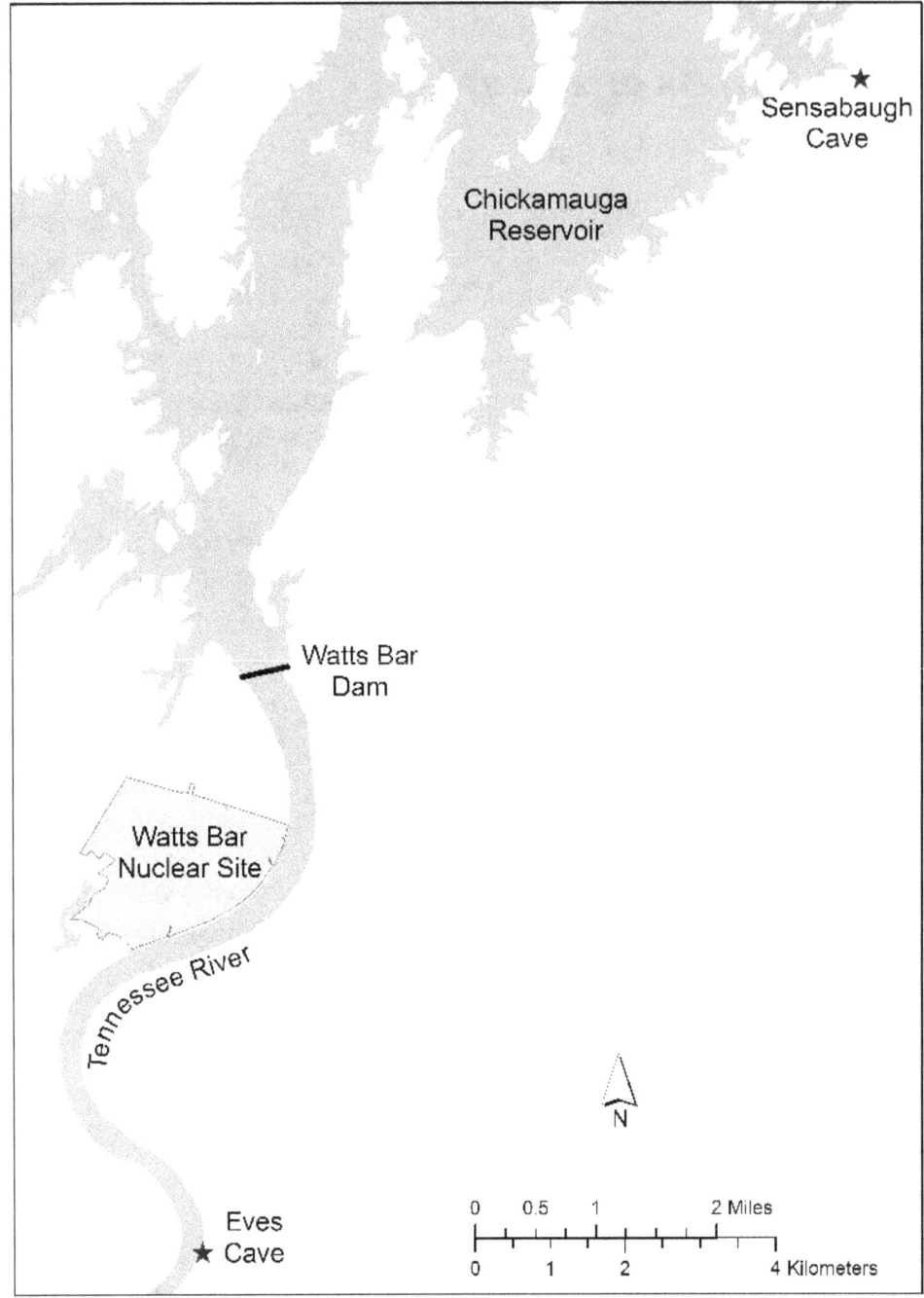

Figure 4-1. Known Caves Occupied by Gray Bats in the Vicinity of the WBN Site

4.2.1 Life History

The life histories of the freshwater mussels and the snail darter are discussed separately.

4.2.1.1 Life History of Freshwater Mussels

Mussels spend their entire juvenile and adult lives buried either partially or completely in the substrate. Although mussels are able to change their position and location, they rarely move more than a few hundred yards during their lifetime unless dislodged. Native freshwater mussels have an unusual reproductive cycle. Although some species are hermaphroditic, the species discussed in this BA have separate sexes. The eggs of female mussels move from the ovaries to the gills where fertilization occurs. Sperm is released to the water by male mussels and is carried into the female's body through the incurrent aperture. The gills, or a portion of the gills, serve as brood pouches, called marsupia. The fertilized eggs develop into small larvae, called glochidia, which release into the water. At the time of their release from the marsupia, the glochidia possess only the embryonic stages of a mouth, intestines, a foot, and a heart. If the glochidia do not encounter a passing fish and attach to its gills, skin, or fins then they fall to the bottom and die a short time later. The glochidia usually remain on the fish from one to six weeks (sometimes longer) and then fall off and begin their growth into adulthood. Each mussel species has specific species of fish that serve as a host fish for the glochidia (Parmalee and Bogan 1998). The survival of freshwater mussel species depends not only on the environmental conditions for the mussel, but on the survival and health of the host fish populations.

Pink mucket mussel – Pink muckets prefer free-flowing reaches of large rivers, typically in silt-free and gravel substrates. Fishes that reportedly serve as hosts for glochidia include the smallmouth bass (*Micropterus dolomieu*), spotted bass (*M. punctulatus*), and largemouth bass (*M. salmoides*) as well as freshwater drum (*Aplodinotus grunniens*), and possibly sauger (*Sander canadensis*) (Mirarchi et al. 2004).

Eastern fanshell pearlymussel – Fanshells are usually found on coarse sand and gravel less than 0.9 m (3 ft) deep (Parmalee and Bogan 1998). The glochidial hosts have been reported to be banded sculpin (*Cyprogenia stegaria*), mottled sculpin (*Cottus bairdi*), greenside darter (*Etheostoma blennioides*), Tennessee snubnose darter (*E. simoterum*), banded darter (*E. zonale*), tangerine darter (*Percina aurantiaca*), blotchside logperch (*P. burtoni*), logperch (*P. caprodes*), and the Roanoke darter (*P. roanoka*).

Rough pigtoe – The rough pigtoe is found primarily in large rivers inhabiting a mixture of sand and gravel in areas kept free of silt by moderate to strong currents. A fish host for the glochidia has not been identified (Mirarchi et al 2004).

Dromedary pearlymussel – The dromedary pearly mussel inhabits small-to-medium, low-turbidity, high-to-moderate-gradient streams. In recent studies, FWS has identified the fantail darter (*Etheostoma flabellare*) as the host species. Other potential hosts include the banded darter, tangerine darter (*Percina aurantiaca*), logperch, gilt darter (*P. evides*), black sculpin (*Cottus baileyi*), greenside darter, Tennessee snubnose darter, blotchside logperch, channel darter (*P. copelandi*), and the Roanoke darter (FWS 2010a).

Orangefoot pimpleback – The orangefoot pimpleback is primarily a big river species found in silt-free areas in a mixture of sand and gravel. The species still survives in the tailwaters of some Tennessee River dams, such as Pickwick Dam. A glochidial host has not been identified (Mirarchi et al. 2004).

4.2.1.2 Snail Darter

Snail darters inhabit larger creeks where they frequent sand and gravel shoal areas in low-turbidity water. They also inhabit deeper portions of rivers and reservoirs in areas where there is a current. Snail darters are known to burrow beneath the substrate, possibly for concealment or to conserve energy. Snail darters spawn early with their spawning season extending from February to mid-April in shoal areas. Females contain an average of 600 mature eggs and may mate with several males during the mating season. Eggs hatch in 15 to 20 days depending on the water temperature. The larvae of snail darters may drift considerable distances to deeper water areas downstream, although by late summer they have migrated upstream again toward the spawning habitat. Snail darters prefer small pleurocerid river snails although they may also feed on caddis fly larvae, midge, and blackfly larvae (Etnier and Starnes 1993).

4.2.2 Status of Listed Species

Federally listed aquatic species include freshwater mussels and the snail darter.

4.2.2.1 Freshwater Mussels

The Tennessee River is home to both introduced and native mussel and clam species. Approximately 130 of nearly 300 species of freshwater mussels in the United States live or have lived in waters within Tennessee (Parmalee and Bogan 1998). The numbers of native mussels in the Tennessee River have been declining since the early 1940s when TVA filled the Chickamauga and Watts Bar reservoirs. Based on studies of shell midden material and evaluations conducted before the impoundments were built, ecologists believe a total of 64 freshwater mussel species occurred near the WBN site prior to impoundment of the river (TVA 1986). Surveys conducted by TVA between 1983 and 1997 identified only 30 native mussel species (TVA 1998b).

Because of the loss of diversity in mussel species, the State of Tennessee created a freshwater mussel sanctuary in Chickamauga Reservoir in the vicinity of the WBN site. The freshwater mussel sanctuary, in which harvesting mussels is illegal, currently extends 16 km (10 river mi from TRM 520.0 to TRM 529.9) (TVA 1998a). Figure 4-2 shows the extent of the freshwater mussel sanctuary, as well as the approximate locations of the mussel beds and the locations of TVA's mussel sampling stations.

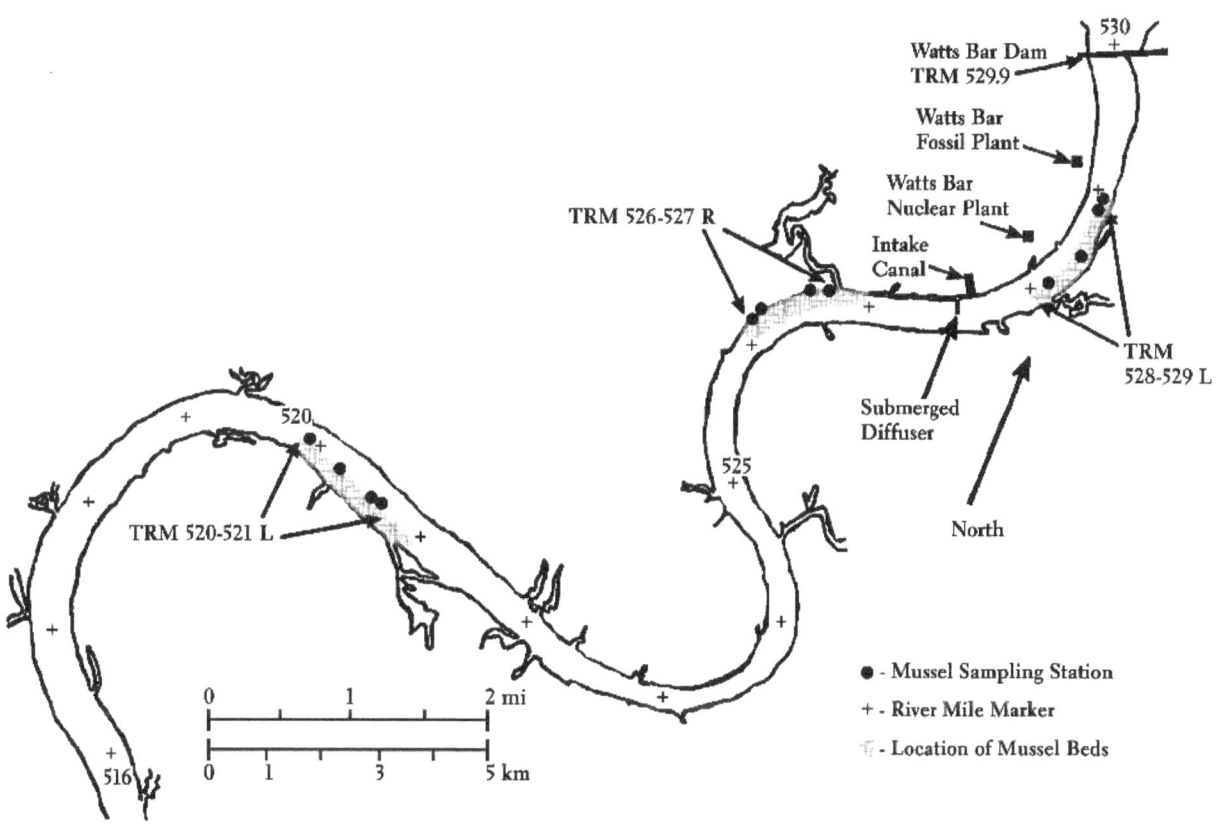

Figure 4-2. Mussel Beds and Monitoring Stations (TVA 1998b)

TVA has monitored three known concentrations of mussels (mussel beds) within this sanctuary since 1983. The beds are all located on submerged gravel and cobble bars in water 2.7 m to 6.4 m (9 ft to 21 ft) deep (TVA 2010b). The furthest bed downstream is located at TRM 520 to TRM 521 on the left descending bank of the river. This bed is 10 km (6 mi) downstream of the WBN site and on the opposite side of the river. A second bed is roughly from TRM 526 to TRM 527 on the right descending bank, and the third from TRM 528 to TRM 529 on the left descending bank (TVA 1998b). The most recent data reported is from surveys from 2010 (TVA 2011a, 2011c).

Table 4-1 provides the results of 15 mussel surveys over a period of 14 years (1983-1997) adjacent to or downstream of the site between TRM 520 and TRM 529.2. The table includes only those species considered in this BA. TVA sampled the same locations in 2010 but did not observe any of the listed species (TVA 2011a), with the exception of a single pink mucket mussel.

Table 4-1. Results of 15 Native Mussel Surveys from TRM 520 to TRM 528.9 (includes one survey from TRM 529.2)

Species	Common Name	1983 (Sep/Nov)	1984 (Jul/Nov)	1985 (Jul/Oct)	1986 (Jul/Oct)	1988 (July)	1990 (July)	1992	1994	1996 (July)	1997 (July)	1997 at TRM 529.2 (TVA 1998a)
Lampsilis abrupta	Pink mucket	3/7	6/2	1/7	6/2	12	4	6	2	4	0	1
Cyprogenia stegaria	Eastern fanshell	2/1	0/1	1/0	0/0	0	0	0	0	0	0	0
Dromus dromas	Dromedary pearlymussel	1/0	0/0	0/0	0/0	0	0	0	0	0	0	0
Pleurobema plenum	Rough pigtoe	1/1	2/0	1/0	0/0	0	0	0	0	0	0	0
Plethobasus cooperianus	Orangefoot pimpleback	0/0	0/0	0/0	0/0	0	0	0	0	0	0	0

Source: Adapted from TVA 1998b, and TVA 1998a

Pink mucket mussel – The FWS designated the pink mucket mussel as endangered in 1976 (41 FR 24062) and wrote a recovery plan in 1985 (FWS 1985). Historically, this species inhabited the entire reach of the Tennessee River across northern Alabama. Currently, it occurs only in the riverine reaches downstream of Wilson Dam in Tennessee and Guntersville Dam in Alabama. However, FWS considers the species to be uncommon to rare. Researchers report specimens younger than 10 years of age as rare in the Wilson and Guntersville dam tailwaters. TVA found the pink mucket in the vicinity of the WBN site during every mussel survey from 1986 to 1997, although the number of specimens was never more than 10 (1988) in the surveys from TRM 528.2 to TRM 528.9 (TVA 1998b) as shown in Table 4-1. The occurrence data provided by TVA (TVA 2010a) indicated that nine specimens were found in the 1990 survey, six specimens in the 1992 survey as well as two specimens in the vicinity of the SCCW discharge (TRM 529.2). The most recent sighting was of a single individual located between TRM 526 and 527 during the most recent survey conducted in 2010 (TVA 2011a).

Eastern Fanshell Pearlymussel – The FWS has listed the Eastern fanshell pearlymussel, also known simply as the fanshell, as endangered since 1990 (55 FR 25591). According to the Fanshell Recovery Plan (FWS 1991), the species is known from only three reproducing populations. The closest population to the WBN Site is in the Clinch River in Tennessee, although it also inhabits the Green and Licking rivers in Kentucky. This species generally is distributed in the Tennessee and Cumberland river systems. The fanshell is generally considered a big river species, though it also may be found inhabiting shallow, unimpounded upper stretches of the Clinch River, and in unimpounded portions of the Tennessee and Cumberland rivers. Researchers think fanshells may be reproducing below Pickwick Landing Dam on the Tennessee River (Parmalee and Bogan 1998). Many factors have caused the decline of this species, including impoundment, navigation projects, water quality degradation, and other forms of habitat alternation such as gravel and sand dredging. These habitat modifications either directly affected the species or reduced or eliminated the fish hosts (55 FR 25591). TVA last found the fanshell in 1985 in the mussel bed nearest the WBN site (TRM 528.2 to TRM 528.9) (TVA 1998b). In addition, three specimens were observed in 1983 and a single specimen in 1984. The occurrence data provided by TVA (TVA 2010a) indicated that a single individual was reported from survey years 1983 to 1984 and that two individuals were confirmed from a survey in 1983.

Dromedary Pearlymussel – The FWS listed the dromedary pearlymussel as endangered in 1976 throughout its entire range in Kentucky, Tennessee, and Virginia (41 FR 24062), and its recovery plan was published in 1983 (FWS 1983a). This species was historically widespread in the Cumberland and Tennessee river systems. The dromedary pearlymussel commonly is found near riffles on sand and gravel substrates with stable rubble. Individuals also have been found in slower waters and up to a depth of 5.5 m (18 ft). Most historic populations apparently were lost when the river sections they inhabited were impounded. The more than 50 impoundments on the Tennessee and Cumberland Rivers eliminated the majority of riverine habitat for this species in its historic range. TVA did not find the dromedary pearlymussel in the bed closest to the WBN site (TRM 528.2 to TRM 528.9) in surveys conducted between 1983 and 1997, but it did find one specimen in the bed located at TRM 520.0 to TRM 520.8 once in 1983 (TVA 1998a). The occurrence data provided by TVA did not show this siting (TVA 2010a).

Rough Pigtoe – The FWS listed the rough pigtoe as endangered in 1976 (41 FR 24062), and published a recovery plan in 1984 (FWS 1984a). Researchers have identified extant populations in the Tennessee River tailwaters of Wilson Dam, where they are very rare, and possibly in the tailwaters of Guntersville Dam (Mirarchi et al. 2004). During surveys conducted near the WBN site, TVA found a single rough pigtoe in each of two surveys in 1983 and two in the early survey of 1984. TVA reported a single individual rough pigtoe as recently as 1985 in the mussel bed closest to the site (TRM 528.2 to TRM 528.9). The occurrence data from TVA (TVA 2010a) indicated the presence of only one specimen from the surveys conducted between 1983 and 1984 at TRM 528.9.

Orangefoot Pimpleback – The FWS has listed the orangefoot pimpleback, also known as the Cumberland pigtoe (Mirarchi et al. 2004), as endangered since 1976 (41 FR 24062), and a recovery plan was published in 1984 (FWS 1984b). The orangefoot pimpleback is primarily a big river species found in silt-free areas in a mixture of sand and gravel. The species still survives in the tailwaters of some Tennessee River dams, such as Pickwick Dam. TVA has not found the orangefoot pimpleback near the WBN site during surveys conducted in 1983 or since that time (TVA 1998b). The occurrence data provided TVA shows that the nearest occurrence of the orangefoot pimpleback was at TRM 595.0 in Watts Bar Reservoir in 1978 (TVA 2010a).

4.2.2.2 Snail Darter

The snail darter was classified as endangered on October 9, 1975 (40 FR 47506) and was reclassified to threatened on July 5, 1984 (49 FR 27510). The FWS wrote a recovery plan in 1979, and updated it in 1982 (FWS 1983b). FWS believes that snail darters originally inhabited the main stem of the Tennessee River and possibly ranged from the Holston, French Broad, Lower Clinch, and Hiwassee Rivers downstream in the Tennessee drainage to northern Alabama (FWS 1992). Etnier and Starnes (1993) report that it is likely that the snail darter inhabited the main channel of the upper Tennessee River and the lower reaches of its major tributaries; however, impoundments fragmented much of the species' range. In 1973, the snail darter was thought to be restricted to the lower Little Tennessee River, with some additional individuals dispersed into Watts Bar Reservoir below Loudon Dam. In 1975, TVA biologists transplanted snail darters into the Nolichucky River until another jeopardized fish species was found in that vicinity. In 1976, they transplanted snail darters into the lower Hiwassee River and, during 1979 and 1980, into the lower Holston and Middle Elk Rivers. Subsequently, in 1988 and 1989, snail darters were collected from the lower French Broad and lower Holston Rivers, respectively. However, the transplant attempts into the lower Holston and Middle Elk rivers did not appear to be successful (Etnier and Starnes 1993). In 1980, an additional population was discovered (estimated to number between 200 to 400 individuals) in South Chickamauga Creek (between Creek Mile 5.6 in Tennessee [Hamilton County] and Creek Mile 19.3 in Georgia [Catoosa County]) (Etnier and Starnes 1993; TVA 2010a). Biologists also found a few darters in the Tennessee River mainstream just below Chickamauga and Nickajack Dams (FWS 1992). The upper Watts Bar Reservoir contained a population of snail darters, but the population did not appear to be reproducing subsequent to the impoundment of the Tellico Reservoir (Etnier and Starnes 1993). Individuals were found at TRM 591.4 as recently as 1976 and at TRM 597.2 as recently as 1982. They were also found as recently as 1979 in the Little Tennessee River, which empties into the Watts Bar Reservoir. As recently as 1985, snail darters inhabited Sewee Creek (Meigs County), which empties into the Tennessee River just south of the WBN site (TVA 2010a). They were identified as living from Creek Mile 3.2 to Creek Mile 5.7. TVA has not observed snail darters since 1975 in any sampling they have conducted in the upper Chickamauga Reservoir (TVA 1998b; Simmons and Baxter 2009).

5.0 Environmental Effects of WBN Unit 2 on Listed Species

Listed species could potentially be affected by the addition of the second nuclear unit as a result of operational noise, water consumption, entrainment or impingement of fish or fish hosts from the intake or as a result of chemical or thermal discharges to Guntersville Reservoir. The potential environmental impact on the gray bat is discussed separately from that of the aquatic species (freshwater mussels and snail darter).

5.1 Gray Bat

Because gray bats do not occur on the WBN site, the potential affect from WBN Unit 2 operations is minimal. The proximity of caves used by gray bats in summer to the site likely means gray bats forage over the Tennessee River immediately adjacent to the site. In a previous biological opinion for the operation of WBN Units 1 and 2, FWS determined that the discharge of excess heat, chemicals, and radionuclides into the river would likely be the primary threat to this species from the operation of WBN Units 1 and 2 (Widlak 1995). Discharge of radioactive materials, chemicals, and other substances could have detrimental effects on larvae of insect species that make up the gray bats' diet. Standards established within the NPDES permit issued by the State of Tennessee are designed to prevent water quality degradation that would result from unregulated discharge of pollutants into the Tennessee River. The NPDES permit also governs monitoring and testing of discharges to ensure continued compliance with permit requirements.

Operational noise also may preclude use of habitats near the WBN site by gray bats. Gray bats forage while flying over open water, and emit sounds to detect flying insects via echolocation. Bats may avoid noise when foraging (Schaub et al. 2008). Greater mouse-eared bats (*Myotis myotis*) foraged most often in experimental chambers where neither broadband noise, traffic noise, nor noise recorded in a noisy outdoor setting was broadcast. However, unlike gray bats, mouse-eared bats forage by listening for sounds produced by non-flying prey while using echolocation for navigation only. Anthropogenic (i.e., traffic) noise may mask sounds made by ground-dwelling insects, while call frequencies of echolocating bats like the gray bat are above frequencies produced by traffic (Jones 2008). Sound frequencies of operational noise and the degree that operational noise may affect foraging gray bats are not known. However, the portion of the Tennessee River adjacent to the WBN site that might experience operational noise from Units 1 and 2 has not been identified as an especially important foraging area for gray bats. Additionally, the displacement of gray bats from using this portion of the Tennessee River for foraging would not noticeably affect gray bat populations that spend summers in nearby caves. Therefore, the staff concludes, as did FWS in 1995, that although the operation of WBN Unit 2, within the bounds of the NPDES permit, may affect the Tennessee River, it would not jeopardize the continued existence of the gray bat in the vicinity of the WBN site.

5.2 Freshwater Mussels and Snail Darters

Operations at the WBN Unit 2 site have the potential to affect freshwater mussels and fish in the vicinity of the site as a result of water consumption, entrainment, impingement, and thermal and chemical effects.

5.2.1 Water Consumption

As discussed in Section 3, the maximum annual plant consumption rate (amount of water that will be consumed by WBN Unit 2) represents 0.1 percent of the mean annual flow of the Tennessee River at Watts Bar Dam. This is small and will not measurably affect the habitat available for Federally listed species.

5.2.2 Entrainment and Impingement

The SCCW intake pulls water from the reservoir above Watts Bar Dam. As a result, snail darters or freshwater mussels residing below the dam would not be affected by continued operation of the SCCW.

Although adult mussels are not susceptible to entrainment or impingement by the IPS, the fish host on which the glochidia implants could be entrained or impinged. Hosts for the rough pigtoe and the orange pimpleback have not been identified. The hosts for the pink mucket include smallmouth, spotted, and largemouth bass, as well as freshwater drum and sauger. Less than 10 percent of the larval fish in the intake canal were drum, sauger, or bass (see Table 5-1).

Other fish present in the vicinity of the intakes, including any snail darters potentially present in the Watts Bar or Chickamauga Reservoirs, also could be subject to entrainment and impingement. As shown in Table 5-2, very small numbers of fish are impinged overall by the IPS, with the exception of shad impinged between January 2011 and the first week of March (TVA 2011d). As a result, the NRC staff considers the likelihood that entrainment or impingement from operation of WBN Unit 2 would affect the host for pink mucket glochidia would be minimal. A variety of darters and sculpins are hosts for larval Eastern fanshell pearlymussel and the dromedary pearlymussel. Except for the logperch, which is a host for the Eastern fanshell, the other host fish for these two mussel species are not present based on sampling studies as far back as 1975. Snail darters are not known to be present in the vicinity of the WBN site.

Table 5-1. Percent Composition of Dominant Larval Fish Taxa Collected in the CCW Intake Channel during 1984 and 1985 and 1996 and 1997

| | | Percent Composition of Larval Fish Taxa | | | |
| | | Preoperational | | Operational | |
Taxon	Common name	1984	1985	1996	1997
Aplodinotus grunniens	Freshwater drum	0.1	0.2	0.8	0.4
Centrachidae	Sunfish	0.9	12.5	7.7	8.2
Clupeidae	Unidentified shad	97.8	86.4	90.5	84.7
Dorosoma sp.	Threadfin or gizzard shad	0.09	--	0.8	0.2
Morone (not *saxatilis*)	Bass (not striped)	0.6	0.5	0.09	0.9
Morone sp.	Bass	0.5	0.5	0.09	5.6
Source: TVA 1998b					

Table 5-2. Actual and Estimated Numbers of Fish Impinged at WBN Plant during Sample Periods from March 1996 through March 1997, March 1997 through October 7, 1997 and during March 2010 through March 2011

	March 1996 – March 1997 and March 1997 – October 1997						March 2010 – March 2011		
	Actual Number Impinged		Total Annual Estimated Number		Percent Composition		Actual Number Impinged	Total Annual Estimated Number	Percent Composition
	Sampling Period		Sampling Period		Sampling Period				
Common Name	1	2	1	2	1	2			
Gizzard shad	4	0	41	0	25%	0%	1,172	8,204	60.4%
Threadfin shad	2	0	20	0	12.5%	0%	766	5,362	39.5%
Freshwater drum	3	3	30	31	18.7%	75%	0	0	0%
Channel catfish	1	0	10	0	6.3%	6.3%	0	0	0%
Flathead catfish	1	0	10	0	6.3%	0%	0	0	0%
Bluegill	2	0	20	0	12.5%	0%	0	0	0%
Redear sunfish	1	0	10	0	6.2%	0%	0	0	0%
White crappie	2	0	20	0	12.5%	0%	0	0	0%

Table 5-2. (contd)

Common Name	March 1996 – March 1997 and March 1997 – October 1997						March 2010 – March 2011		
	Actual Number Impinged		Total Annual Estimated Number		Percent Composition		Actual Number Impinged	Total Annual Estimated Number	Percent Composition
	Sampling Period		Sampling Period		Sampling Period				
	1	2	1	2	1	2			
Log perch	0	1	0	10	0%	25%	0	10.2	0%
Inland silverside	0	0	0	0	0%	0%	1		0.1%
Total	16	4	161	41	100%	100%	1,939	13,573	100%
Source: TVA 1998a; TVA 2011d									

5.2.3 Thermal and Chemical Effects

The current NPDES permit issued by the State of Tennessee for Unit 1 specifies limits on the amount of thermal effluent the plant may discharge into the Tennessee River, establishes an active mixing zone, and defines in-stream monitoring and reporting requirements necessary to comply with effluent limitations. The additional increment for flow of the SCCW is approximately 14 percent of the current amount of heat discharged. The measurements and model indicate that the plume rises after hitting the concrete pad located at the end of the discharge, allowing room underneath for fish passage and not directly affecting the freshwater mussels.

In an effort to limit the impact to the mussels in the vicinity of the SCCW discharge, a mussel relocation zone was established that extended 46 m (150 ft) from the right bank and 23 m (75 ft) upstream and downstream of the centerline of Outfall 113. The area was surveyed for mussels in 1997. The only Federally protected mussel identified was a single specimen of the pink mucket. The freshwater mussels that were in an area of 46 m by 46 m (150 ft by 150 ft) at the outlet to the SCCW system (23 m [75 ft] upstream and downstream of the centerline of Outfall 113) were relocated before the startup of the SCCW (TVA 1999). TVA moved these mussels in an effort to prevent adverse effects from operation of the SCCW system discharge. In addition, TVA placed a ramp on the invert of the SCCW outfall to deflect the discharge upward, and away from the bottom of the river (TVA 2004a). The analysis of in-stream data collected by TVA for Outfall 113 showed that heat from the SCCW effluent does not reach the bottom in significant amounts (TVA 2004a).

TVA also conducted field studies to confirm the diffuser performance for Outfall 101 (TVA 1998a). To provide adequate dilution of the plant effluent, TVA permits the diffusers to discharge water only when Watts Bar Dam releases at least 99 m^3/s (3,500 cfs). This policy will remain the same when both units are operating. The location and design of the diffuser discharge should not impede fish passage up and down the Tennessee River. Fish (including darters) and other organisms likely would avoid the warmer water, but mussels and benthic organisms would not be able to avoid the elevated temperatures. However, as indicated, the diffuser's plume angles upward at 45 degrees above horizontal in the downstream direction, and as a result, the plume would not have much of an effect on the mussels and other benthic organisms in the area of or immediately downstream of the diffuser.

TVA conducted hydrothermal surveys (combined with ichthyoplankton surveys) in May 2010 to coincide with the period of expected peak abundance of ichthyoplankton and in August 2010 to coincide with the near maximum ambient water temperatures. TVA mapped and tracked the thermal plume from discharge Outfall 113 at a time when there were no releases from the Watts Bar Dam, showing that the plume remained near the surface and spread across the river. During periods of normal release from Watts Bar Dam, the plume remains near the right descending bank. Based on the ichthyoplankton taxa collected, thermal tolerance data, river temperatures, and exposure times, TVA concluded, "there is essentially no risk of thermal damage to ichthyoplankton during no-flow conditions" from the dam (TVA 2011e).

According to NPDES permit requirements, TVA conducts biotoxicity tests (i.e., 3-brood *Ceriodaphnia dubia* survival and reproduction tests and 7-day fathead minnow (*Pimephales promelas*) larval survival and growth tests) on samples of final effluent from Outfalls 101, 102, 112, and 113. The NRC staff reviewed 12 years of toxicity testing data provided in the NPDES permit request (TVA 2009c). The data showed that percentage survival in the highest concentration tested for 96-hour survival was a mean of 92.8 percent for Outfall 101 and 99 percent survival for Outfall 113. Based on the results of these tests and the lack of changes from the quantity of chemicals that would be discharged, the NRC staff determined that the aquatic biota of Chickamauga Reservoir would not be affected by chemical discharges resulting from the additional operation of WBN Unit 2.

5.3 Summary

Based on the information provided in this section of the BA, the staff determines that there would be no adverse impact to threatened and endangered species from noise, cooling tower operation, water consumption, entrainment, impingement, and thermal, and chemical discharge operations of WBN Unit 2.

6.0 Cumulative Impacts

The NRC staff considered potential past, present, and reasonably foreseeable activities that could have cumulative effects on Federally protected species in conjunction with operating another nuclear unit at the WBN site.

6.1 Terrestrial Species (Gray Bat)

For this analysis, the geographic area of interest includes all of Rhea and Meigs Counties and lands of Hamilton, Bradley, McMinn, Roane, Anderson, Knox, Blount, and Loudon Counties that occur within 0.8 km (0.5 mi) of the transmission line system that would support WBN Unit 2. Based on the nature of the potential impacts and attributes of the affected terrestrial resources, these counties would bound the area expected to be affected by the operation of WBN Unit 2.

WBN Unit 2 is co-located with WBN Unit 1. Operation of Unit 1 produces a visible vapor plume and operational noise. However, because of the nature of the effects from operating Unit 2, the synergistic effect of operating both units is not expected to affect the gray bat any more than the operation of a single unit.

Little is known about a phenomenon known as white-nose syndrome that has caused massive mortality of many bat species in the northeastern United States. (Cohn 2008). The name comes from a white *Geomyces* fungus that grows on affected bats' muzzles. White-nose syndrome has affected at least six species of bats and has been confirmed in at least eight U.S. states, including Tennessee, and three Canadian provinces (FWS 2010b). The mortality rate of affected bats is high, with bat colony reductions in infected caves over 90 percent. White-nose syndrome afflicts at least six bat species, and it may be affecting gray bats (FWS 2010c). Because little is known about white-nose syndrome, the extent that it may affect the gray bat population is still unknown.

6.2 Aquatic Species

Historically, the Tennessee River was free flowing and flooded annually. Before 1936, the few power dams that obstructed streams in Tennessee backed up relatively small impoundments. In 1936, TVA completed Norris Reservoir, its first reservoir on the Tennessee River. Currently, TVA operates nine dams on the Tennessee River. The dams have fragmented the watershed, and the isolation and stress dams have imposed on tributaries of the river have caused and will continue to cause extirpation of fish (such as the snail darter) and freshwater mussels. Historically, species introduced after building the dams, over fishing of species such as paddlefish, harvesting of mussels, toxic spills, mining, and agriculture have affected the fish fauna.

Impacts on aquatic biota from operations at both WBN Unit 1 and Unit 2 are difficult for NRC staff to separate, because both units share the same intake and discharge systems. The makeup flow rate through the IPS would be almost twice that for the single unit operation. The intake flow rate for the SCCW when both units are operating would be less than that for operating a single unit. The volume of water returned to the river through the SCCW discharge would be less because of greater amounts of water evaporation. WBN Units 1 and 2 together would consume 1.7 m^3/s (61 cfs) of water, which is approximately 0.2 percent of the mean flow past the WBN site. This would result in an increase of less than 10 percent from the current consumptive use of WBN Unit 1 (see Table 3.1).

Other facilities also have adverse impacts on the aquatic biota of Watts Bar and Chickamauga Reservoirs by entrainment, impingement, or thermal, chemical, or physical discharges. These facilities include Watts Bar Dam (TRM 529.9), which is immediately upstream of the facility (the SCCW intake is located on the dam); Sequoyah Nuclear Plant, which is located on the Chickamauga Reservoir (TRM 484.5); the Kingston Fossil Plant, which is located at the junction of Emory River and Clinch River (approximately 69 river kilometers [42 river miles]); and Oak Ridge National Laboratory, which is located on the Clinch River (approximately 89 river kilometers [55 river mile]) upstream of Watts Bar Dam. The facility that has the greatest effect on the freshwater mussels would be the Watts Bar Dam. Watters (1999) points to impoundments, dredging, snagging, and channelization as having long-term detrimental effects on freshwater mussels. The impoundments result in silt accumulation, loss of shallow-water habitat, stagnation, pollutant accumulation, and nutrient-poor water.

7.0 Conclusions

The potential impacts of the operation of WBN Unit 2 on Federally protected species near the site have been evaluated. This BA considers the known distributions and records of those species, and the potential ecological impacts of facility operations on those species. Based on this review, the NRC staff reached the following conclusions:

- Operation of proposed Unit 2 at the WBN site may affect foraging for a small number of gray bats. However, the portion of the Tennessee River adjacent to the WBN site that may receive operational noise has not been identified as an especially important foraging area for gray bats. Gray bat avoidance of this portion of the Tennessee River for foraging would not noticeably affect populations that spend summers in nearby caves. Therefore, the staff concludes, as did FWS in 1995, that although the operation of WBN Unit 2 may affect the Tennessee River, it would not jeopardize the continued existence of the gray bat in the vicinity of the WBN Site. Therefore, the NRC staff concludes that direct, indirect,or cumulative impacts from the operation of WBN Unit 2 are not likely to adversely affect the gray bat.

- Operation of the proposed Unit 2 may affect the pink mucket mussel that is known to potentially be present in the vicinity of the WBN site. The impact of entrainment or impingement is not likely to affect the survival of the pink mucket because of the low fraction of water withdrawn and the low demonstrated rates of entrainment and impingement from the intake in the Chickamauga Reservoir. Although thermal discharges may affect the pink mucket, this is unlikely from the discharge of the SCCW as a result of the relocation of freshwater mussels near the outlet of the SCCW discharge system. It is also unlikely at the IPS discharge because of mitigative strategies enacted by the applicant, such as the use of diffusers only when Watts Bar Dam releases at least 99 m^3/s (3,500 cfs) and the orientation of the diffuser plume (45 degrees above horizontal in the downstream direction). Further, based on a review of 12 years of toxicity testing data provided in the NPDES permit request (TVA 2009b) it is unlikely that chemical discharges will affect the pink mucket mussel. Thus, the NRC staff concludes that operation of the proposed Unit 2, even in addition to the operation of Unit 1, is not likely to adversely affect the pink mucket.

- Operation of the proposed Unit 2 is not likely to affect the Eastern fanshell mussel because they are likely no longer present in the vicinity of the WBN site. The last Eastern fanshell was found in 1985 in the mussel bed nearest the WBN site (TRM 528.2 to TRM 528.9). It was not seen in any of the following 10 surveys that were conducted in the vicinity or downstream of the WBN site between 1985 and 1997, or in the survey conducted in 2010. Therefore, the NRC staff concludes that operation of WBN Unit 2, even in addition to the operation of Unit 1, will have no effect on the Eastern fanshell mussel.

- Operation of the proposed Unit 2 is not likely to affect the rough pigtoe because the species probably is no longer present in the vicinity of the WBN site. The last rough pigtoe was observed in Chickamauga Reservoir near the site in 1985. TVA conducted seven additional surveys of the mussel beds downstream of the WBN site between 1985 and 1997, and one in 2010 without observing a live rough pigtoe mussel. Therefore, the NRC staff concludes that operation of WBN Unit 2 will have no effect on the rough pigtoe mussel.

- Operation of the proposed Unit 2 is not likely to affect the dromedary pearly mussel because they probably are no longer present in the vicinity of the WBN site. The most recent observation of a dromedary pearly mussel occurred in 1983. Additional surveys were conducted annually over the next 14 years, with an additional survey in 2010 and no specimens of the dromedary pearly mussel were identified. Therefore, the NRC staff concludes that operation of WBN Unit 2 will have no effect on the dromedary pearly mussel.

- The orangefoot pimpleback mussel has not been reported from the vicinity of the proposed Unit 2 during any of the surveys conducted since 1983. Therefore, the NRC staff concludes that operation of WBN Unit 2 will have no effect on the orangefoot pimpleback mussel.

- Operation of the proposed Unit 2 is unlikely to affect the snail darter because they have not been observed in Chickamauga Reservoir in the vicinity of the WBN site. The population that was identified as recently as 1985 as living in Sewee Creek from Creek Mile 3.2 to Creek Mile 5.7 could possibly still be located in the creek since no additional studies were found to have been conducted since that time. However, operation of the proposed Unit 2 would be unlikely to affect a population located in Sewee Creek. Therefore, the NRC staff concludes that operation of WBN Unit 2 will have no effect on the snail darter.

8.0 References

40 FR 47506. October 9, 1975. "Amendment Listing the Snail Darter as an Endangered Species." *Federal Register.* U.S. Department of the Interior.

41 FR 17736. April 28, 1976. "Determination that Two Species of Butterflies are Threatened Species and Two Species of Mammals are Endangered Species." *Federal Register.* U.S. Department of the Interior.

41 FR 24062. June 14, 1976. "Endangered and Threatened Wildlife and Plants; Endangered Status for 159 Taxa of Animals." *Federal Register.* U.S. Department of the Interior.

49 FR 27510. November 10, 1975. "Endangered and Threatened Wildlife and Plants: Final Rule Reclassifying the Snail Darter (*Percina tanasi*) from an Endangered Species to a Threatened Species and Rescinding Critical Habitat Designation." *Federal Register.* U.S. Department of the Interior.

55 FR 25591. June 21, 1990. "Endangered and Threatened Wildlife and Plants; Designation of the Freshwater Mussel, the Fanshell as an Endangered Species." *Federal Register.* U.S. Department of the Interior.

76 FR 3392. January 19, 2011. "Endangered and Threatened Wildlife and Plants; Endangered Status for the Sheepnose and Spectaclecase Mussels. Proposed Rule." *Federal Register.* U.S. Department of the Interior.

Brady, J., T. Kunz, M.D. Tuttle, and D. Wilson. 1982. *Gray Bat Recovery Plan.* Fish and Wildlife Reference Service, Denver, Colorado.

Cohn, J.P. 2008. "White-nose Syndrome Threatens Bats." *BioScience* 58(11).

Endangered Species Act, as amended. 16 USC 1531 et seq.

Etnier, D.A. and W.C. Starnes. 1993. *The Fishes of Tennessee.* University of Tennessee Press, Knoxville, Tennessee.

Harvey, M.J. and E.R. Britzke. 2002. *Distribution and Status of Endangered Bats in Tennessee.* Tennessee Wildlife Resources Agency, Nashville, Tennessee.

Jones, G. 2008. "Sensory Ecology: Noise Annoys Foraging Bats." *Current Biology* 18(23):R1099.

Mirarchi, R., J. Garner, M. Mettee, and P. O'Neil, eds. 2004. *Alabama Wildlife, Volume 2, Imperiled Aquatic Mollusks and Fishes.* University of Alabama Press, Tuscaloosa, Alabama.

NatureServe. 2010. *NatureServe Explorer: An Online Encyclopedia of Life.* Version 7.1. NatureServe, Arlington, Virginia. Accessed July 1, 2010 at http://www.natureserve.org/explorer.

Pacific Northwest National Laboratory (PNNL). 2009. Letter from J.A. Stegen (Pacific Northwest National Laboratory) to U.S. Nuclear Regulatory Commission dated November 12, 2009. "Final Watts Bar Site Audit Trip Report for JCN J-4261, 'Technical Assistance for Operating License Environmental Review – Watts Bar Unit 2 Reactivation.'" Accession No. ML100220022.

Parmalee, P.W. and A.E. Bogan. 1998. *Freshwater Mussels of Tennessee.* University of Tennessee Press, Knoxville, Tennessee.

Schaub, A., J. Ostwald, and B.M. Siemers. 2008. "Foraging Bats Avoid Noise." *The Journal of Experimental Biology* 211:3174-3180.

Simmons, J.W. and D.S. Baxter. 2009. *Biological Monitoring of the Tennessee River Near Watts Bar Nuclear Plant Discharge, 2008.* Tennessee Valley Authority, Aquatic Monitoring and Management, Chattanooga, Tennessee. Accession No. ML073510313.

Tennessee Department of Environment and Conservation (TDEC). 2011. *State of Tennessee NPDES Permit No. TN0020168.* Nashville, Tennessee. Accession No. ML11215A099.

Tennessee Valley Authority (TVA). 1972. *Final Environmental Statement, Watts Bar Nuclear Plant Units 1 and 2.* TVA-OHES-EIS-72-9, Office of Health and Environmental Science, Chattanooga, Tennessee. Accession No. ML073470580.

Tennessee Valley Authority (TVA). 1976. *Tennessee Valley Authority Environmental Information Watts Bar Nuclear Plant Units 1 and 2, November 18, 1976.* Chattanooga, Tennessee. Accession No. ML073230776.

Tennessee Valley Authority (TVA). 1986. *Preoperational Assessment of Water Quality and Biological Resources of Chickamauga Reservoir, Watts Bar Nuclear Plant, 1973-1985.* TVA/ONRED/WRF-87/1a. Knoxville, Tennessee. Accession No. ML073510313.

Tennessee Valley Authority (TVA). 1997. *Watts Bar Nuclear Plant Supplemental Condenser Cooling Water Project, Thermal Plume Modeling.* Norris, Tennessee.

Tennessee Valley Authority (TVA). 1998a. *Watts Bar Nuclear Plant Supplemental Condenser Cooling Water Project, Environmental Assessment.* Knoxville, Tennessee.

Tennessee Valley Authority (TVA). 1998b. *Aquatic Environmental Conditions in the Vicinity of Watts Bar Nuclear Plant During Two Years of Operation, 1996-1997.* Norris, Tennessee. Accession No. ML073510313.

Tennessee Valley Authority (TVA). 1999. *July 1999 Verification Study of Thermal Discharge for Watts Bar Nuclear Plant Supplemental Condenser Cooling Water System.* WR99-2-85-143. Norris, Tennessee. Accession No. ML073510313.

Tennessee Valley Authority (TVA). 2004. *Proposed Modifications to Water Temperature Effluent Requirements for Watts Bar Nuclear Plant Outfall 113.* WR2004-3-85-149. Knoxville, Tennessee. Accession No. ML073510313.

Tennessee Valley Authority (TVA). 2005a. Letter from Betsy M. Eiford (Tennessee Valley Authority, Environmental Support Manager) to Ed Polk (Tennessee Department of Environmental Conservation, Manager) dated July 22, 2005, "Tennessee Valley Authority (TVA) - Watts Bar Nuclear Plant (WBN) - NPDES Permit No. TN0020168 - Phase 11 316(b) Rule for Existing Facilities - Proposal for Information Collection for Supplemental Condenser Cooling Water (SCCW) System." Accession No. ML052150033.

Tennessee Valley Authority (TVA). 2005b. *Winter 2005 Compliance Survey for Watts Bar Nuclear Plant Outfall 113 Passive Mixing Zone.* WR2005-2-85-151. Knoxville, Tennessee. Accession No. ML073510313.

Tennessee Valley Authority (TVA). 2006. *Summer 2005 Compliance Survey for Watts Bar Nuclear Plant Outfall 113 Passive Mixing Zone.* WR2005-85-152. Knoxville, Tennessee. Accession No. ML073510313.

Tennessee Valley Authority (TVA). 2007a. Letter from William R. McCollum (Tennessee Valley Authority, Chief Operating Officer) to U.S. Nuclear Regulatory Agency dated August 3, 2007, "Watts Bar Nuclear Plant (WBN) – Unit 2 – Reactivation of Construction Activities." Accession No. ML072190047.

Tennessee Valley Authority (TVA). 2007b. *Summer 2006 Compliance Survey for Watts Bar Nuclear Plant Outfall 113 Passive Mixing Zone.* Knoxville, Tennessee. Accession No. ML073510313.

Tennessee Valley Authority (TVA). 2007c. *Winter 2006 Compliance Survey for Watts Bar Nuclear Plant Outfall 113 Passive Mixing Zone.* Knoxville, Tennessee. Accession No. ML073510313.

Tennessee Valley Authority (TVA). 2008. *Final Supplemental Environmental Impact Statement; Completion and Operation of Watts Bar Nuclear Plant Unit 2, Rhea County, Tennessee*, submitted to NRC as the TVA Environmental Report for an Operating License, Knoxville, Tennessee.

Tennessee Valley Authority (TVA). 2009a. Letter from Masoud Bajestani (Watts Bar Unit 2, Vice President) to U.S. Nuclear Regulatory Commission dated March 4, 2009, "Watts Bar Nuclear Plant (WBN) Unit 2 – Operating License Application Update." Accession No. ML090700378.

Tennessee Valley Authority (TVA). 2009b. *Watts Bar Nuclear Plant (WBN) - Unit 2 - Final Safety Analysis Report (FSAR)*. Amendment 94, Spring City, Tennessee.

Tennessee Valley Authority (TVA). 2009c. Letter from Darrin Hutchison (Tennessee Valley Authority, Environmental Technical Support Manager) to Vijin Janjic (Tennessee Department of Environment & Conservation, Permit Writer) dated April 15, 2009, "Watts Bar Nuclear Plant (WBN) – National Pollutant Discharge Elimination System (NPDES) Permit No. TN0020168 – Request for Raw Water Treatment Modification." Accession No. ML091190175.

Tennessee Valley Authority (TVA). 2010a. Letter from Masoud Bajestani (Watts Bar Unit 2, Vice President) to U.S. Nuclear Regulatory Commission dated February 25, 2010 in response to NRC letter dated December 3, 2009 and TVA letters dated July 2, 2008, January 27, 2009, and December 23, 2009, "Watts Bar Nuclear Plant (WBN) Unit 2 – Additional Information Regarding Environmental Review (TAC No. MD8203)." Accession No. ML100630115.

Tennessee Valley Authority (TVA). 2010b. Letter from Masoud Bajestani (Watts Bar Unit 2) to U.S. Nuclear Regulatory Commission dated July 2, 2010, "Watts Bar Nuclear Plant (WBN) Unit 2 – Submittal of Additional Information Requested During May 12, 2010, Request for Additional Information (RAI) Clarification Teleconference Regarding Environmental Review (TAC No. MD8203)." Accession No. ML101930470.

Tennessee Valley Authority (TVA). 2010c. Letter from Lindy Johnson (Tennessee Valley Authority, Water Permits and Compliance) to Richard Urban (Tennessee Department of Environment & Conservation) dated August 17, 2010, "Tennessee Valley Authority (TVA) - Watts Bar Nuclear Plant (WBN) – NPDES Permit No. TN0020168 – Permit Modification Request – Addition of Unit 2 Operation."

Tennessee Valley Authority (TVA). 2010d. Letter from R.M. Krich (Nuclear Licensing, Tennessee Valley Authority) to U.S. Nuclear Regulatory Commission dated July 6, 2010, "Notification of National Pollutant Discharge Elimination System Permit Renewal." Accession No. ML101890069.

Tennessee Valley Authority (TVA). 2010e. *2009 Annual Non-Radiological Environmental Operating Report*. Chattagnooga, Tennessee. Accession No. ML101310204.

Tennessee Valley Authority (TVA). 2011a. Letter from Edwin E. Freeman (Watts Bar Unit 2, Engineering) to U.S. Nuclear Regulatory Commission dated January 4, 2011, "Watts Bar Nuclear Plant (WBN) Unit 2 – Additional Information Related to U.S. Nuclear Regulatory Commission (NRC) Request for Additional Information (RAI) Regarding Environmental Review." Accession No. ML110060510.

Tennessee Valley Authority (TVA). 2011b. Letter from David Stinson (Watts Bar Unit 2) to U.S. 11 Nuclear Regulatory Commission dated May 19, 2011, "Watts Bar Nuclear Plant (WBN) Unit 2 –Intake Pumping Station Water Velocity – Response to Request for Additional Information." Accession No. ML11143A083.

Tennessee Valley Authority (TVA). 2011c. Letter from David Stinson (Watts Bar Unit 2, Vice President) to U.S. Nuclear Regulatory Commission dated March 24, 2011, "Watts Bar Nuclear Plant (WBN) Unit 2 – Additional Information Related to U.S. Nuclear Regulatory Commission (NRC) Regarding Environmental Review." Accession No. ML110871475.

Tennessee Valley Authority (TVA). 2011d. Letter from Edwin E. Freeman (Watts Bar Unit 2 Engineering Manager) to U.S. Nuclear Regulatory Commission dated March 28, 2011, "Watts Bar Nuclear Plant (WBN) Unit 2 – Additional Information Related to U.S. Nuclear Regulatory commission (NRC) Environmental Review." Accession No. ML110890472.

Tennessee Valley Authority (TVA). 2011e. Letter from Marie Gillman (Watts Bar Unit 2, Vice President, Acting) to U.S. Nuclear Regulatory Commission dated February 7, 2011, "Submittal of 'Hydrothermal Effects on the Ichthyoplankton from the Watts Bar Nuclear Plant Supplemental Condenser Cooling Water outfall in Upper Chickamauga Reservoir' Report." Accession Nos. ML110400383, ML110400384.

U.S. Fish and Wildlife Service (FWS). 1983a. *Dromedary Pearly Mussel Recovery Plan*. Atlanta, Georgia.

U.S. Fish and Wildlife Service (FWS). 1983b. *Snail Darter Recovery Plan*. Atlanta, Georgia.

U.S. Fish and Wildlife Service (FWS). 1984a. Rough *Pigtoe Pearly Mussel Recovery Plan*. Atlanta, Georgia.

U.S. Fish and Wildlife Service (FWS). 1984b. *Orange-footed Pearly Mussel Recovery Plan*. Atlanta, Georgia.

U.S. Fish and Wildlife Service (FWS). 1985. *Recovery Plan for the Pink Mucket Pearly Mussel, Lampsilis obiculata (Hildreth, 1828)*. Atlanta, Georgia.

U.S. Fish and Wildlife Service (FWS). 1991. *Recovery Plan for Fanshell (Cyprogenia stegaria (= C. irrorata))*. Atlanta, Georgia.

U.S. Fish and Wildlife Service (FWS). 1992. "Species Accounts, Snail Darter (Percina (Imostoma) tanasi)." In *Endangered and Threatened Species of the Southeastern United States: The Red Book*. Accessed December 22, 2009 at http://www.fws.gov/cookeville/docs/endspec/snldtrsa.html.

U.S. Fish and Wildlife Service (FWS). 2009. Letter from Mary E. Jennings (Field Supervisor) to U.S. Nuclear Regulatory Commission dated October 29, 2009, "FWS #2009-SL-0885. Watts Bar Nuclear Plant, Unit 2 – Request for list of Protected Species, Rhea County, Tennessee." Accession No. ML0929301820.

U.S. Fish and Wildlife Service (FWS). 2010a. "Dromedary Pearlymussel - *Dromus dromas*." Accessed May 17, 2010 at http://ecos.fws.gov/docs/life_histories/F00K.html.

U.S. Fish and Wildlife Service (FWS). 2010b. *White-Nose Syndrome*. Accessed May 26, 2010 at http://www.fws.gov/WhiteNoseSyndrome/about.html.

U.S. Fish and Wildlife Service (FWS). 2010c. *The White-nose Syndrome Mystery-Something is Killing Our Bats*. Available at http://www.fws.gov/WhiteNoseSyndrome/pdf/White-nose_mystery.pdf.

U.S. Fish and Wildlife Service (FWS). 2011. "Sheepnose (a freshwater mussel) – *Plethobasus cyphyus*)." Accessed March 28, 2011 at http://www.fws.gov/midwest/endangered/clams/sheepnose/SheepnoseFactSheetJan2011.html.

U.S. Nuclear Regulatory Commission (NRC). 1978. *Final Environmental Statement Related to Operation of Watts Bar Nuclear Plant Units Nos. 1 and 2*. NUREG-0498, Washington, D.C. Accession No. ML082540803, ML082560293, ML082560292, ML082560291, ML082560289.

U.S. Nuclear Regulatory Commission (NRC). 1995. *Final Environmental Statement Related to the Operation of Watts Bar Nuclear Plant, Units 1 and 2*. NUREG-0498 Supplement No. 1, Washington, D.C. Accession No. ML081430592.

U.S. Nuclear Regulatory Commission (NRC). 2009. Letter from U.S. Nuclear Regulatory Commission to M.E. Jennings (Field Supervisor, U.S. Fish and Wildlife Service) dated September 2, 2009. "Watts Bar Nuclear Plant, Unit 2- Request for List of Protected Species within the Area under Evaluation for the Operating License Application Environmental Review." Accession No. ML092100088.

Watters, G.T. 1999. "Freshwater Mussels and Water Quality: A Review of the Effects of Hydrologic and Instream Habitat Alterations." In *Proceedings of the First Freshwater Mollusk Conservation Society Symposium*. Ohio Biological Survey, Columbus, Ohio.

Widlak, J.C. 1995. *Biological Opinion for the Proposed Operation of the Watts Bar Nuclear Plant-Rhea County, Tennessee*. U.S. Fish and Wildlife Service. Ecological Services Field Office, Cookeville, Tennessee.

Appendix G

List of Authorizations, Permits, and Certifications

Table G-1. Federal, State, and Local Authorizations

Agency	Authority	Phase/Requirement/Status	Activity Covered
U.S. Nuclear Regulatory Commission (NRC)	Title 10 of the Code of Federal Regulations (CFR) Part 50	Preconstruction. Construction Permit CPPR-92 EXP: 31MAR2013.	Permit for construction of a utilization facility. TVA submitted a request to NRC dated May 17, 2012 to extend the construction permit expiration date to September 30, 2016.
NRC	10 CFR Part 50	OL Submittal. Updated license application filed 04MAR2009.	Operation of a utilization facility for commercial purposes.
U.S. Fish and Wildlife Service (FWS)	16 U.S.C. §§ 1531 et seq.	SFES. Concurrence. 1995 consultation with FWS, cited in SFES Appendix D, applied to WBN Unit 1 and WBN Unit 2. 2007 SFES also found no impacts.	Consultation concerning potential impacts to Federal threatened and endangered species.
U.S. Department of the Interior (DOI)	42 U.S.C. § 1996; 25 U.S.C. § 3001 et seq.	SFES. Consultation. Consultation not required as SFES did not identify any items of cultural significance to Native American tribes.	Identification, protection, and repatriation of items of cultural significance to Native American tribes.
Federal Aviation Administration (FAA)	14 CFR Part 77	Preconstruction. Notification not required as no activities affect structures over 60 m (200 ft).	Preconstruction letter of notification to FAA results in a written response certifying that no hazards exist or recommending project modification.
U.S Coast Guard	14 U.S.C. §§ 81, 83, 85, 633; 49 U.S.C. § 1655(b).	Preconstruction. Authorization not required as no activities affect navigation.	Navigation markers authorization to protect river navigation from hazards connected with temporary construction activities in a river.
Tennessee Department of Environment and Conservation (TDEC)	Water Quality Control Act, TCA §§ 69-3-101 et seq.	Preoperation. Certification. TVA will seek any required certification from TDEC prior to issuance of the OL.	Aquatic resource alteration permit for any alteration of the properties of State waters. This permit also serves as a Section 401 water quality certification, which is required prior to seeking a Federal permit or license, including an operating license from the NRC.

Table G-1. (contd)

Agency	Authority	Phase/Requirement/Status	Activity Covered
TDEC	CWA Sections 316(a), 316(b) and 402(p)	National Pollutant Discharge Elimination System (NPDES) Permit. TDEC issued NPDES Permit TN0020168 on June 30, 2011; effective August 1, 2011 to June 30, 2016; modifications made November 28, 2011.	Facility permit for point source discharges of wastewater to surface waters and in-stream monitoring. Limits liquid pollutant discharges to surface water from WBN Units 1 and 2.
U.S. Army Corps of Engineers (USACE)	33 U.S.C. § 1344; 33 U.S.C. §§ 1341	Preconstruction. Permit. USACE stated, as listed in SFES Appendix D, that a Section 404 permit is not required as no work requires discharge of dredged or fill material.	Section 404 permit required for discharge of dredged and fill material. A Section 401 certification that the action does not violate state water quality standards is required prior to obtaining a Section 404 permit.
TDEC Air Division	Tennessee Air Quality Act, TCA §§ 68-201-101 et seq. 42 U.S.C. §§ 7401 et seq.	Preconstruction. Construction permit. Permit 957606P held by TVA. EXP: 01JAN2007 Renewal pending. Requested update and consolidation with operating permit 448529 on 23JAN2007.	Construction permit for prevention of significant deterioration of air quality required to construct an air contaminant source.
TDEC Air Division	TCA §§ 68-201-101 42 U.S.C. §§ 7401 et seq.	Preoperation. Operating permit. Permit 448529 held by TVA. EXP: 01SEP2010. Continued operation is permissible since TVA submitted a 2010 Air Permit application on June 29, 2010 (with July 29, 2011 update), which met the 60-day requirement prior to expiration.	This permit covers emissions from the WBN site for both Unit 1 and Unit 2 equipment. TVA - WBN opted out of major source - Not a Title V Permit.
TDEC Water Division	33 U.S.C. §1342, TCA §§ 69-3-101 et seq.	Continuing permit requirement. Industrial Storm Water Multi-Sector General Permit TNR050000 held by TVA. EXP: 14MAY2014.	Permit for discharge of storm water associated with land disturbance and industrial activity.

Table G-1. (contd)

Agency	Authority	Phase/Requirement/Status	Activity Covered
TDEC Water Division	33 U.S.C. §1342; TCA §§ 69-3-101 et seq.	Preconstruction. Permit. Not required, as no construction activities planned that would result in storm water discharge.	Permit for discharge of storm water associated with construction involving clearing, grading or excavation that result in an area of disturbance of one or more acres, and activities that result in the disturbance of less than one acre if it is part of a larger common plan of development.
TDEC Division of Solid and Hazardous Waste Management (SHW)	Tennessee Solid Waste Disposal Act, TCA §§ 68-211-101 et seq.	Preoperation. Permit. Permit number DML72-103-0025 held by TVA. EXP: N/A.	Site Permit for operation of a Class IV disposal facility (onsite construction and demolition landfill).
TDEC Division of SHW	TCA §§ 68-212	EPA Facility ID TN2640030035 Construction Demolition Landfill Permit Number DML 721030025 EXP: N/A.	Transportation of waste.
Alabama Department of Environmental Management (ADEM)	ADEM Admin. Code R. 335-14	Ongoing. Permit. Operation Permit AL2-640-090-005 held by TVA. EXP: 06MAY2011.	Storage of hazardous waste at the hazardous waste storage facility in Muscle Shoals, Alabama.
TDEC Division UST or Solid and Hazardous Waste	TCA §§ 68-212	Preconstruction/operation. Permit. Not required as no underground storage tanks as defined by TDEC.	Installation/operation of underground storage tanks that store regulated substances.
Tennessee Historical Commission (THC) (State Historic Preservation Officer)	16 U.S.C. §§ 470 et seq. 36 CFR Part 800	Preoperation. Consultation. Consultation with THC completed and documented in SFES Appendix D (TVA ER 2008).	Review and analysis of cultural and historic resources, including completion of National Historic Preservation Act of 1966, as amended, Section 106 consultation.

Table G-1. (contd)

Agency	Authority	Phase/Requirement/Status	Activity Covered
Tennessee Public Service Commission		Operation. Certification not required.	Certificate of public convenience and necessity.
TVA	Executive Order 11514 (Protection and Enhancement of Environmental Quality) 40 CFR Parts 1500-1508	SFES. Completed.	Protect and enhance the quality of the environment; develop procedures to ensure the fullest practicable provision of timely public information and understanding of Federal plans and programs that may have potential environmental impacts that the views of interested parties can be obtained.
TVA	Executive Order 11988 (Floodplain Management) TVA Procedure for Compliance With NEPA, Section 5.7	SFES. Completed.	Floodplain impacts to be avoided to the extent practicable.
TVA	Executive Order 11990 (Protection of Wetlands) TVA Procedure for Compliance With NEPA, Section 5.7	SFES. Completed.	Requires Federal agencies to avoid any short- and long-term adverse impacts on wetlands wherever there is a practicable alternative.

Appendix H

Severe Accident Mitigation Design Alternatives

Appendix H

Severe Accident Mitigation Design Alternatives

H.1 Introduction

Tennessee Valley Authority (TVA) submitted an initial assessment of severe accident mitigation design alternatives (SAMDAs)[a] for Watts Bar Nuclear (WBN) Unit 2 as part of its final supplemental environmental impact statement (EIS) (TVA 2009a). This assessment was based on the most recent WBN Unit 1 probabilistic risk assessment (PRA) available at the time modified to reflect expected two unit operation. Subsequently, TVA submitted an updated SAMDA assessment using the latest Computer Aided Fault Tree Analysis- (CAFTA-) based dual unit PRA (TVA 2010a). In addition to these plant-specific PRAs, the SAMDA assessments were based on a plant-specific offsite consequence analysis using the MELCOR Accident Consequence Code System 2 (MACCS2) computer code (Chanin and Young 1998), insights from the WBN Unit 1 individual plant examination (IPE) (TVA 1992), the WBN Unit 1 individual plant examination of external events (IPEEE) (TVA 1998), and, in the updated assessment, the WBN Unit 2 IPE (TVA 2010b). In identifying and evaluating potential SAMDAs, TVA considered SAMDA candidates that addressed the major contributors to core damage frequency (CDF) and large early release frequency (LERF) at WBN, as well as severe accident mitigation alternative (SAMA) candidates for operating plants which have submitted license renewal applications. TVA initially identified 283 potential SAMDAs, followed by an additional 24 in the updated submittal, all of which were reduced to 38 by eliminating those inapplicable to WBN Unit 2 due to (1) design differences; (2) prior implementation at WBN Unit 2; (3) similarity in nature so as to be combined with another SAMDA candidate; (4) excessive implementation cost such that the estimated cost would exceed the dollar value associated with completely eliminating all severe accident risk at WBN Unit 2; or (5) determined to provide very low benefit. TVA assessed the costs and benefits associated with each potential SAMDA, and concluded in the EIS that several are potentially cost-beneficial.

Based on a review of the SAMDA assessments, the U.S. Nuclear Regulatory Commission (NRC) issued requests for additional information (RAIs) to TVA by letters dated November 30,

---

(a) While TVA submittals generally refer to potential enhancements at WBN Unit 2 as severe accident mitigation alternatives (SAMAs) this appendix refers to the potential enhancements as severe accident mitigation design alternatives (SAMDAs). In general usage, SAMAs include mitigation measures such as training and procedures as well as design changes.

2009 (NRC 2009); January 11, 2011 (NRC 2011a); March 30, 2011 (NRC 2011b); and June 13, 2011 (NRC 2011c). Key questions concerned the following:

- major plant and modeling changes incorporated within each evolution of the PRA model

- justification for the multiplier used for external events

- binning and structure of the Level 2 and Level 3 analyses

- resolution of peer review findings

- basis for the source term used for the release categories

- incorporation of computer code corrections into the Level 3 analysis

- process for identifying plant-specific SAMDAs to address internal event risk

- identification of SAMDAs to mitigate fire risk

- further information on several specific candidate SAMDAs and low cost alternatives.

TVA submitted additional information in letters dated July 23, 2010 (TVA 2010c); September 17, 2010 (TVA 2010d); January 31, 2011 (TVA 2011a); May 13, 2001 (TVA 2011b); May 25, 2011 (TVA 2011c); June 17, 2011 (TVA 2011d); June 17, 2011 (TVA 2011e); September 16, 2011 (TVA 2011f); and October 17, 2011 (TVA 2011g). In response to the RAIs, TVA provided the following:

- additional information regarding the PRA model development and resultant changes to dominant risk contributors to CDF

- additional justification for the treatment of external events; a more detailed description of the Level 2 and Level 3 analyses

- justification for the release categories and associated consequence development

- information on resolution of peer review findings

- information on 31 additional SAMDA candidates

- additional information regarding several specific SAMDAs

- the impact of uncertainty on the Phase I screening of SAMDAs.

The responses also included revised results of the SAMDA analysis (incorporating updated computer codes, revised release category characterization, a revised external events multiplier and a revised Level 3 consequence analysis) based on several corrections/changes to the SAMDA analysis contained in the original and updated submittal (TVA 2011a, f, g). The TVA responses and revised SAMDA analysis addressed the NRC staff's concerns.

An assessment of the SAMDAs for WBN Unit 2 is presented below.

H.2 Risk Estimates for Watts Bar Nuclear Plant Unit 2

The TVA estimates of offsite risk at WBN Unit 2 are summarized in Section H.2.1. The summary is followed by the NRC staff's review of the TVA risk estimates in Section H.2.2.

H.2.1 TVA Risk Estimates

TVA combined two distinct analyses to form the basis for the risk estimates used in the SAMDA analysis: (1) the WBN Level 1 and 2 dual unit PRA model, which is updated from the WBN Unit 2 IPE, and (2) a supplemental analysis of offsite consequences and economic impacts (essentially a Level 3 PRA model) developed specifically for the SAMDA analysis. The updated SAMDA analysis is based on the most recent WBN Level 1 and Level 2 PRA models available at the time of the assessment, referred to as the WBN_U1_U2_FLOOD_SAMA model (TVA 2010a). This model is referred to as the SAMDA model throughout this appendix. The scope of the WBN PRA does not include external events.

The WBN Unit 2 CDF is approximately 1.7×10^{-5} per year for internal events as determined from quantification of the Level 1 PRA model. The CDF is based on the risk assessment for internally initiated events, which includes internal flooding. TVA did not include the contribution from external events in the WBN risk estimates; however, it did account for the potential risk-reduction benefits associated with external events by multiplying the estimated benefits for internal events by a factor of 2. This factor was subsequently increased to 2.28 in response to an NRC staff RAI (TVA 2011a). This is discussed further in Sections H.2.2 and H.6.2.

The breakdown of CDF by initiating event is provided in Table H-1. As shown in this table, events initiated by loss of offsite power (LOOP) and internal floods are the dominant contributors to CDF. Station blackout (SBO) sequences, a subset of sequences initiated by LOOP, make up 27 percent (4.7×10^{-6} per year) of the CDF while anticipated transient without scram (ATWS) sequences make up approximately 4 percent (6.2×10^{-7} per year) of the CDF (TVA 2011a).

The Level 2 portion of the SAMDA model represents an updated version of the WBN Unit 2 IPE Level 2 model. The IPE model was based on enhancements to NUREG/CR-6595 (NRC 2004a) and included quantification of containment threats resulting from high pressure failure of the reactor vessel and hydrogen deflagrations/detonations as well as additional detail on the treatment of interfacing system loss-of-coolant accidents (ISLOCAs) and induced steam generator tube rupture (SGTR). Two large containment event trees (CETs) were developed: one for SBO and one for non-SBO sequences.

Table H-1. WBN Unit 2 Core Damage Frequency for Internal Events[a]

Initiating Event	CDF (Per Year)	% Contribution to CDF[b]
Loss of Offsite Power (Grid Related)	3.2×10^{-6}	19
Loss of Offsite Power (Plant Centered)	2.8×10^{-6}	16
Total Loss of Component Cooling Unit 2	1.6×10^{-6}	10
Loss of Offsite Power (Weather Induced)	1.1×10^{-6}	6
Flood Event Induced by Rupture of Raw Cooling Water (RCW) Line in Room 772 0 – A8	1.1×10^{-6}	6
Flood Event Induced by Rupture of RCW Line in Room 772 0 – A9	1.1×10^{-6}	6
Total Loss of Essential RCW (ERCW) Cooling	9.6×10^{-7}	6
Small Loss-of-Coolant Accident (LOCA) Stuck-Open Safety Relief Valve	6.5×10^{-7}	4
Flood Event Induced by Rupture of High Pressure Fire Protection in Common Areas of the Auxiliary Building	3.2×10^{-7}	2
Turbine Trip	3.0×10^{-7}	2
Others (each 1% or less)	4.1×10^{-6}	24
Total CDF (internal events)	1.72×10^{-5}	100

(a) Information provided in response to the NRC staff RAIs (TVA 2010a, 2011a).
(b) May not total to 100 percent due to round off.

The result of the Level 2 model is a set of four release categories [I – Early Containment Failure (LERF), II – Containment Bypass (BYPASS), III – Late Containment Failure (LATE) and IV – Small Preexisting Leak (ISERF)] with their respective frequency and one category for intact containment, which is considered to have a negligible release. The frequency of each release category was obtained by summing the frequency of the individual Level 2 sequences assigned to each release category. The results of this analysis for WBN Unit 2 are provided in Table 2.a.iv-1 of the January 31, 2011 RAI responses (TVA 2011a). The four release categories were characterized by a total of 11 representative scenarios with their associated release parameters (i.e., source terms, release heights, release times, and release energies). The source terms for the representative scenarios were based on a SEQSOR (NRC 1990a) emulation spreadsheet methodology. The release parameters for the 11 representative sequences are provided in Table 2.a.iv-4 of the September 16, 2011 submittal (TVA 2011f) and Table 2.a.iv-5 of the June 17, 2011 RAI response (TVA 2011d).

The offsite consequences and economic impact analyses use the WinMACCS code, the current version of the MACCS2 code, to determine the offsite risk impacts on the surrounding environment and public. Inputs for these analyses include plant-specific and site-specific input values for core radionuclide inventory, source term and release characteristics, site meteorological data, projected population distribution (within an 80-km [50-mi] radius) for the year 2040, emergency response evacuation modeling, and economic data. The magnitude of the onsite impacts (in terms of cleanup and decontamination costs and occupational dose) is based on information provided in NUREG/BR-0184 (NRC 1997a).

The consequence analysis was performed for each of the 11 representative scenarios and the results (in terms of person-rem and offsite economic consequence) weighted by the contribution each representative scenario makes to the release category to obtain the consequences for the four release categories. This average consequence for the release category times the frequency of the release category yields the annual population dose and offsite economic consequence for each release category.

TVA estimated the dose to the population within 80 km (50 mi) of the WBN site to be approximately 0.20 person-Sv (20.0 person-rem) per year (TVA 2011f). The breakdown of the total population dose by release category is summarized in Table H-2. Late containment overpressure failure is the dominant contributor to population dose risk at WBN Unit 2.

Table H-2. Breakdown of Population Dose by Containment Release Category

Containment Release Category	Population Dose (Person-Rem[a] Per Year)	Percent Contribution
I – Early Containment Failure (LERF)	3.7	19
II – Containment Bypass (BYPASS)	0.8	4
III – Late Containment Failure (LATE)	14.1	71
IV – Small Preexisting Leak (ISERF)	1.2	6
V – Intact Containment (Intact)	negligible	negligible
Total	20.0[b]	100

(a) One person-rem = 0.01 person-Sv.
(b) Total is not equal to the sum of the above due to round off.

H.2.2 Review of TVA Risk Estimates

The TVA determination of offsite risk at WBN Unit 2 is based on the following major elements of analysis:

- the Level 1 and 2 risk models that form the bases for the WBN Unit 2 IPE submittal (TVA 2010b), the external event analyses of the WBN Unit 2 IPEEE submittal (TVA 2010e), and the modifications to the IPE model that have been incorporated into the latest WBN Unit 2 model—WBN_U1_U2_FLOOD_SAMA

- the WinMACCS analyses performed to translate fission product source terms and release frequencies from the Level 2 PRA model into offsite consequence measures (essentially this equates to a Level 3 PRA).

Appendix H

Each of these analyses was reviewed to determine the acceptability of the TVA risk estimates for the SAMDA analysis, as summarized below.

The NRC staff's review of the WBN Unit 2 IPE is described in an NRC report dated August 12, 2011 (NRC 2011d). Based on a review of the IPE submittal and responses to RAIs, the NRC staff found that the TVA definition of vulnerability and its conclusion that no severe accident vulnerabilities exist at WBN Unit 2 to be reasonable. Consequently the NRC staff found the WBN Unit 2 IPE to be consistent with the intent of Generic Letter (GL) 88-20 "Initiation of the Individual Plant Examination for Severe Accident Vulnerabilities – 10 CFR [Title 10, *Code of Federal Regulations*] 50.54(f)" (NRC 1988), subject to the completion of the applicable commitments and the TVA plan to confirm that, prior to WBN Unit 2 startup, the WBN Unit 2 PRA model matches the as-built, as-operated plant. Although no severe accident vulnerabilities were identified in the WBN Unit 2 IPE, the IPE did cite cost-beneficial SAMDAs identified in the original final supplemental EIS submittal (TVA 2009a). Each of these improvements is addressed by a SAMDA in the current evaluation and discussed further in Section H.3.2.

The CDF value from the WBN Unit 2 IPE submittal (3.3×10^{-5} per year) is near the average of the CDF values reported in the IPEs for other Westinghouse four-loop plants. Figure 11.6 of NUREG-1560 shows that the IPE-based total internal events CDF for these plants range from 3×10^{-6} to 2×10^{-4} per year, with an average CDF for the group of 6×10^{-5} per year (NRC 1997b). It is recognized that other plants have updated the values for CDF subsequent to the IPE submittals to reflect modeling and hardware changes. The current internal event CDF result for WBN Unit 2 (1.7×10^{-5} per year, including internal flooding) is comparable to that for other plants of similar vintage and characteristics.

Since WBN Units 1 and 2 are essentially identical, the history of both units' PRA models is relevant to this evaluation. There have been eight revisions to the WBN PRA model since the 1992 WBN Unit 1 IPE submittal, including the 2009 dual unit model, which used the CAFTA PRA software, whereas earlier versions utilized the RISKMAN® PRA software (PLG 1992). A description of the most significant changes made to each revision was provided by TVA in the original and updated assessments and in response to NRC staff RAIs (TVA 2009b, c; TVA 2010a, c; TVA 2011a, b), and is summarized in Table H-3.

A comparison of internal events CDF between the 1994 WBN Unit 1 IPE update and the initial WBN Unit 2 PRA model (referred to as WBN4SAMA) indicates a decrease of approximately 80 percent (from 8.0×10^{-5} per year to 1.5×10^{-5} per year). This reduction is attributed to the resolution of various 2001 peer review F&Os. The approximate factor of two increase for the WBN Unit 2 IPE (from 1.5×10^{-5} per year to 3.3×10^{-5} per year) is attributed primarily to the removal of credit for LOOP recovery factors from switchyard cross ties which are no longer feasible. The factor of two reduction in CDF from the WBN Unit 2 IPE (3.3×10^{-5} per year) to the SAMDA model (1.7×10^{-5} per year) is attributed to taking credit for cross-tying WBN Unit 1 and Unit 2 shutdown boards and recovery of total loss of ERCW by use of a portable diesel-driven fire pump (TVA 2011a). A comparison of the contributors to the total CDF between the WBN4SAMA model and the current SAMDA model indicates that some have increased while others have decreased. A summary listing of those changes that resulted in the greatest impact on the internal events CDF and, in particular, the changed risk profile was provided in response to an NRC staff RAI and is included in Table H-3 (TVA 2011a, b).

Table H-3. WBN PRA Historical Summary

PRA Version	Summary of Changes from Prior Model	CDF (per year)
WBN Unit 1 IPE (1992)	• WBN Unit 1 IPE Submittal	3.3×10^{-4}
WBN Unit 1 IPE Revision 1 (1994)	• Revised success criteria for the component cooling water system to one of two pumps being successful • Provided for the use of nitrogen bottles for steam generator power-operated relief valves (PORVs) and auxiliary feedwater (AFW) flow control valves under SBO conditions • Revised human reliability analysis to reflect updated procedures and training	8.0×10^{-5}
WBN Unit 1 Revision 2 (1997)	• Enhanced recovery from LOOP through use of WBN Unit 2 equipment and changing certain emergency diesel generator (EDG) cooling valves from normally closed to locked open • Credited improved operator actions resulting from changes in emergency operating procedures associated with high pressure recirculation • Revised component cooling water system (CCS) model to credit WBN Unit 2 pump and reduced loss of CCS initiating event frequency	4.4×10^{-5}
WBN Unit 1 Revision 3 (2001)	• Integrated Level 1 and 2 models to allow calculation of LERF • Revised seal LOCA model to reflect new high temperature seals • Incorporated plant-specific data	4.5×10^{-5}
WBN Unit 1 Revision 4	• Updated initiating event data • Incorporated latest maintenance rule data • Incorporated comments by WBN system engineers	1.3×10^{-5}
WBN Unit 2 WBN4SAMA (2008)	• Revised core damage arrest model in Level 2 to be consistent with Level 1 model • Revised bleed and feed success criteria to indicate two PORVs required with one safety injection pump • Added loss of plant compressed air initiating event • Revised ventilation system recovery modeling • Accounted for dual unit operation by removing credit for WBN Unit 2 component cooling water pumps from WBN Unit 1 model and changing ERCW success criteria	1.5×10^{-5}

Table H-3. (contd)

PRA Version	Summary of Changes from Prior Model	CDF (per year)
Dual Unit WBN_U1_U2_FLOOD (2009)	• Updated PRA model to be dual unit and to use CAFTA software package • Developed from WBN4SAMA event trees and fault trees • Removed LOOP recovery factors from switchyard cross ties no longer feasible • Replaced previous CETs with updated models	2.9×10^{-5}
WBN Unit 2 IPE WBN_U1_U2_FLOOD (2010)	• Resolved selected findings from Westinghouse Owner's Group (WOG) 2009 review • Updated LOOP model so that all batteries were not failed at time zero • Corrected basic event coding for turbine-driven (TD) AFW pump failure to start • Revised the linking of steam generator condition after core damage to the correct plant damage state	3.3×10^{-5}
Dual Unit WBN_U1_U2_FLOOD_SAMA (2010)	• Prepared for the updated WBN Unit 2 SAMDA analysis • Added cross tie of WBN Unit 1 and 2 shutdown boards • Credited recovery of total loss of ERCW by use of portable diesel-driven fire pump	1.7×10^{-5}

The WBN Unit 2 IPE is stated to be based on the WBN Unit 1 design and operation as of April 1, 2008. In response to an NRC staff RAI, TVA discussed the design and procedural changes since the IPE 2008 freeze date with potential PRA significance (TVA 2011a). A significant number of mainly procedural changes that were identified in the initial WBN Unit 2 SAMDA assessment have been implemented and incorporated in the current SAMDA PRA. The NRC staff concludes that those changes that have not been incorporated into the PRA will tend to reduce the CDF and thus make the current results conservative.

The NRC staff considered the peer reviews performed for the WBN PRA and the potential impact of the review findings on the SAMDA evaluation. The most relevant review is that performed by the WOG in November 2009. A summary of the results of this peer review is provided in the WBN Unit 2 IPE submittal along with a listing of the peer review findings (TVA 2010b). Of the 326 supporting requirements of the ASME PRA standard, 9 were judged to not be applicable, 272 were judged to meet Capability Category I/II or greater, 19 met Capability Category I, and 26 were judged as not met. In response to a NRC staff RAI, TVA discussed the status of the peer review findings relative to the SAMDA model (TVA 2011a). While most of the findings have been resolved as part of the updated SAMDA model, a significant number of findings remain open. These are in two categories: those considered by TVA to be documentation-only issues and those pertaining to internal flooding.

In response to an NRC staff RAI, TVA discussed three specific findings considered by TVA to be documentation related. With regard to the finding concerning the diesel generator load sequencer modeling, additional information from the finding was presented including the reviewer's conclusion that the missing failure modes would be expected to have minimal impact on the PRA results. Also, the peer review indicated that Capability Category II was met. With regard to the finding concerning the lack of simulator observations to support the human reliability analysis (HRA), TVA pointed out that the PRA standard required either simulator observations or talk-throughs with the operators. The talk-throughs with several members of the operations staff conducted for the WBN PRA model are considered by TVA to meet the PRA standards requirements. With regard to the finding concerning optimistic mission times used for room heatup calculations, TVA identified those systems or functions for which a mission time of less than 24 hours was assumed. These include emergency boration, electric power equipment related to 4-hour battery coping time, residual heat removal, reactor protection system (RPS), and emergency safety features actuation system (ESFAS). The NRC staff considers the mission time used for each of these systems acceptable for determining the need for room cooling.[a] TVA stated that, except for two cases, for those areas where the heatup analysis showed that the temperatures for affected components/functions exceeded the equipment qualification temperature prior to the desired mission time, cooling was considered necessary and included in the model. For Room 757-01, Auxiliary Control Room, the room temperature reaches the equipment qualification temperature of 104°F at just over 21 hours into the event and peaks at 105.8°F. Room 757-24, 6.9 kV Shutdown Board Room B reaches 104°F at 23 hours and exceed this by less than one degree. TVA concludes that these small differences will not lead to failure of the associated equipment in the 24-hour mission time (TVA 2011c, d). The NRC staff finds this conclusion reasonable since the peak temperature was reached within the 24-hour mission time.

With regard to the findings concerning the internal flooding analysis, TVA provided, in response to an NRC staff RAI, a general discussion of the open findings including identifying those that could be resolved by additional documentation or supporting analysis and those that represent conservatisms in the flooding CDF (TVA 2011a). Based on the exclusion of recovery actions for many of the important flooding contributors, TVA concludes that the flooding analysis is conservatively bounding for the present application. The NRC staff review of the available information indicates that it is not clear if the overall impact of the resolution of the findings will increase or decrease the flooding CDF. Several findings indicate the results are clearly conservative, while for others the impact of the finding resolution is not clear. Further, additional credit for recovery actions would reduce the flood CDF for those scenarios where recovery can

---

(a) In the June 17 TVA submittal (TVA 2011d), it is stated that the mission time for the RPS, ESFAS, and instrumentation is 12 hours. Subsequently TVA clarified that this reference to instrumentation was meant to refer to instruments associated with a reactor trip but are not specifically classified as RPS or ESFAS. They further stated that instrumentation necessary and sufficient to monitor safe shutdown is considered to have a mission time of 24 hours (TVA 2011e).

Appendix H

be credited. As discussed below, TVA has addressed the two important flood induced sequences by committing to add flood detection equipment in rooms where floods are important contributors (Rooms 772.0-A8 and 772.0-A9). In addition, as discussed below for SAMDAs 70 and 339, TVA has committed to provide the capability to transfer normal compressed air supply to the station nitrogen system. This addresses flooding sequences that cause loss of the station air system. Further, SAMDA identification has considered and adequately disposed of other less important flooding risk contributors. Based on this, the NRC staff concludes that updating the internal flooding analysis to resolve the remaining peer review findings is unlikely to result in any additional cost beneficial SAMDAs.

TVA also indicated that the changes between the WBN Unit 2 IPE model and the SAMDA model were independently reviewed internally by the contractor making the changes, by an independent contractor, and by TVA.

In response to NRC staff RAIs, TVA identified the systems shared between WBN Units 1 and 2, and described the modeling of these systems and how WBN Unit 1 outages potentially impacting these systems are accounted for in the WBN Unit 2 model. WBN Units 1 and 2 share the electric power; ERCW; CCS; and plant and control air, and heating, ventilation and air conditioning systems (TVA 2010c). The WBN CAFTA model is described as a single fault tree constructed with systems and components for each unit and common systems modeled. The impact of the unavailability of WBN Unit 1 components/systems with respect to mitigation and to initiating events (unit-specific and dual unit) is incorporated in the model for WBN Unit 2. Model quantification for each unit accurately tracks the dependent failure for each unit (TVA 2011a). The test and maintenance unavailability of WBN Unit 1 components impacting the WBN Unit 2 CDF includes the unavailability of these components when WBN Unit 1 is operating as well as when it is shutdown. Testing and maintenance unavailability is based on WBN Unit 1 experience data. Since it is TVA practice to perform testing and maintenance with the unit on line, the unavailability data include routine testing as well as infrequent but more extensive maintenance activities (TVA 2011b, c, d).

Given that the WBN internal events PRA model has been peer-reviewed and the peer review findings were all addressed, and that TVA has satisfactorily addressed the NRC staff's questions regarding the PRA, the NRC staff concludes that the internal events Level 1 PRA model, WBN_U1_U2_ FLOOD_SAMA, is of sufficient quality to support the SAMDA evaluation.

As indicated above, the WBN PRA does not include external events. The SAMDA submittals cite the WBN Unit 1 IPEEE which indicates that the only vulnerability found has been corrected and no longer impacts either unit. The WBN Unit 1 IPEEE was submitted in November 1998 (TVA 1998), in response to Supplement 4 of GL 88-20 (NRC 1991). The Unit 2 IPEEE was submitted in April 2010 (TVA 2010e).

The WBN Unit 2 IPEEE uses the same methodology and to a large extent the same assessment as the WBN Unit 1 IPEEE, subject to validation that the WBN Unit 1 assessments are applicable to the as-built WBN Unit 2. This submittal included a summary of the seismic margin analysis, the fire-induced vulnerability evaluation (FIVE), and the screening analysis for other external events, all subject to validation for WBN Unit 2 when construction is complete. No fundamental weaknesses or vulnerabilities to severe accident risk in regard to external events were identified in the WBN Unit 1 IPEEE with the exception of one item related to tornado missiles discussed below. No seismic, fire, high winds, external floods, or other external hazard improvements were identified. In a letter dated May 19, 2000, the NRC staff concluded that the licensee's WBN Unit 1 IPEEE process is capable of identifying the most likely severe accidents and severe accident vulnerabilities, and therefore, that the WBN IPEEE has met the intent of Supplement 4 to GL 88-20 (NRC 2000). The NRC staff's review of the WBN Unit 2 IPEEE is described in an NRC letter dated September 20, 2011 (NRC 2011e). Based on a review of the IPEEE submittal and responses to RAIs, the NRC staff found that the TVA definition of vulnerability and its conclusion that no severe accident vulnerabilities exist at WBN Unit 2 to be reasonable. Consequently the NRC staff found the WBN Unit 2 IPEEE to be consistent with the intent of GL 88-20 (NRC 1988), subject to the completion of validation activities to confirm that the assumptions concerning the WBN Unit 2 design are valid for the as-built, as-operated plant.

The WBN Unit 1 IPEEE used a focused-scope Electric Power Research Institute (EPRI) seismic margins analysis (EPRI 1991). This method is qualitative and does not provide numerical estimates of the CDF contributions from seismic initiators. For this assessment, the seismic walkdown took advantage of the extensive walkdowns performed prior to the issuance of the WBN Unit 1 low power operating license including: the corrective action program in which the plant structures were reevaluated against more recent seismic criteria, the hanger and analysis update program, the integrated interaction program, and the equipment seismic qualification program. The components in the safe shutdown equipment list were screened using an overall high confidence of low probability of failure (HCLPF) capacity of 0.3 g, the review level earthquake value for the plant, and the screening level that would be used for a focused-scope plant. No significant seismic concerns were identified. A small number of maintenance and housekeeping items were noted and corrected (TVA 1998, TVA 2010e).

While the WBN Unit 2 seismic assessment makes considerable use of the WBN Unit 1 assessment, individual aspects are repeated and/or the WBN Unit 1 results were reviewed to confirm that they are applicable to WBN Unit 2. TVA considered this an acceptable approach since the designs of the units are nearly identical and use the same design criteria The NRC staff finds this approach reasonable.

The WBN Unit 2 IPEEE did not identify any vulnerabilities due to seismic events or any improvements to reduce seismic risk.

Appendix H

To provide insight into the appropriate estimate of the seismic CDF to use for the SAMDA evaluation, the NRC staff noted that, in the attachments to NRC Information Notice 2010-18, Generic Issue (GI) 199 (NRC 2010), the NRC staff estimated a "weakest link model" seismic CDF for WBN Unit 1 of 3.6×10^{-5} per year using updated seismic hazard curves developed by the U.S. Geological Survey (USGS) in 2008 (USGS 2008) and requested TVA to provide an assessment of the impact of the updated USGS seismic hazard curves on the SAMDA evaluation (NRC 2011a). The NRC Information Notice referenced the August 2010 NRC document, "Safety/Risk Assessment Results for Generic Issue 199, Implications of Updated Probabilistic Seismic Hazard Estimates in Central and Eastern United States on Existing Plants" (ADAMS Accession No. ML100270582 [package]), that discusses recent updates to estimates of the seismic hazard in the central and eastern United States. Appendix A of that document describes how the seismic CDF estimate can be acceptably derived using various approaches; including a maximum estimate, averaging estimates, and the weakest link estimate. All these approaches use the plant-specific ground motion characterization (i.e., spectral accelerations at various frequencies and/or peak ground accelerations). For WBN Unit 1, the peak ground acceleration estimate is greater than the spectral acceleration estimates derived at 1 hertz (Hz), 5 Hz, and 10 Hz. As a result, the peak ground acceleration estimate is equal to the maximum estimate and dominates the weakest link model estimate at 3.6×10^{-5} per year.

In response to the NRC staff request, TVA noted that the WBN site was used as the test case for closure of GI-194, "Implications of Updated Probabilistic Seismic Hazard Estimates" (NRC 2003a; TVA 2011a). In the NRC staff evaluation supporting the closure of GI-194, the NRC staff used new seismic hazard curves for the East Tennessee Seismic Zone, which includes the WBN site, to develop the updated seismic CDF estimate for the WBN site. Initially, the NRC staff estimated the seismic CDF using the updated peak ground acceleration and derived a value similar to the latest updated value. However, the NRC staff noted that the WBN site's updated seismic spectral acceleration values differed significantly from the design safe shutdown earthquake uniform hazard spectrum. In order to account for the difference in the uniform hazard spectrum shape of the new hazard curves, the WBN plant HCLPF capacity of 0.3 g was scaled to the spectral acceleration values at 5 Hz and 10 Hz, based on the natural frequency range for most structures and equipment in nuclear power plants being below 10 Hz (NRC 2003a). The average of the seismic CDF for these two seismic acceleration values resulted in the seismic CDF of 1.8×10^{-5} per reactor-year for the WBN site. Based on the GI-194 staff analysis, TVA concluded that 1.8×10^{-5} per year is an appropriate estimate of the seismic CDF for use in the WBN Unit 2 SAMDA evaluation.

The seismic CDF estimated by the NRC staff for WBN Unit 1 for soil (vs. bedrock – NRC GI-199 estimates CDF for both) used the 2008 USGS seismic hazard curves, for spectral acceleration values of 5 and 10 Hz is 1.3×10^{-5} per year and 2.8×10^{-5} per year, respectively (NRC 2010). The average of the seismic CDF for these two acceleration values is 2×10^{-5} per year. Based on the spectral-averaged seismic CDF of 2×10^{-5} per year using the 2008 USGS data being essentially the same as the spectral-average seismic CDF of 1.8×10^{-5} per year determined for closure of GI 194, the NRC staff concludes that 1.8×10^{-5} per year is an acceptable estimate of the seismic CDF for use in the WBN Unit 2 SAMDA evaluation.

For the analysis of plant vulnerability to fire, the WBN Unit 2 IPEEE used the FIVE (EPRI 1992) methodology and modified versions of the WBN Unit 2 IPE. The methodology consists of a series of progressive screens. In the first phase of screening, fire areas are screened based on area fire boundary integrity, the absence of safe shutdown components, and the lack of plant trip initiators. The second phase of screening consists of an initial, bounding quantitative analysis followed by a more detailed quantitative evaluation for those areas not screened out based on a fire CDF of 1×10^{-6} per year. The initial quantification consisted of generating an area specific fire ignition frequency and a conditional core damage probability from the IPE model assuming all components in the fire area were damaged. In the detailed quantification further evaluation was performed which included consideration of fire severity, zones of fire influence, and fire suppression probability. The NRC staff review of the WBN Unit 2 IPEEE fire analysis notes that while the approach was somewhat unique and deviated from the traditional FIVE methodology in some respects, the methods used and the implementation of those methods are adequate to meet the IPEEE objectives. The NRC staff also concluded that the fire analysis incorporated a degree of conservatism (NRC 2011e). The WBN Unit 2 IPEEE did not identify any vulnerabilities due to fire events or any improvements to reduce fire risk.

In response to an NRC staff RAI, TVA provided more information concerning the modified PRA used in the fire evaluation (TVA 2011a). These modifications consisted mainly of changes to conform to the FIVE analysis screening assumptions. The changes were unique to the fire analysis or had a minimal affect on non-fire scenarios. Also, the changes made to the WBN Unit 2 model to create the SAMDA model were not incorporated in the modification used in the fire analysis. This results in a degree of conservatism in the FIVE results and is discussed further below.

The dominant fire areas, defined as those having a fire CDF greater than 3×10^{-7} per year, and their contributions to the fire CDF are listed in Table H-4. The total fire CDF is not given in the IPEEE submittal, but the total for those subjected to the final stage of screening is stated to be 9.3×10^{-6} per year (TVA 2011a).

In response to an NRC staff RAI, TVA identified 15 conservatisms and 8 non-conservatisms in the WBN fire analysis (TVA 2011a). The conservative items included the following:

- most fire ignition frequencies
- triple counting fires in one area
- conservative fire severity factors in the control room
- conservative fire suppression failure probabilities

- conservative treatment of core damage given control room evacuation
- the assignment of fire impacts to the individual fire scenarios
- not incorporating model changes credited in the SAMDA model, principally recovery actions for station blackout and loss of ERCW.

Table H-4. Dominant Fire Areas and Their Contribution to Fire CDF[a]

Fire Area Description	CDF (per year)
Main Control Room	9.7×10^{-7}
Corridor in Auxiliary Building (713.0-A1 & A2)	9.3×10^{-7}
125V Vital Battery Board Room IV	8.4×10^{-7}
Refueling Room	7.5×10^{-7}
Auxiliary Instrument Room 2	6.8×10^{-7}
Turbine Building	5.9×10^{-7}
Corridor (737.0-A1B)	5.1×10^{-7}
Corridor (737.0-A1A)	4.2×10^{-7}
Auxiliary Building Roof	3.1×10^{-7}

(a) Information provided in responses to NRC staff RAIs (TVA 2010e and TVA 2011a).

Non-conservatisms included the following:

- not modeling fire propagation between analysis volumes
- not modeling some spurious equipment actuations
- assumption that reactor coolant pump (RCP) seal return valves would not transfer closed as a result of fires
- not including the increased probability of the 182 gpm per pump seal LOCA on loss of seal cooling.

TVA concludes that the conservatisms outweigh the non-conservatisms so that the fire contribution to risk is less than that given by the sum of the final fire screen results.

TVA used a fire CDF of 4.1×10^{-6} per year for the SAMDA evaluation. This value is based on the sum of the detailed CDF quantification of the previously unscreened fire areas as given in the IPEEE and updated in response to an NRC staff RAI, or 9.3×10^{-6} per year as discussed previously (TVA 2010e, TVA 2011a), which is reduced by a factor of 2.29 to account for the conservatisms in the fire analysis. This factor only accounts for the conservatisms in the PRA model used to evaluate the conditional core damage probabilities compared to the SAMDA model and does not account for conservatisms in the FIVE methodology or those conservatisms discussed above. This factor is the ratio of the internal events CDF of 2.68×10^{-5} per year given by the modified PRA used for the fire analysis with no fire-induced failures nor flood failures to the CDF of 1.17×10^{-5} per year given by the October 2010 SAMDA PRA for internal events only, excluding floods (TVA 2011a). Based on the conservatisms in the fire analysis, the NRC staff concludes that a fire CDF of 4.1×10^{-6} per year is reasonable for the SAMDA analysis.[a]

The IPEEE analysis of high winds, floods, and other (HFO) external events followed the screening and evaluation approaches described in Supplement 4 of GL 88-20 (NRC 1991) and focused on demonstrating that the design and construction of the plant in the HFO areas met the 1975 Standard Review Plan Criteria (NRC 1975). During the HFO walkdown, it was noted that there was an opening on the WBN Unit 2 side of the Auxiliary Building that had the potential for allowing tornado missiles to penetrate into the auxiliary building and damage safety-related equipment. A corrective action to design and install a steel shield to eliminate this concern for both WBN Units 1 and 2 was completed. TVA did not identify other vulnerabilities or improvements. Based on this result, TVA did not consider specific SAMDAs for these events.

In the original and updated SAMDA submittals, TVA assumed that the estimated benefits from external events was equivalent to the estimated benefits from internal events. This was based on the SAMA submittals for license renewal applications for several other four-loop Westinghouse plants. Accordingly, TVA applied an external events multiplier of 2 to the internal events results to account for the additional benefits from fire, seismic, and other external events. In an RAI, the NRC staff questioned the basis for this multiplier and asked TVA to use a higher multiplier that is supported by WBN-specific information relative to the seismic and fire contributors to CDF (NRC 2011a). In response to the RAI, TVA developed a new external events multiplier of 2.28 based on the aforementioned results (based on a seismic CDF of 1.8×10^{-5} per year, a fire CDF of 4.1×10^{-6} per year, and an internal events CDF of 1.7×10^{-5} per year) (TVA 2011a). In a revised SAMDA analysis submitted in response to the RAI, TVA multiplied the benefit that was derived from the internal events model by a factor of 2.28 to account for the combined contribution from internal and external events. The NRC staff

---

(a) An alternative method of assessing the potential conservatism in the TVA fire CDF due to the PRA model used for the SAMDA analysis is to ratio the total internal events CDF of the PRA that was the basis for the WBN Unit 2 fire analysis to the internal event CDF of the PRA used for the SAMDA analysis. Using the IPE CDF of 3.3E-05 per year (TVA 2010b) as the internal events CDF on which the fire CDF was based and the WBN_U1_U2_FLOOD_SAMA CDF of 1.7E-05 per year results in a fire CDF reduction factor of 3.3E-05/1.7E-05 = 1.94. This yields an adjusted WBN Unit 2 fire CDF of 9.3E-06/1.94 = 4.8E-06 per year versus the TVA value of 4.1E-06 per year. While higher than the TVA value, the difference is well within the approximate nature of the adjustment methodology. Use of the higher value of fire CDF would increase the external events multiplier (and, therefore, the potential benefit), but would not change the ultimate disposition of the various SAMDAs. The NRC staff considers the TVA result, but not necessarily its approach, as acceptable for use in the SAMDA evaluations.

finds with the applicant's overall conclusion concerning the impact of external events and concludes that the applicant's use of a multiplier of 2.28 to account for external events is reasonable for the purposes of the SAMDA evaluation.

The NRC staff reviewed the general process used by TVA to translate the results of the Level 1 PRA into containment releases, as well as the results of the Level 2 analysis, as described in the SAMDA submittal and in response to NRC staff RAIs (TVA 2011a, b, c, d). As indicated above, the Level 2 portion of the SAMDA model represents an updated version of the WBN Unit 2 IPE Level 2 model. The IPE model was based on enhancements to NUREG/CR-6595 (NRC 2004a) and included quantification of containment threats resulting from high pressure failure of the reactor vessel and hydrogen deflagrations/detonations as well as additional detail on the treatment of ISLOCA and induced SGTR. The accident progression was modeled using a 32-node containment model in MAAP4.0.7. Two large CETs were developed; one for SBO and one for non-SBO sequences (TVA 2010b).

In response to NRC staff RAIs, TVA provided additional information on the linking of the Level 1 and 2 models, the binning of CET end states and their assignment to release categories, the dominant sequences for each release category, and the determination of the release characteristics for each release category. Each of the Level 1 core damage sequences is assigned to one of eight plant damage state (PDS) bins, based on characteristics such as bypass containment or not, the type of bypass, and high or low reactor coolant pressure. Each core damage sequence is linked to the Level 2 CET in accord with the PDS bin. The CETs consisted of 18 questions (or events or nodes), the first 5 of which link each PDS to the appropriate portion of the CET. The remainder of the questions determine the appropriate containment failure type, the CET end states, and release category. This process results in 11 CET end state groups (plus the INTACT end state), which are then assigned to the release categories used in the Level 3 consequence analysis. It is noted that some changes/corrections were made to the IPE PDS bin assignments for the SAMDA model. The CET end states are binned into release categories that represent similar containment failure modes and release timing. The frequency of each release category was obtained by summing the frequency of the individual Level 2 sequences assigned to each release category.

The NRC staff noted that the sum of the frequencies of release categories I through IV of 1.85×10^{-5} per year is greater than the CDF of 1.72×10^{-5} per year even without the inclusion of release category V for containment intact sequences. In response to an NRC staff RAI to explain this greater value and to provide the frequency for release category V, TVA explained the reason for the sum of release category I through IV frequencies being greater than the CDF as due to not excluding release category IV (a small preexisting containment leak) from the other branches of the containment event tree that are subsequently assigned to the other release categories. While the NRC staff finds that this leads to double counting of release category IV frequency, the degree of conservatism in risk is not clear since the presence of a small preexisting containment leak would not necessarily preclude the other containment failure modes, particularly those early failures due to high pressure reactor vessel failures. TVA further indicated that the CAFTA based Level 2 model did not calculate the intact containment frequency correctly. TVA stated that a proportion of the late containment failure sequences (release category III) was inadvertently included in the intact category and thus a correct value for the intact containment frequency is not available. TVA indicated that this does not impact the risk results since the containment intact category has negligible impact on the results (TVA 2011b).

TVA stated that the LERF for the latest SAMDA model is 1.70×10^{-6} per year (TVA 2011a). In an NRC staff RAI, the NRC staff noted that this is different from the sum of the frequencies of release categories I and II of 1.61×10^{-6} per year (NRC 2011b). In response, TVA indicated that the correct value for LERF is the sum of the frequencies of release categories I and II or 1.61×10^{-6} per year (TVA 2011b).

Source terms for use in the Level 3 consequence analysis are based on representative accident scenarios that reflect the post core damage behavior for the dominant sequence or sequences within a PDS that contribute to each release category. The release fractions were determined for each representative scenario using a spreadsheet version of the SEQSOR computer code (NRC 1990a). SEQSOR was used to calculate the release fractions for the NUREG-1150 analysis of the Sequoyah Nuclear Plant (NRC 1990b). The SEQSOR methodology determines release fractions using a parametric approach with probabilistic data blocks based on supporting first principle analyses as well as expert panel judgments. SEQSOR determines the mean release fractions for each representative sequence that makes up each release category using input release characteristics describing the representative scenario and parametric data included in the code. The release characteristics used for the WBN Unit 2 analysis are reported in Table 2.a.iv-3 of NRC staff RAI (TVA 2011a). Since the SEQSOR data blocks were developed for the Sequoyah Nuclear Plant, a sister plant to WBN Unit 2, the use of SEQSOR was considered appropriate by TVA for the WBN Unit 2 SAMDA analysis. TVA indicated that the same data blocks and data were used in the SEQSOR emulator used for the WBN Unit 2 analysis as in the NUREG-1150 SEQSOR code except where process or equipment that needed to be considered in the WBN Unit 2 analysis were not included in the NUREG-1150 analysis. TVA states that the SEQSOR emulator was independently reviewed prior to its use in the SAMDA analysis (TVA 2011c, d).

Based on the NRC staff's review of the Level 2 methodology and the fact that (1) the LERF model was reviewed by the WOG and the review findings have all been addressed in the SAMDA Level 2 model, (2) the updated Level 2 model was reviewed by an external contractor and independently reviewed by the TVA PRA team, and (3) TVA has adequately addressed NRC staff RAIs concerning the Level 2 model, the NRC staff concludes that the Level 2 PRA provides an acceptable basis for evaluating the benefits associated with various SAMDAs.

The reactor core radionuclide inventory used in the consequence analysis contained in the EIS is for 5 percent enrichment and a burnup of 1,000 effective full power days for WBN Unit 2 at 3,565 megawatts thermal (MW(t)) as evaluated using the ORIGEN code. TVA states that these conditions bound those expected for the WBN Unit 2 fuel management program for the license period (TVA 2010c).

The NRC staff reviewed the process used by TVA to extend the containment performance (Level 2) portion of the PRA to an assessment of offsite consequences (essentially a Level 3 PRA). This included consideration of the source terms used to characterize fission product releases for the applicable containment release categories and the major input assumptions used in the offsite consequence analyses. The WinMACCS code, the current version of the MACCS2 code, was used to estimate offsite consequences. Plant-specific input to the code includes the source terms for each release category and the reactor core radionuclide inventory (both discussed above), site-specific meteorological data, projected population distribution within an 80-km (50-mi) radius for the year 2040, emergency evacuation modeling, and economic data. This information is provided in Section 4.6 of the SAMDA submittal (TVA 2010a).

TVA used site-specific meteorological data for the 2002 calendar year as input to the WinMACCS code. The data were collected from the onsite meteorological tower. Data from 2001 through 2005 were also considered, but the 2002 data were chosen because they were found to give the largest risk based on sampling the population dose consequences for each year with a reference set of fission product releases. Missing data were obtained by linear interpolation from the recorded data. The NRC staff notes that previous SAMA analyses results for WBN have shown little sensitivity to year-to-year differences in meteorological data and concludes that the use of the 2002 meteorological data in the SAMDA analysis is reasonable. In response to an NRC staff RAI, TVA stated that the WinMACCS evaluation for WBN Unit 2 applied a large rainfall boundary condition that results in conservative deposition of radionuclides in the last spatial interval (40 to 50 mi [64 to 80 km]) (TVA 2010c). The NRC staff notes that previous SAMA analyses have indicated that this assumption results in a relatively substantial increase in offsite consequences.

All releases were modeled as occurring at a height of 10 m (33 ft), and buoyant plume rise appropriate to the release category was modeled. TVA did not perform a sensitivity analysis on these assumptions, instead citing previous SAMA analyses as indicating relatively small changes in overall risk. Of the two SAMA analyses cited (i.e., Vogtle and Wolf Creek), one shows an increase in population dose risk of up to 10 percent while the other shows a decrease of 17 percent in population dose risk with an increase in release elevation from ground level to the top of the containment building and/or heat release rates of 1 to 10 MW (SNC 2007; WCNOC 2006). The NRC staff notes that previous SAMA analyses have shown only minor sensitivities to release height and buoyancy. The NRC staff concludes that the release parameters TVA used are acceptable for the purposes of the SAMDA evaluation.

The population distribution the licensee used as input to the WinMACCS analysis was estimated for the year 2040, based on the U.S. Census Bureau population data for 2000. A map was prepared displaying county and census tract boundaries partly or entirely within the 80-km (50-mi) boundary. A grid of concentric circles and radii were overlaid on this map to display the 160 zones or sectors needed. County population data for 2000 were allocated to the appropriate sectors, using census tracts to the extent feasible. Block groups were used where census tracts crossed the sector boundaries. For sectors near the plant, aerial photographs and local knowledge were used. Projected county growth rates to the year 2030 were then used and extended using linear trend lines to the years 2040, 2050, and 2060 and the results applied to each sector. Transient population was included based on peak recreation visitation estimates at the various sites around the Tennessee River system. The numbers were estimated from TVA recreational facility information and extrapolated to the year 2040 using population projections for an 11 county region around the site. The NRC staff considers the methods and assumptions for estimating population reasonable and acceptable for purposes of the SAMDA evaluation.

The emergency evacuation model was modeled as a single sheltering and evacuation zone extending out 16 km (10 mi) from the plant. It was assumed that 99.5 percent of the population would evacuate. This assumption is consistent with the NUREG-1150 study (NRC 1990b). The NRC staff notes that previous SAMA analyses have shown only minor sensitivities to the percent of population evacuated within the range of 95 and 100 percent. The evacuation speed used in the SAMDA analysis of 2.2 mph (1.0 m/s) was selected considering average evacuation speeds under adverse weather conditions using evacuation data from the multi-jurisdictional emergency response plan for WBN (TVA 2006). The evacuation was assumed to start after a sheltering and evacuation delay time of 45 minutes and 2.5 hours respectively. These values were obtained from the multi-jurisdictional emergency response plan and the NUREG-1150 model.

A sensitivity analysis was performed in which the evacuation speed was decreased from 2.2 mph (1 m/s) to 1.6 mph (0.7 m/s) and then increased to 3.4 mph (1.5 m/s). The result was a small change in the population dose risk (a maximum of approximately an 8 percent increase when the speed is reduced to 1.6 mph [0.7 m/s]) and no change in the offsite economic cost risk for each release category for the baseline consequence analysis (TVA 2011f). The NRC staff estimates that this increases the maximum averted cost risk (MACR) by approximately 0.5 percent. The NRC staff notes that the TVA analysis did not account for the reduced evacuation speed that would result from population growth from the time the evacuation time estimates were made to 2040. While this may decrease the speed below that considered in the sensitivity study, the NRC staff concludes that the impact on the risk results will be small in light of the above result and since other studies have shown that population dose is not very sensitive to evacuation speed and population dose risk contributes only a relatively small part of the total maximum benefit. The NRC staff concludes that the evacuation assumptions and analysis are reasonable and acceptable for the purposes of the SAMDA evaluation.

The site-specific economic data input to WinMACCS2 code used for the SAMDA analysis was provided from SECPOP2000 (NRC 2003b) by specifying the data for each of the counties surrounding the plant to a distance of 80 km (50 mi). The original SAMDA submittal used SECPOP2000 version 3.12 which was found to have several problems that could lead to erroneous data being used. In response to an NRC staff RAI, TVA corrected the earlier analysis using the results from SECPOP2000 version 3.13.1 which corrected the previous version's errors (TVA 2010d). TVA used the corrected version in the updated and subsequent SAMDA analysis. SECPOP2000 version 3.13.1 used data from the 2000 census and the 2002 Census of Agriculture to determine the population, land fractions and region index, and associated economic data for each sector around the plant for use in WinMACCS. The dollar values were increased by a factor of 1.15 to account for inflation from 2002 to 2007. This was determined from the United States Bureau of Labor Statistics Consumer Price Index (CPI) Inflation Calculator (BLS 2010). The sector population from SECPOP2000 was replaced with the TVA-generated values discussed above. The NRC staff concludes that the approach taken for determining the site-specific economic data is appropriate for the SAMDA analysis.

Since SECPOP2000 provides only generic values for the number of watersheds, the watershed index, the watershed definition and the crop seasons and share, definitions and values more appropriate for the WBN site were used. TVA described these changes and their basis (TVA 2010a) and the NRC staff concludes that their use is appropriate for the SAMDA analysis.

The MACCS2 analysis described above was performed for each of the representative scenarios that contribute to each of the four release categories. The consequences in terms of person-rem and offsite economic costs were then combined using the relative contribution each representative scenario makes to its release category, as provided in Table 2.a.iv-3 of the January 31, 2011 TVA submittal (TVA 2011a). There are 11 representative scenarios: 4 for release category I, 1 for release category II, 4 for release category III, and 2 for release category IV. The combined consequences for each release category were then used to assess the risk associated with each SAMDA. In response to an NRC staff RAI, TVA provided the source terms, other release parameters, and the person-rem and offsite economic results for each of the 11 representative scenarios. The revised consequences given in Table 2.a.iv-6 of the September 16, 2011 submittal (TVA 2011f) for the 11 representative scenarios were averaged using the relative contributions cited above to yield the consequences for the four release categories used in the cost-benefit analysis of each SAMDA.

The NRC staff noted that this approach, while valid for the base case, may not be valid for the determination of the risk reduction for a given SAMDA. The methodology assumes that the relative contribution of the representative scenarios for a SAMDA remains the same as for the base case. If a given SAMDA decreases the relative importance of a high-consequence scenario while not impacting other lower consequence scenarios, then the benefit of the SAMDA would be underestimated. To assess the impact of this assumption on the determination of a SAMDA cost-benefit TVA performed a sensitivity study which applied the worst accident scenario consequences of the representative scenarios making up a release category to the entire release category to the evaluation of each SAMDA (TVA 2011c). The impact of this sensitivity study is discussed in Section H.6.2.

Subsequent to TVA responses to all NRC staff RAIs, TVA determined that all prior consequence analyses were based on a misinterpretation of the consequence model (MACCS2) output for the total person-rem for each of the several release categories (and representative scenarios) and on two less significant source term errors (TVA 2011f). This misinterpretation led to the underestimation of the total person-rem for the base case and all SAMDAs. Revised consequence and cost-benefit results and a summary the impact of these changes on the SAMDA evaluations and the responses to the prior RAI responses are provided in the TVA September 16, 2011 (TVA 2011f) and October 17, 2011 (TVA 2011g) submittals. The revised consequence analyses are cited in the above discussions.

The NRC staff concludes that the corrected methodology used by TVA to estimate the offsite consequences for WBN provides an acceptable basis from which to proceed with an assessment of risk-reduction potential for candidate SAMDAs. Accordingly, the NRC staff based its assessment of offsite risk on the CDF and revised offsite doses reported by TVA.

H.3 Potential Plant Improvements

The process for identifying potential plant improvements, an evaluation of that process, and the improvements evaluated in detail by TVA are discussed in this section.

H.3.1 Process for Identifying Potential Plant Improvements

The TVA process for identifying potential plant improvements (SAMDAs) consisted of the following elements:

- Review of other industry documentation discussing potential plant improvements as developed in NEI 05-01 (NEI 2005)
- Review of Phase II SAMAs from license renewal applications for five other U.S. nuclear sites
- Review of potential plant improvements identified in the WBN IPE and IPEEE
- Review of the most significant basic events and systems from the WBN Unit 2 PRA submitted in support of the original WBN Unit 2 SAMDA assessment (TVA 2009a)

- In response to NRC staff RAIs, review of the most significant basic events from the WBN Unit 2 IPE based PRA submitted in support of the updated SAMDA assessment (TVA 2010a).

Based on this process, an initial set of 307 candidate SAMDAs, referred to as Phase I SAMDAs, was identified. In Phase I of the evaluation, TVA performed a qualitative screening of the initial list of SAMDAs and eliminated SAMDAs from further consideration using the following criteria:

- The SAMDA is not applicable to the WBN design

- The SAMDA or its equivalent has already been implemented at WBN

- The SAMDA is similar in nature and can be combined with another SAMDA

- The SAMDA has estimated costs that would exceed the dollar value associated with completely eliminating all severe accident risk at WBN

- The SAMDA is related to a non-risk significant system known to have negligible impact on risk.

Based on this screening, 269 SAMDAs were eliminated leaving 38 for further evaluation. The remaining SAMDAs, referred to as Phase II SAMDAs, are discussed in Section 8 and listed in Table 16 of the updated SAMDA submittal (TVA 2010a). In Phase II, a detailed evaluation was performed for each of the 38 remaining SAMDA candidates, as discussed in Sections H.4 and H.6 below. To account for the potential impact of external events, the estimated benefits based on internal events were initially multiplied by a factor of 2. As discussed above, in response to an NRC staff RAI, an external events multiplier of 2.28 was used in a subsequent reassessment (TVA 2011a).

In response to NRC staff RAIs, TVA addressed the potential for SAMDAs resulting from: the enhancements identified in the WBN Unit 1 SAMDA analysis (TVA 1994a), the review of the WBN Unit 2 PRA down to a lower value of risk reduction worth (RRW), and the dominant fire zones as identified in the IPEEE. In this process, 31 additional candidate SAMDAs were identified. All were, however, screened from detailed analysis (TVA 2011a). In assessing the impact of the corrected consequence analysis on the SAMDA identification process, TVA identified one additional candidate SAMDA.[a] This SAMDA was screened out (TVA 2011f). These additional SAMDA candidates are discussed further below.

(a) The September 16, 2011 submittal (TVA 2011f) designated this new SAMDA candidate as SAMDA 340. In the May 25, 2011 RAI responses (TVA 2011c), a different SAMDA candidate was designated as SAMDA 340. The May 25, 2011 SAMDA candidate, to install flood detection in areas 772.0-A8 and 772.0-A9, is referred to in this report as SAMDA 340. The new September 16, 2011 candidate SAMDA is not referred to by a number in this review to avoid confusion.

H.3.2　Review of the TVA Process

TVA efforts to identify potential SAMDAs focused primarily on areas associated with internal initiating events, but also, in response to an RAI, included explicit consideration of potential SAMDAs for fire events. The initial list of SAMDAs generally addressed the systems and basic events considered to be important to CDF and LERF from an RRW perspective at WBN, and included selected SAMDAs from prior SAMA analyses for other plants.

TVA provided a tabular listing of PRA basic events sorted according to their CDF RRW (TVA 2011a, f). TVA stated SAMDAs impacting these basic events would have the greatest potential for reducing risk. TVA reviewed the list down to an RRW cutoff of 1.006 (after accounting for the impact of revised consequence on the MACR) for potential SAMDAs that could reduce operator error (e.g., enhanced procedures, training). An RRW of 1.006, which corresponds to about a 0.6 percent change in CDF given 100 percent reliability of the SAMDA in eliminating the risk due to the basic event, was selected as the threshold because it is approximately equivalent to a benefit of $27,000 (the minimum cost for the types of enhancements mentioned above) using the revised MACR and an external event multiplier of 2.28. TVA reviewed the RRW list down to 1.0227 (a 2.27 percent change in CDF) for hardware-based modifications. This corresponds to a $100,000 maximum benefit (TVA 2011a, f).

TVA also provided and reviewed the LERF-based RRW basic events down to a RRW of 1.029. This was determined using the definition of LERF as the sum of release categories I and II and corresponds to a benefit of $27,000 assuming that all changes in frequency occurred in release category II since this release category has the greatest consequences (TVA 2011b). It is noted that with the revised consequence analysis, this screening value increases to 1.044 due to changes in offsite economic costs since the original assessment (TVA 2011f).

Based on the review of these basic events and the important basic events in the WBN Unit 2 IPE (TVA 2010b), TVA identified 49 new WBN Unit 2-specific SAMDAs. This number does not include the SAMDAs previously identified from the generic SAMAs, from the review of other SAMA assessments, or from the WBN IPE or IPEEE insights and enhancements that also addressed the WBN Unit 2 important basic events. In response to a NRC staff RAI, TVA correlated the Phase I SAMDAs with the basic events having the highest risk importance in the Level 1 and 2 PRA. With a few exceptions, all of the significant basic events (excluding the basic events which represent "tag events" – not actual equipment or operator failure events – or physical parameters) are addressed by one or more SAMDAs. Of the basic events of high risk importance that are not addressed by SAMDAs, each is closely tied to other basic events that had been addressed by one or more SAMDAs. The NRC staff noted that no SAMDAs were identified that directly address two EDG sequencer failures which contribute a total of about 2.3 percent to the CDF. In response to the RAI, TVA discusses the failure modes and modeling of the sequencers and states that no credit is taken for EDG recovery in the WBN PRA. TVA also discusses the existence of plant procedures and training that address some of the sequencer failure modes (TVA 2011b). Because the modeling is conservative and procedures not credited in the PRA address sequencer failure, the NRC staff concludes that no additional SAMDA to address sequencer failures is likely to be cost-beneficial.

The WBN Unit 2 IPE did not result in the identification of any vulnerability. The IPE submittal did cite three SAMDAs from the original WBN Unit 2 SAMDA assessment as providing a risk reduction for internal events (TVA 2010b). The NRC staff noted that the WBN Unit 2 IPE submittal cited two sets of sensitivity studies concerning internal flooding, including one which was intended to evaluate alternative design/procedural changes that would significantly impact the flood related CDF. In response to an RAI, TVA described these sensitivity studies which supported the decision that, in addition to replacing carbon steel piping on the RCW piping in certain plant areas with stainless steel piping, TVA planned to install leak detection instrumentation in these areas (TVA 2011c). See the discussion concerning SAMDAs 293 and 294 below.

The review of the generic list of SAMAs as developed in NEI 05-01 (NEI 2005) led to 153 Phase I SAMDAs for WBN Unit 2. The review of Phase I SAMAs contained in the license renewal applications for Cook, Catawba, McGuire, Wolf Creek, and Vogtle led to the identification of 105 additional Phase I SAMDAs for WBN Unit 2.

Although no vulnerabilities were identified in the WBN Unit 1 IPE, 12 procedural and hardware enhancements and additional insights and recommendations were identified (TVA 1992). These 12 were included in the SAMDA Phase I list. The Unit 1 IPE update identified 13 additional insights and recommendations (TVA 1994a). In response to an NRC staff RAI, TVA discusses each of these by indicating the Phase I SAMDA which addresses the item or stating that it had been implemented at WBN (TVA 2010c).

In 1994, TVA performed a SAMDA analysis for WBN Unit 1 (TVA 1994b). A number of potential enhancements were identified in this analysis and in the NRC's review of the analysis (NRC 1995). In response to an NRC staff RAI, TVA indicated that all of these enhancements have been either implemented at WBN or included in the current Phase I SAMDA list (TVA 2010c).

Based on this information, the NRC staff concludes that the set of SAMDAs evaluated in the EIS, together with those identified in response to NRC staff RAIs, address the major contributors to internal event CDF.

Although several Phase I SAMDAs were identified based on the generic and other plant SAMA reviews, the WBN Unit 2 SAMDA assessments did not include any WBN-specific SAMDAs that addressed external events. As discussed above, the WBN Unit 1 IPEEE did not identify any vulnerability to external events (except one tornado missile issue, which has been addressed). The WBN Unit 2 IPEEE also did not identify any vulnerabilities, although validation and finalization of this assessment will not be completed until plant construction is finished. The

NRC staff concludes that the availability of information and status of construction for WBN Unit 2 is sufficient for the purposes of the SAMDA assessment to indicate that no vulnerabilities will be identified. If vulnerabilities are identified, they will be addressed under the IPE/IPEEE program.

NRC requested that TVA consider potential SAMDAs for the dominant fire areas and scenarios as identified in the WBN Unit 2 IPEEE. In response to this RAI, TVA provided a listing of the 18 fire scenarios that contribute to more than 90 percent of the screening fire CDF in the dominant fire areas (see Table H-4 above). From the review of this list, TVA identified 20 additional Phase I SAMDAs (TVA 2011a).

As indicated above, the WBN Unit 2 IPEEE includes a seismic margins assessment performed in accordance with the requirements of Supplement 4 to GL 88-20 (NRC 2000). The lowest value of HCLPF (0.36 g) is greater than the review level earthquake of 0.3 g. TVA defined a seismic vulnerability as any component on the safe shutdown equipment list for which the HCLPF capacity is less than 0.3 g. Thus, there is some margin to this definition of a vulnerability. The NRC staff's review of the Unit 2 IPEEE analysis concluded that the TVA definition of vulnerability and its conclusion that no severe accident vulnerabilities exist at WBN Unit 2 are reasonable (NRC 2011e).

Based on the TVA IPEEE and the expected cost associated with further risk analysis and potential plant modifications, the NRC staff concludes that the opportunity for seismic and fire-related SAMDAs has been adequately explored and that it is unlikely that there are any additional cost-beneficial seismic or fire-related SAMDA candidates.

The NRC staff questioned TVA about other lower cost alternatives to some of the SAMDAs evaluated (NRC 2009), including the following:

- purchasing or manufacturing a "gagging device" that could be used to close a stuck-open steam generator safety valve for a SGTR event prior to core damage

- using the spare 5th diesel generator mentioned in the disposition of SAMDA 261 without going through the expense of complete refurbishing and licensing

- providing procedures and cabling to enable the use of the trailer-mounted 2 MW diesel generator provided in response to GSI-189 to be used to power selected equipment such as battery chargers and/or individual pumps

- purchasing and installing a permanent diesel generator to supply power to the normal charging pump.

In response to the RAI, TVA addressed the suggested lower cost alternatives and determined that they are not feasible or had been implemented at WBN (TVA 2009b, TVA 2011a). This is discussed further in Section H.6.2.

In response to NRC staff RAIs concerning the screening of Phase I SAMDAs, TVA provided additional information on a number of SAMDAs to support the screening disposition. For those WBN Unit 2-specific SAMDAs identified through the original RRW review that were screened as "already implemented," TVA described the status of implementation at the plant and of incorporation into the WBN PRA used in the review. A group of SAMDAs were screened on the basis of design changes that were in progress or other actions to be taken in the future. TVA discussed each, providing information on the status and schedule for the change or action. For several SAMDAs screened out on the basis of "low benefit," TVA provided additional information supporting this conclusion.

SAMDA 29, which is to provide capability for alternate injection via diesel-driven fire pump, was identified from the list of generic SAMAs provided in NEI 05-01 rather than from WBN plant-specific PRA results. In response to NRC staff RAIs, TVA provided a discussion of the sequences in which this SAMDA would potentially be a benefit, the existing procedures and guidelines that would address these conditions, and the feasibility of implementing this SAMDA (TVA 2011b). The sequences potentially benefitted by this SAMDA all involve failure of the RCP seal cooling with some involving loss of steam generator cooling. Existing procedures, some of which involve use of the diesel-driven fire pump to prevent seal failure or loss of steam generator cooling address these sequences. Further, the conditions under which reactor coolant system (RCS) depressurization to a pressure low enough to allow fire pump injection would be called for occur only after core damage has already occurred. TVA also points out that other SAMDAs address the sequences of concern here and TVA has committed to follow the installation of a new seal design that addresses the RCP seal failure and, if favorable, to install these new seals. Based on this discussion, the NRC staff finds that further pursuit of SAMDA 29 is not likely to result in a cost-beneficial SAMDA.

SAMDA 58, which is to install improved RCP seals, was initially screened as not being applicable based on the cost for a new design by Westinghouse not being available and, hence, since this SAMDA is not under TVA control, the inability to perform a cost-benefit analysis. Subsequently, TVA indicated that a cost estimate is available and that, while not cost-beneficial in the baseline analysis, it would be cost-beneficial if the benefit analysis used the 95th percentile CDF. While TVA states that this SAMDA would not be considered further for implementation, TVA does commit to following the initial experience with the new seal design and, if proven reliable during operation, it would be installed at the earliest refueling outage following startup during normal seal replacements (TVA 2011a).

SAMDA 80, which is to provide a redundant train or means of ventilation, was originally screened on the basis of having a very low benefit. In response to NRC staff RAI concerning the use of temporary fans and ducting to mitigate room cooling failures in the centrifugal charging pump (CCP) area, the TD AFW pump room, and the EDG switchgear rooms, TVA indicated that, since such equipment was relatively inexpensive and easy to use, additional equipment will be made available and procedures will be written for the use of such equipment in these areas (TVA 2011b).

SAMDA 183, which is to implement internal flood prevention and mitigation enhancements, was screened as being of very low benefit. Subsequently, two flood related SAMDAs, SAMDAs 293 and 294, associated with RCW failures were added based on the review of the RRW values for the October 2010 SAMDA PRA. These two SAMDAs were screened as having already been identified as implementation commitments to the NRC. In response to an NRC staff RAI, TVA indicated that they had committed to replacing the existing piping with stainless steel piping. While credit was taken for the lower pipe leak frequencies in the SAMDA PRA, floods from this piping in specific plant areas (*772.0-A8* and *772.0-A9*) still contributed significantly to the CDF. Since rerouting the RCW piping was considered impractical, TVA committed to install flood detection instrumentation in these rooms (TVA 2011a). This was supported by sensitivity studies cited in the IPE and described in the TVA response to an RAI (TVA 2011c). This was designated by TVA as SAMDA 340 (TVA 2011c).

Subsequently, TVA revised its commitment to install flood detection and change piping from carbon to stainless steel in the WBN Auxiliary Building areas, 772.0-A8 and 772.0-A9, to only install flood detection in those areas, consistent with SAMDA 340 (TVA 2012). The rationale for this revision came from a bounding evaluation of the change in CDF for multiple electrical board areas including the specified areas showing that changing both the pipe material and installing flood detection provided a very small increase in benefit as compared to the change in risk from either modification by itself. The piping change alone would be expected to reduce the CDF by about 1.5E-5/yr. The installation of flood detection alone would be expected to reduce the CDF by about 1.6E-5/yr. The expected CDF reduction from taking both actions is only slightly greater, about 1.7E-5/yr. Therefore, TVA determined that installation of flood detection by itself sufficiently reduced CDF without replacement of the existing piping.

SAMDA 242, which is to provide a permanent dedicated generator for the normal charging pump with local operation of the TD AFW pump after 125V battery depletion, was screened out as having excessive implementation cost. In an NRC staff RAI, it was noted that this SAMDA was similar to SAMDA 255 (provide a permanent dedicated generator for the normal charging pump, one motor driven AFW pump, and a battery charger), except that SAMDA 242 had a smaller scope and therefore would be expected to have a smaller cost. TVA provided additional information on the cost-benefit for SAMDA 242 that supported the screening.

SAMDA 314, which is to enhance training for local control of AFW given station blackout, loss of control air, or fires affecting AFW level control valves (LCVs), identified from the review of the fire FIVE assessment, was screened out as already implemented; citing SAMDAs 285 and 299.

The NRC staff noted that while a number of SAMDAs cite enhancements to training in a general sense, none appear to specifically address the training enhancement needed for SAMDA 314. In response to an NRC staff RAI to provide a specific citation which incorporates the requirements of SAMDA 314, TVA provided a discussion of the existing training and cited the procedures that deal with failures due to the fires in the key fire areas and specifically the local manual operation of the AFW LCVs. In addition, TVA, as part of its response to questions on the Appendix R analysis, has committed to provide a new capability to allow the operators, from the control room, to transfer from normal compressed air supply to the station nitrogen system for control of the LCVs. This would be expected to have a greater benefit for these fire scenarios then the enhanced training of SAMDA 314 (TVA 2011b). The NRC staff concludes that the existing training and the new commitment for use of the nitrogen supply adequately addresses the mitigation of fire scenarios originally addressed by SAMDA 314.

The NRC staff notes that one SAMDA (SAMDA 273, which is to provide a redundant path for emergency core cooling system (ECCS) suction from the refueling water storage tank (RWST) around check valve 62-504), originally identified and included as a Phase II SAMDA in the January 2009 submittal (TVA 2009a) was subsequently screened in the updated submittal (TVA 2010a). Check valve 62-504 appeared in the original list of important components but did not have a RRW of 1.007 or greater in the revised analysis. Based on the NRC staff review, the NRC staff considers this screening to be appropriate.

The NRC staff notes that the set of SAMDAs submitted is not all inclusive, because additional, possibly even less expensive, design alternatives can always be postulated. However, the NRC staff concludes that the benefits of any additional modifications are unlikely to exceed the benefits of the modifications evaluated and that the alternative improvements would not likely cost less than the least expensive alternatives evaluated, when the subsidiary costs associated with maintenance, procedures, and training are considered.

The NRC staff concludes that TVA used a systematic and comprehensive process for identifying potential plant improvements for WBN, and that the set of potential plant improvements identified by TVA is reasonably comprehensive and therefore acceptable. This search included reviewing insights from the plant-specific risk studies, and reviewing plant improvements considered in previous SAMDA analyses. While explicit treatment of external events in the SAMDA identification process was limited, the absence of external event vulnerabilities reasonably justifies examining primarily the internal events risk results for this purpose.

H.4 Risk-Reduction Potential of Plant Improvements

TVA evaluated the risk-reduction potential of the 38 remaining SAMDAs applicable to WBN. The majority of the SAMDA evaluations were performed in a bounding fashion in that the SAMDA was assumed to completely eliminate the risk associated with the proposed enhancement. Such bounding calculations overestimate the benefit and are conservative.

TVA used model requantification to determine the potential benefits. The CDF and population dose reductions were estimated using the WBN Unit 2 PRA (version WBN_U1_U2_FLOOD_SAMA) model and the Level 3 consequence analysis. The changes made to the model to quantify the impact of SAMDAs are detailed in Section 8 of the updated SAMDA assessment (TVA 2010a). Table H-5 lists the assumptions considered to estimate the risk reduction for each of the evaluated SAMDAs, the estimated risk reduction in terms of percent reduction in CDF and population dose, and the estimated total benefit (present value) of the averted risk. The estimated benefits reported in Table H-5 reflect the combined benefit in both internal and external events, as well as a number of changes to the analysis methodology subsequent to the above referenced submission. The determination of the benefits for the various SAMDAs is further discussed in Section H.6.

The NRC staff questioned the assumptions used in evaluating the benefits or risk-reduction estimates of certain SAMDAs provided in the SAMDA assessment (TVA 2011a).

For SAMDA 45, which is to enhance procedural guidance for the use of cross-tied component cooling or service water pumps, TVA clarified that the model requantification assumed that the cross tie provided backup cooling not only to the charging pumps but to the component cooling system and all of its loads.

For SAMDA 70, which is to install accumulators for TD AFW pump flow control valves, the risk benefit is stated to be bounded by eliminating the cognitive portion of human error to restore AFW control following loss of instrument air. TVA supported the assumption on the basis that the feasible accumulator size would only provide enough air for a few cycles of the flow control valves; hence operator action would ultimately be required. The additional time for operator response provided by the accumulators is assumed to eliminate the cognitive portion of the human error but not the action portion. In response to an NRC staff RAI, TVA confirmed that the assessment eliminated the relevant cognitive portion of the human error in both the independent and dependent human error contributors. Subsequently, as discussed above, TVA has committed to provide a new capability to allow the operators from the control room to transfer from normal compressed air supply to the station nitrogen system for control of the LCVs. This new capability, identified by TVA as SAMDA 339, will have a greater benefit than that associated with SAMDA 70 and thus supersedes it (TVA 2011b). The NRC staff finds that this commitment adequately dispositions SAMDA 70.

Table H-5. SAMDA Cost/Benefit Screening Analysis for WBN[a]

SAMDA	Assumptions	% Risk Reduction		Total Benefit ($)		Cost ($)
		CDF[b]	Population Dose[b]	Baseline (Internal + External)[c]	Baseline with Uncertainty[d]	
4 – Improve DC bus load shedding	AC power always recovered prior to battery failure	1.1	1.2	40K	110K	32K
8 – Increase training on response to loss of two 120V AC buses which causes inadvertent actuation signal	Eliminate the contributions of the loss of 120V bus initiators	0.8	~0	12K	350K	27K
26 – Provide an additional high pressure injection (HPI) pump with independent diesel	Added a new basic event in parallel with existing HPI pump without any power dependency	1.4	1.4	65K	180K	3.6M
32 – Add the ability to automatically align ECCS to recirculation mode upon RWST depletion	Swap over to high pressure recirculation is always successful	7.4	12	400K	1.1M	2.1M
45 – Enhance procedural guidance for use of cross-tied component cooling or service water pumps	Cross-tying ERCW headers is always successful	0.3	~0	5K	14K	32K
46 – Add service water pump	ERCW pump 1A-A is always successful	7.0	3.7	150K	410K	1.0M
56 – Install an independent RCP seal injection system, without dedicated diesel	RCP seal injection is always successful when AC power available	24	29	1.1M	3.2M	8.2M
70 – Install accumulators for TD AFW pump flow control valves [Superseded by SAMDA 339][e]	Operator cognitive error to manually operate the valves to control level set to zero	2.5	2.2	100K	280K[e]	260K
71 – Install a new condensate storage tank (AFW storage tank)	New tank would require same operator actions as current design	~0[f]	~0[f]	~0	~0	1.7M
87 – Replace service and instrument air compressors with more reliable compressors which have self-contained air cooling by shaft driven fans	Normal plant air system is always successful	0.2	~0	2.2K	6.0K	890K
93 – Install an unfiltered hardened containment vent to eliminate the containment overpressure failure	LATE release category revised to be half the release for a late containment rupture from the SEQSOR release methodology.	0	38	1.2M	3.5M	3.1M

Table H-5. (contd)

SAMDA	Assumptions	% Risk Reduction		Total Benefit ($)		Cost ($)
		CDF[b]	Population Dose[b]	Baseline (Internal + External)[c]	Baseline with Uncertainty[d]	
93 – Install an unfiltered hardened containment vent to eliminate the containment overpressure failure	**LATE release category revised to be half the release for a late containment rupture from the SEQSOR release methodology.**	0	38	1.2M	3.5M	3.1M
101 – Provide a reactor exterior cooling system to cool a molten core before vessel failure	Removed the rocket mode and ex-vessel steam explosion failure modes from the containment failure probability	0	8.5	210K	580K	2.5M
103 – Institute simulator training for severe accident scenarios	Eliminated human action failure to arrest the severe accidents	33	32	1.4M	3.9M	8.0M
109 – Install a passive hydrogen control system	Assumed hydrogen igniters always successful	0	12	300K	840K	3.7M
110 – Erect a barrier that would provide enhanced protection of the containment walls (shell) from ejected core debris following a core melt scenario at high pressure.	Set rocket mode early containment failure probability (0.05) to zero for high pressure vessel breach sequences	0	4.0	100K	290K	1.2M
112 – Add redundant and diverse limit switches to each containment isolation valve	Completely eliminate all ISLOCA events.	<0.1	~0	3.2K	8.9	690K
136 – Install motor generator set trip breakers in the control room	Operator action to trip reactor is always successful	0.9	~0	13K	37K	240K
156 – Eliminate RCP thermal barrier dependence on condenser cooling water (CCW), such that loss of CCW does not result directly in core damage (Enhance procedural guidance for use of ERCW for RCP thermal barrier cooling)[a]	**RCP seal injection is always successful when AC power is available.**	13	20	780K[h]	2.2M[h]	32K
176 – Provide a connection to alternate offsite power source	Remove grid related failures from frequency of loss of offsite power	19	17	780K	2.2M	9.1M
191 – Provide self-cooled ECCS seals	Eliminate seal cooling failures	~0[f]	~0	~0	~0	1.0M

Table H-5. (contd)

SAMDA	Assumptions	% Risk Reduction CDF[b]	% Risk Reduction Population Dose[b]	Total Benefit ($) Baseline (Internal + External)[c]	Total Benefit ($) Baseline with Uncertainty[d]	Cost ($)
215 – Provide a means to ensure RCP seal cooling so that RCP seals LOCAs are precluded for SBO events	Assume for SBO that 21 GPM seal event always occurs and other seal LOCAs never occur	26	31	1.3M	3.7M	1.5M
226 – Provide permanent self-powered pump to back up normal charging pump	Assume guaranteed success of seal injection system	26	31	1.3M	3.7M	2.7M
255 – Install a permanent, dedicated generator for the normal charging pump, one Motor Driven AFW Pump and a Battery Charger	Added an additional diesel generator to the power inputs of one charging pump, one AFW pump and one battery charger	18	20	840K	2.3M	3.2M
256 – Install fire barriers around cables or reroute the cables away from fire sources (*Enhance procedure for controlling temporary alterations to reduce fire risk from temporary cables*)[g]	Reduce CDF and consequences of all release categories except SGTR by 25%	25	25	1.1M	3.1M	20K
276 – Provide an auto start signal for the AFW on loss of standby feedwater pump	Beneficial only for startup accidents which are assumed to have approximately same risk as at-power accidents. Reduce risk for all initiators except SGTR by 1/365 assuming that standby feedwater pump in use for a total time of one day per year	0.7	0.6	25K	70K	620K
279 – Provide a permanent tie-in to the construction air compressor	Assume air compressor D is always successful given success of power supply and no flood events.	1.8	1.6	72K	200K	910K
280 – Add new Unit 2 air compressor similar to the Unit 1 D compressor	Assume air compressor D is always successful given success of power supply and no flood events.	1.8	1.6	72K	200K	810K
282 – Provide cross tie to Unit 1 RWST	Assume operator actions involving makeup to RWST set to success and cognitive error to sump swap over eliminated	1.3	~0	21K	58K	10M

Table H-5. (contd)

SAMDA	Assumptions	% Risk Reduction		Total Benefit ($)		Cost ($)
		CDF[b]	Population Dose[b]	Baseline (Internal + External)[c]	Baseline with Uncertainty[d]	
285 – Improve training to establish feed and bleed cooling given no centrifuge charging pumps (CCPs) are running or a vital instrument board fails	Set human action to initiate bleed and feed cooling and associated dependent events to guaranteed success	6.4	0.3	140K	144K	27K
292 – Improve training to reduce failure probability to terminate inadvertent safety injection prior to water challenge to PORVs	Set human action to terminate safety injection to prevent PORV water challenge to guaranteed success	4.2	13	400K	1.1M	27K
295 – Increase frequency of containment leak rate testing	Set containment small and preexisting leak frequency to zero	0	6.1	144K	400K	2.5M
299 – Initiate frequent awareness training for plant operators/ maintenance/testing staff on key human actions for plant risk (Initiate frequent awareness training for maintenance and testing staff as on key human actions for plant risk)[g]	Reduced key human actions for CDF and release category	4.6	6.6	290K	793K	27K
300 – Revise procedure FR-H.1 to eliminate and/or simplify complex decision logic for establishing feed and bleed cooling and to improve operator recovery from initial mistakes	Reduce human error rate of operator action to initiate bleed and feed cooling to just the cognitive part	3.4	0.2	57K	160K	100K
303 – Move indicator/operator interface for starting igniters to front main control room panel	Set the cognitive portion of human action to place igniters in service as success	0	~0	1.7K	4.8K	50K
304 – Add annunciator or alarm signaling parameters to initiate hydrogen igniters to front panel on MCR	Set the cognitive portion of human action to place igniters in service as success	0	~0	1.7K	4.8K	50K

Table H-5. (contd)

SAMDA	Assumptions	% Risk Reduction		Total Benefit ($)		
		CDF[b]	Population Dose[b]	Baseline (Internal + External)[c]	Baseline with Uncertainty[d]	Cost ($)
305 – Revise procedure E-1 to include recovery steps for failure to initiate hydrogen igniters	Set human action to place igniters in service as success	0	6.2	150K	420K	100K
306 – Improve operator performance by enhancing likelihood of recovery from execution errors	Reduce joint probability of dependent action involved in important recovery action	2.4	5.3	170K	470K	100K
307 – Make provisions for connecting ERCW to CCP 2B-B	Added potential for using ERCW for CCP 2B-B similar to that for CCP 2A-A	0.1	0.0	0.6K	1.7K	99K
339 – Provide a capability to allow the operators from the control room to transfer from normal compressed air supply to the station nitrogen system for control of the AFW LCVs.	Not explicitly evaluated.[e]	NA[e]	NA[e]	NA[e]	NA[e]	NA[e]
340 – Install flood detection in areas 772.0-A8 and 772.0-A9	Not explicitly evaluated.[i]	NA[i]	NA[i]	NA[i]	NA[i]	NA[i]

NA – Not available

(a) SAMDAs in bold are potentially cost beneficial, and have either been committed to be further considered for implementation by TVA and/or have already been implemented. See the discussion in Section H.6.2.

(b) Determined by NRC staff from values given by TVA (TVA 2010a, 2011a, f, g).

(c) Using an external events multiplier of 2.28.

(d) Determined from baseline benefit times a 95th percentile to point estimate ratio of 2.78. Note that this is different from the 2.70 value used by TVA. See the discussion in Section H.6.2 below.

(e) While SAMDA 70 is slightly cost-beneficial at the 95th percentile CDF uncertainty, it has been superseded by SAMDA 339 to which TVA has committed and should have a greater benefit. See the discussion below.

(f) TVA states that the risk benefit of the SAMDA is zero. While the NRC staff believes the risk benefit may not be zero it does conclude that it is negligible compared to the estimated cost.

(g) SAMDA title given in parentheses is considered a more accurate description of the actual SAMDA.

(h) Due to time constraints, procedure change envisioned for SAMDA 156 is now considered not to be effective; hence benefit would be essentially negligible. Hardware change considered in SAMDA 215.

(i) This SAMDA captures the previous commitment by TVA to install this flood detection equipment (TVA 2011c).

For SAMDA 93, which is to install an unfiltered hardened containment vent to eliminate the containment overpressure failure, TVA provided additional information on the adjustments made to the LATE release category to evaluate the benefit. The early portion of the category for the releases from the reactor coolant system remained unchanged, while the later portion for the releases for the core-concrete interaction phase were taken to be half of those from the SEQSOR methodology for late rupture. The NRC staff noted that the TVA assessment indicated there is no reduction in CDF for SAMDA 93. The usual purpose of containment venting is to prevent core damage for loss of containment heat removal sequences where the functioning core injection systems would fail upon containment overpressure failure. In response to an RAI to discuss the reason why there is no CDF reduction for this SAMDA, TVA provided a discussion of core damage modeling in the Sequoyah Nuclear Plant NUREG-1150 analysis (NRC 1990c) and the specific sequences at WBN that might lead to a CDF reduction for containment venting (TVA 2011b). The WBN CAFTA model adopts a similar approach as that found in NUREG-1150 modeling of Sequoyah Nuclear Plant. This approach evaluates the frequency of core damage as independent of containment heat removal and thus venting (which is equivalent in impact to containment heat removal) would not affect the CDF. TVA provides a discussion and bounding analysis of the WBN sequences for which containment venting might reduce the CDF. The result indicates a maximum potential CDF reduction of approximately 6×10^{-9} per year, which is equivalent to an added cost-benefit of $1,400. This is very small contribution to the estimated benefit due to release category changes alone of $1,100,000. While the above cost-benefit values increase slightly (10 to 15 percent) as a result of the revised consequence analysis, the conclusion remains valid. The NRC staff finds that the TVA updated assessment of the benefit of SAMDA 93 is acceptable for the SAMDA analysis.

For SAMDA 110, which is to erect a barrier that would provide enhanced protection of the containment walls (shell) from ejected core debris following a core melt scenario at high pressure, the NRC staff questioned the basis for the benefit—which was originally estimated by removing the rocket mode and ex-vessel steam explosion failure modes from the containment event tree. TVA revised the assessment to only eliminate the rocket mode from the model since core debris would only be expected to reach the containment wall for high pressure reactor pressure vessel failures such as those that lead to the rocket mode but not ex-vessel steam explosions. The rocket mode was used since the Level 2 risk model did not explicitly include the debris impingement failure mode due to the assumption that the seal table would prevent this impingement. In addition, debris impingement would only be possible for station blackout sequences because, with AC power available, containment spray injection would be expected to flood the reactor cavity. The probability of rocket mode failure of 0.05 was set to zero for a number of containment event tree split fractions. Estimation of the risk reduction using this value is considered adequate by TVA to represent the debris impingement mode based on information from NUREG-1150 (NRC 1990a). TVA reported the results of two sensitivity studies making different assumptions concerning the Level 2 model. Both resulted in smaller risk benefits than elimination of the rocket mode of failure (TVA 2011a, b). Based on these results and the fact that this SAMDA potentially reduces the risk of only a small portion of the CDF, the NRC staff concludes that SAMDA 110 would not be cost-beneficial.

For SAMDA 215, which is to provide a means to ensure RCP seal cooling so that RCP seal LOCAs are precluded for SBO events, the benefit was assessed by modifying RCP seal LOCA probabilities. The NRC staff questioned limiting the benefit to only SBO scenarios. TVA confirmed that the modified seal LOCA probabilities were made for all sequences including SBO, loss of ERCW, or loss of component cooling system.

For SAMDA 299, which is to initiate frequent awareness training for plant operators/maintenance/testing staff on key human actions for plant risk, the NRC staff questioned the calculated reduction in CDF compared to the similar but apparently more limited SAMDA 300. TVA indicated that the title of SAMDA 299 is slightly misleading in that the training for operators had already been implemented and this SAMDA should have been described as additional training for maintenance and testing staff as appropriate to address key actions they perform.

For SAMDA 300, which is to revise procedure FR-H.1 to eliminate and/or simplify complex decision logic for establishing feed and bleed cooling and to improve operator recovery from initial mistakes both of which involve reducing human errors associated with CDF and release categories, TVA revised the benefit analysis to be based on eliminating only the action portion of the human error.

For SAMDAs 303 and 305, both of which involve actions to reduce operator error to initiate hydrogen igniters, the risk benefit for both was stated to be determined by setting the human action to place igniters in service as success. The NRC staff questioned the assessment since the net benefit of the two SAMDAs is significantly different. TVA corrected the assessment of SAMDA 303 by indicating that the SAMDA would only reduce the cognitive portion of the human error and thus the benefit was based on eliminating this portion of the human action. For SAMDA 305, TVA indicated that the procedure change would have a greater impact on the human error than SAMDA 303 and the benefit would be bounded by the elimination of all human error to initiate the igniters.

The NRC staff has reviewed the TVA bases for calculating the risk reduction for the various plant improvements as described in the SAMDA assessments and in response to NRC staff RAIs and concludes that the rationale and assumptions for estimating risk reduction are reasonable and generally conservative (i.e., the estimated risk reduction is higher than what would actually be realized). Accordingly, the NRC staff based its estimates of averted risk for the various SAMDAs on the TVA risk reduction estimates.

H.5 Cost Impacts of Candidate Plant Improvements

TVA estimated the costs of implementing the 38 Phase II SAMAs by focusing on labor (e.g., craft, engineering) and component cost related to installing the proposed physical change. Costs do not include lifetime operation, testing, or maintenance costs or contingency for unforeseen obstacles or inflation (TVA 2010c). Procedure development and training associated with the physical changes were also not included, except for those SAMDAs which were solely procedural and/or training activities (TVA 2011a).

The NRC staff reviewed the bases for the applicant's cost estimates as described above. In response to an NRC staff RAI concerning per unit cost savings associated with implementing the changes to both WBN units, TVA stated that the cost of procedural or training module development is only marginally increased to apply to a second unit and that for physical unit design changes the costs are for the affected unit only (TVA 2011a). While TVA states that dividing the cost of procedure and training SAMDAs by a factor of two would not be appropriate, the NRC staff concludes that the per unit cost of physical changes (for the scope of the cost estimate as described above) would be less than that given by TVA. The scope of TVAs cost estimate, however, does not include lifetime costs associated with the procedure and training and hence is conservative. This is borne out by comparison with similar costs given in license renewal SAMA submittals. With regard to physical changes, the NRC staff concludes that while there may be some savings with respect to sharing engineering costs between units to the extent that these cost are included in the cost estimate, other factors such as lifetime costs and procedure and training associated with the change that are not included in the TVA estimate result in a conservative estimate for use in the SAMDA assessments. Further, in response to a specific RAI concerning SAMDA 70, TVA stated that engineering and design cost were not considered (TVA 2011b). This is the only SAMDA where such cost savings would impact the cost-benefit conclusions.

For a number of the Phase II SAMDAs evaluated by TVA, the information provided did not sufficiently describe the associated modifications and what is included in the cost estimate. In response to an NRC staff RAI, TVA provided a more detailed description of both the modification and the cost estimate for these SAMDAs (TVA 2010c). This information resolved the NRC staff concerns. In addition, conflicting information was provided for the costs associated with several SAMDAs. In response to an NRC staff RAI, TVA discussed the reasons for these differences and indicated the correct value to be used in the cost-benefit analysis (TVA 2010c).

The NRC staff concludes that the cost estimates provided by TVA are sufficient and appropriate for use in the SAMDA evaluation.

H.6 Cost-Benefit Comparison

The TVA cost-benefit analysis and the NRC staff's review are described in the following sections.

H.6.1   The TVA Evaluation

The methodology used by TVA was based on NEI 05-01, *Severe Accident Mitigation Alternatives (SAMA) Analysis Guidance Document* (NEI 2005),which in turn is based on NRC's guidance for performing cost-benefit analysis, i.e., NUREG/BR-0184, *Regulatory Analysis Technical Evaluation Handbook* (NRC 1997b). NEI 05-01 was endorsed for use in the license renewal application by the NRC (NRC 2007). The guidance involves determining the net value for each SAMA (or SAMDA) according to the following formula:

$$\text{Net Value} = (APE + AOC + AOE + AOSC) - COE$$

where   APE   =   present value of averted public exposure ($)
            AOC   =   present value of averted offsite property damage costs ($)
            AOE   =   present value of averted occupational exposure costs ($)
            AOSC =   present value of averted onsite costs ($)
            COE   =   cost of enhancement ($).

If the net value of a SAMDA is negative, the cost of implementing the SAMDA is larger than the benefit associated with the SAMDA and it is not considered cost-beneficial. The TVA derivation of each of the associated costs is summarized below.

NUREG/BR-0058 has recently been revised to reflect the agency's policy on discount rates. Revision 4 of NUREG/BR-0058 states that two sets of estimates should be developed—one at 3 percent and one at 7 percent (NRC 2004b). TVA performed the SAMDA analysis using 7 percent and provided a sensitivity analysis using the 3 percent discount rate in order to capture SAMDAs that may be cost-effective using the lower discount rate, as well as the higher, baseline rate (TVA 2011a). This analysis is sufficient to satisfy NRC policy in Revision 4 of NUREG/BR-0058.

H.6.1.1   Averted Public Exposure (APE) Costs

The APE costs were calculated using the following formula:

APE = Annual reduction in public exposure (Δ person-rem per year)
× monetary equivalent of unit dose ($2,000 per person-rem)
× present value conversion factor (13.42 based on a 40-year period with a
7 percent discount rate).

As stated in NUREG/BR-0184 (NRC 1997a), it is important to note that the monetary value of the public health risk after discounting does not represent the expected reduction in public health risk due to a single accident. Rather, it is the present value of a stream of potential losses extending over the remaining lifetime (in this case, the operating license period) of the facility. Thus, it reflects the expected annual loss due to a single accident, the possibility that such an accident could occur at any time over the time period, and the effect of discounting these potential future losses to present value. For the purposes of initial screening, which assumes elimination of all severe accidents due to internal events, TVA calculated an APE of approximately $536,000 for the 40-year license period (TVA 2011f).

H.6.1.2   Averted Offsite Property Damage Costs (AOC)

The AOCs were calculated using the following formula:

$$\text{AOC} = \text{Annual CDF reduction}$$
$$\times \text{ offsite economic costs associated with a severe accident (on a per-event basis)}$$
$$\times \text{ present value conversion factor.}$$

For the purposes of initial screening, which assumes all severe accidents due to internal events are eliminated, TVA calculated an annual offsite economic risk of about $53,700 based on the Level 3 risk analysis. This results in a discounted value of approximately $720,000 for the 40-year license period (TVA 2011f).

H.6.1.3   Averted Occupational Exposure (AOE) Costs

The AOE costs were calculated using the following formula:

$$\text{AOE} = \text{Annual CDF reduction}$$
$$\times \text{ occupational exposure per core damage event}$$
$$\times \text{ monetary equivalent of unit dose}$$
$$\times \text{ present value conversion factor.}$$

TVA derived the values for AOE from information provided in Section 5.7.3 of the regulatory analysis handbook (NRC 1997a). Best estimate values provided for immediate occupational dose (3,300 person-rem) and long-term occupational dose (20,000 person-rem over a 10-year cleanup period) were used. The present value of these doses was calculated using the equations provided in the handbook in conjunction with a monetary equivalent of unit dose of $2,000 per person-rem, a real discount rate of 7 percent, and a time period of 40 years to represent the license period. For the purposes of initial screening, which assumes all severe accidents due to internal events are eliminated, TVA calculated an AOE of approximately $8,150 for the 40-year license period.

H.6.1.4   Averted Onsite Costs

Averted onsite costs (AOSC) include averted cleanup and decontamination costs and averted power replacement costs. Repair and refurbishment costs are considered for recoverable accidents only and not for severe accidents. TVA derived the values for AOSC based on information provided in Section 5.7.6 of NUREG/BR-0184, the regulatory analysis handbook (NRC 1997a).

$$\text{AOSC} =$$
[(present value of cleanup costs per core damage event × present value conversion factor) + (present value of replacement power for a single event × factor to account for remaining service years for which replacement power is required × reactor power scaling factor)] × annual CDF reduction

The total cost of cleanup and decontamination subsequent to a severe accident is estimated in the regulatory analysis handbook to be $\$1.5 \times 10^9$ (undiscounted). This value was converted to present costs over a 10-year cleanup period and integrated over the 40-year license period to give 1.45×10^{10} \$-years.

TVA based its calculations on the value of 1,160 megawatts electric (MW(e)). Therefore, TVA applied a power scaling factor of 1160/910 to determine the replacement power costs. Using the methodology of NUREG/BR-0184 for a 7 percent discount rate the resulting net present value of replacement power integrated over the 40-year license period is 2.43×10^{10} \$-years.

For the purposes of initial screening, which assumes all severe accidents due to internal events are eliminated, TVA calculated an AOSC of approximately $666,000 for the 40-year license period.

Using the above equations, TVA estimated the total present dollar value equivalent associated with completely eliminating severe accidents from internal events at WBN to be about $1,930,000. Use of a multiplier of 2.28 to account for external events increases the value to $4,401,000 and represents the dollar value associated with completely eliminating all internal and external event severe accident risk at WBN Unit 2, also referred to as the Modified MACR (TVA 2011f).

H.6.1.5   The TVA Results

If the implementation costs for a candidate SAMDA exceeded the calculated benefit, the SAMDA was considered not to be cost-beneficial. In the TVA SAMDA submittal, this is expressed, not as a negative net value (SAMDA benefit less than cost), but as a benefit-to-cost ratio for the SAMDA that is less than 1.0. The benefit, cost, and benefit-to-cost ratio for the Phase II SAMDAs are given in the revised Table 2.a.iv-8 of the TVA September 16, 2011 submittal (TVA 2011f). This table incorporates revised analysis taking into account the responses to NRC staff RAIs on the prior results as well as the results of the corrected consequence analysis.

In the baseline analysis contained in the January 31, 2011, submittal (using a 7 percent discount rate and an external events multiplier of 2.28), TVA identified eight potentially cost-beneficial SAMDAs. The potentially cost-beneficial SAMDAs are:

- SAMDA 4 – Improve DC bus Load shedding

- SAMDA 156 – Eliminate RCP thermal barrier dependence on CCW, such that loss of CCW does not result directly in core damage *(Enhance procedural guidance for use of ERCW for RCP thermal barrier cooling)*[a]

- SAMDA 256– Install fire barriers around cables or reroute the cables away from fire sources *(Enhance procedure for controlling temporary alterations to reduce fire risk from temporary cables)**

- SAMDA 285 – Improve training to establish feed and bleed cooling given no CCPs are running or a vital instrument board fails

- SAMDA 292 – Improve training to reduce failure probability to terminate inadvertent safety injection prior to water challenge to PORVs

- SAMDA 299 – Initiate frequent awareness training for plant operators/maintenance/testing staff on key human actions for plant risk *(Initiate frequent awareness training for maintenance and testing staff as on key human actions for plant risk)*[a]

- SAMDA 305 – Revise procedure E-1 to include recovery steps for failure to initiate hydrogen igniters

- SAMDA 306 – Improve operator performance by enhancing likelihood of recovery from execution errors.

It was subsequently determined that, due to time constraints, the procedural enhancements of SAMDA 156 would not be effective and hence this SAMDA would not have the benefit originally estimated. Also, it was determined that, relative to SAMDAs 305 and 306, the HRA in the PRA had not credited recovery steps in an existing procedure (SAG-6 "Containment Control Conditions") and hence these SAMDAs have already been implemented.

_____

(a) SAMDA title given in parentheses is as given in Section 10, Conclusions, of the submittals and is a more accurate description of the actual SAMDA.

TVA performed additional analyses to evaluate the impact of discount rate, CDF uncertainties and parameter choices on the results of the SAMDA assessment (TVA 2011f). If the benefits are calculated for a 3 percent discount rate or increased by a factor of 2.7 to account for uncertainties, six additional SAMDA candidates were determined to be potentially cost-beneficial:

- SAMDA 8 – Increase training on response to loss of two 120V AC buses which causes inadvertent actuation signal

- SAMDA 70 – Install accumulators for TD AFW pump flow control valves

- SAMDA 93 – Install an unfiltered hardened containment vent to eliminate the containment overpressure failure

- SAMDA 215 – Provide a means to ensure RCP seal cooling so that RCP seals LOCAs are precluded for SBO events

- SAMDA 226 – Provide permanent self-powered pump to back up normal charging pump

- SAMDA 300 – Revise procedure FR-H.1 to eliminate of simplify complex (and/or) decision logic for establishing feed and bleed cooling and to improve operator recovery from initial mistakes.

SAMDA 215, which is to provide a means to ensure RCP seal cooling so that RCP seals LOCAs are precluded for SBO events, is considered by TVA to be essentially the replacement of the RCP seals with a new design which eliminates the high leakage seal failure mode. This is the same as SAMDA 58 and is discussed further in Sections H.6.2 and H.7.

H.6.2   Review of the TVA Cost-Benefit Evaluation

The cost-benefit analysis performed by TVA was based primarily on NUREG/BR-0184 (NRC 1997a) and was executed consistent with this guidance.

To account for external events, TVA initially multiplied the internal event benefits by a factor of 2 for each SAMDA. As discussed above in Section H.2.2, in response to an NRC staff RAI, TVA increased this to 2.28, and this value was used for the results discussed above and included in the results in Table H-4. As a result of the TVA baseline analysis, eight SAMDAs (SAMDAs 4, 156, 256, 285, 292, 299, 305, and 306, as described above) were identified as potentially cost-beneficial.

As indicated above, TVA considered the impact of discount rate, CDF uncertainties and parameter choices on the results of the SAMDA assessment (TVA 2011f). The results of the discount rate assessment are provided in the updated Table 2.a.iv-9 of the September 16, 2011 submittal (TVA 2011f). The change in discount rate from 7 percent used in the baseline case to 3 percent used in the sensitivity analysis increases the assessed benefit of all SAMDAs but only

Appendix H

changed the conclusion concerning the cost-benefit of SAMDAs 215 and 300. The disposition of these SAMDAs is discussed below in Section H.7. Moreover, these results indicated that the impact of the 3 percent discount rate was less than that of the CDF uncertainty. Hence, the SAMDAs that are cost-beneficial based on the CDF uncertainty incorporate those that are cost-beneficial considering the 3 percent discount rate.

TVA provided the results of an additional sensitivity analysis of evacuation speed, a WinMACCS input parameter. This analysis did not identify any additional potentially cost-beneficial SAMDAs. This is as expected since evacuation speed has only a small impact on offsite exposure and no impact on offsite economic consequence and offsite exposure makes up only a small portion of the total maximum benefit.

TVA considered the impact that possible increases in benefits from analysis uncertainties would have on the results of the SAMDA assessment. Since no uncertainty distributions on CDF were available for the CAFTA based SAMDA model, TVA used the results of the uncertainty analysis of the RISKMAN WBN4SAMDA PRA model (PLG 1992) to establish the uncertainty multiplier to be used. From this information TVA chose the ratio of the 95th percentile CDF to the mean CDF or 2.70. The results of the analysis uncertainty assessment are provided in the updated Table 2.a.iv-10 of the September, 16, 2011 submittal (TVA 2011f). Based on this uncertainty consideration, TVA determined that six additional SAMDAs (SAMDAs 8, 70, 93, 215, 226, and 300, as described above) were potentially cost-beneficial.

The NRC staff notes that the CAFTA results are point estimates, not mean values, and hence the ratio of the 95th percentile CDF to the point estimate CDF of 2.78 should be used in the CDF uncertainty analysis instead of 2.7. This difference is small and in the revised analysis of September 16, 2011 (TVA 2011f) did not impact the cost-benefit analysis of any SAMDAs.

SAMDA 70, which is to install accumulators for TD AFW pump flow control valves, was originally assessed by TVA to have a benefit-to-cost ratio of 0.99 (TVA 2010a), but was determined to have a ratio just slightly above 1.0 using 2.78 in the corrected consequence analysis. In response to an NRC staff RAI and as discussed above, TVA, as part of its response to questions on the Appendix R analysis, has committed to provide a new capability to allow the operators from the control room to transfer from normal compressed air supply to the station nitrogen system for control of the LCVs. This new capability, identified as SAMDA 339, will have a greater benefit then that associated with SAMDA 70 and thus supersedes it (TVA 2011b).

As discussed above, the methodology TVA used to determine the benefit of each SAMDA could lead to an underestimate of the benefit. In response to an NRC staff RAI, TVA performed a sensitivity study reevaluating the benefit of each Phase II SAMDA basing the consequences for each release category on the maximum consequence for the scenarios that make up each release category rather than the average consequence. TVA indicated that, for the uncorrected consequence analysis (TVA 2011c), with one exception, the sensitivity study indicated that no additional SAMDAs would be cost-beneficial using the 95th percentile uncertainty factor of 2.78. The one exception is SAMDA 93, which is to install an unfiltered hardened containment vent to eliminate the containment overpressure failure. While use of the maximum consequences increases the benefit-to-cost ratio from slightly less than 1.0 to slightly more than 1.0, TVA argues that use of the average LATE release category (release category III) is appropriate for this SAMDA. In addition, TVA points out that 40 percent of the LATE release category is due to RCP seal LOCAs while 10 percent is due to scenarios involving the loss of control air and operators failing to control AFW manually. Both of these situations are addressed by other SAMDAs, SAMDA 58 for RCP seal failure and SAMDA 339 (replacing SAMDA 70) for loss of control air. TVA further commits to reevaluating SAMDA 93 if the new RCP seal package proves to not be reliable (TVA 2011c).

The revised cost-benefit analysis resulting from the correction to the consequence analysis (TVA 2011f) indicates that SAMDA 93 is cost-beneficial without considering the conservative source terms, and in the submittal, again cites the commitment to re-evaluate SAMDA 93 if the new RCP seal package proves to not be reliable.

The September 16, 2011, submittal does not specifically state that the use of the conservative source terms with the revised consequence analysis will not result in any additional cost-beneficial SAMDAs. TVA does point out that the next largest benefit-cost ratio is 0.70 for SAMADA 255, using the 2.70 uncertainty multiplier, and that this would not be cost-beneficial even if the 2.78 multiplier is used. The NRC staff considers that this SAMDA (and all others which have lower benefit-cost ratios) is sufficiently removed from being cost-beneficial that use of the updated conservative source terms and consequence analysis would not result in it being cost-beneficial.

In response to an NRC staff RAI, TVA reexamined the initial set of SAMDAs to determine if any additional Phase I SAMDAs would be retained for further analysis if the benefits (and Modified MACR) were based on using the 95th percentile CDF. This reexamination used a number of SAMDA maximum benefit cases that represented the possible change in the MACR for a range of assumptions concerning the nature of the impact of the SAMDA on the risk; for example, entire risk changed linearly with the change in CDF, the CDF remained fixed and only individual release categories changed, or combinations of both situations. Using these maximum benefit cases and estimates of the maximum potential reduction in CDF or risk, TVA provided the results of rescreening of all the Phase I SAMDAs originally screened out on the basis of excessive implementation cost or very low benefit. All Phase I SAMDAs screened out remained screened out based on a 95th percentile uncertainty factor of 2.7 (TVA 2011c, f). It is noted, however, while SAMDAs 50, 55 and 242, all impacting RCP seal failure sequences, are screened, TVA has committed to further consider these SAMDAs if the new RCP seal package proves not to be reliable (TVA 2011c). The NRC staff has reviewed the information provided and finds that the conclusion that the Phase I SAMDAs originally screened will remain screened considering the CDF uncertainty to be acceptable. While, as indicated above, the more correct uncertainty factor is believed to be 2.78, the NRC staff concludes use of this higher factor will not change the conclusions.

In RAIs, the NRC staff questioned TVA about other lower cost alternatives to some of the SAMDAs evaluated, as summarized below:

- Purchasing or manufacturing a "gagging device" that could be used to close a stuck-open steam generator safety valve for a SGTR event prior to core damage. In response to the RAI, TVA indicated that using such a device would require access to the stuck-open safety valve. Since the WBN steam generator safety valves do not have tailpipes, the discharge is at the throat of the valve making such access infeasible due to local hazards (TVA 2011a).

- Utilizing the spare 5th diesel generator mentioned in the disposition of SAMDA 261 without going through the expense of complete refurbishing and licensing. In response to the RAI, TVA indicated that the diesel generator has been cannibalized to the point where essentially an entire new unit would be required. In addition, adding to the cost would be the requirement for class IE interfaces to the shutdown boards (TVA 2010c).

- Providing procedures and cabling to enable the use of the trailer-mounted 2 MW diesel generator provided in response to GSI-189 to be used to power selected equipment such as battery chargers, and/or individual pumps. In response to this RAI TVA indicated that this has been implemented at WBN (TVA 2010c).

- Purchasing and installing a permanent diesel generator to supply power to the normal charging pump. In response to this RAI, TVA indicated that such a SAMDA would need to consider power supply arrangements and interfaces with existing power supplies as well as the physical location of the diesel generator. There would be significant cable routing required and procedures and training involved (TVA 2010c).

The NRC staff concludes that, with the exception of the potentially cost-beneficial SAMDAs discussed above (SAMDAs 4, 156, 256, 285, 292, 299, 305, and 306), the costs of the other SAMDAs evaluated would be higher than the associated benefits.

H.7 Conclusions

TVA compiled a list of SAMDAs based on a review of: the most significant basic events from the plant-specific PRA, insights from the plant-specific IPE and IPEEE, Phase I SAMAs from license renewal applications for other plants, and NEI's list of generic SAMAs. An initial screening removed SAMDA candidates that (1) were not applicable to WBN, (2) were already implemented at WBN, (3) were similar to and could be combined with other SAMDAs, (4) had estimated costs that would exceed the dollar value associated with completely eliminating all severe accident risk at WBN, or (5) determined to have negligible impact on risk. Based on this screening, a number of these SAMDAs were eliminated leaving the remaining candidate SAMDAs for Phase II evaluation.

For the remaining SAMDA candidates, more detailed design and cost estimates were developed as shown in Table H-4. The cost-benefit analyses showed that eight of the SAMDA candidates were potentially cost-beneficial in the baseline analysis (SAMDAs 4, 156, 256, 285, 292, 299, 305, and 306). TVA performed additional analyses to evaluate the impact of parameter choices and uncertainties on the results of the SAMDA assessment. As a result, six additional SAMDAs (SAMDAs 8, 70, 93, 215, 226, and 300) were identified as potentially cost-beneficial.

Of these potentially cost-beneficial SAMDAs, SAMDA 156 was found by TVA to not be effective due to time constraints on the operators to perform the action. SAMDAs 305 and 306 are considered by TVA to have been previously implemented in an existing procedure that was not credited in the PRA's HRA.

SAMDAs 93, 215, and 226 (Table H-5) all relate to preventing RCP seal failures, as does SAMDA 58. SAMDA 58 was originally screened due to the unavailability of an approved seal design and associated cost. Subsequent to the publication of this draft SFES, TVA learned that such a seal had been installed at the Farley Nuclear Power Plant. TVA has committed to follow the progress and experience with this seal package design and, if proven reliable during operation, to install it at the earliest refueling outage following startup during normal seal package replacements (TVA 2011b). TVA further committed that if the seal package is not proven reliable, TVA will use the latest PRA model at the time to re-evaluate SAMDAs 93, 215, and 226 as well as 10 CFR 50.55 and 10 CFR 59.56 to determine if an alternate SAMDA is cost-beneficial for implementation and implement the SAMDA accordingly (TVA 2011b). TVA has further committed to similarly re-evaluate other SAMDAs that may be cost-beneficial and/or related to or impacting RCP seal failure sequences including SAMDAs 50, 55, 56, and 242 (TVA 2011c).

SAMDAs 293 and 294, both related to flooding due to RCW pipe failures, were superseded by SAMDA 340 (TVA 2011c), by which TVA has committed to the installation of flood detection instrumentation in the affected areas, 772.0-A8 and 772.0-A9. As discussed above, the originally installed carbon steel piping was to have been replaced with stainless steel piping. This original commitment to include the piping change along with the installation of flood detection has subsequently been found to provide negligible additional benefit in reducing risk relative to the installation of detection alone (TVA 2012). Therefore, as per SAMDA 340, TVA has modified its commitment to one of installing only the flood detection so that operators could take steps to isolate the affected piping.

SAMDA 70, which was found to be cost-beneficial considering uncertainty in CDF, was superseded by SAMDA 339, a new SAMDA to provide in the control room the capability to connect to the station nitrogen system (TVA 2011b).

SAMDA 80 was originally screened on the basis of having a very low benefit. In response to an NRC staff RAI, TVA indicated that this SAMDA will be implemented in the CCP area, the TD AFW pump room, and the EDG switchgear rooms (TVA 2011b).

As stated in the November 1, 2010 submittal, TVA has indicated that the following potentially cost-beneficial SAMDAs will be implemented: SAMDAs 4, 8, 256, 285, 292, 299, and 300.[a] For reasons beyond a cost-benefit analysis, TVA will be implementing SAMDAs 339 and 340 as committed by letters dated May 13 and 25, 2011.

In its September 16, 2011 submittal (TVA 2011f) TVA reaffirms the commitments made in prior SAMDA submittals (TVA 2011a, b, c, d, e).

The NRC staff reviewed the TVA analysis and concludes that the methods used and the implementation of those methods were sound. The treatment of SAMDA benefits and costs support the general conclusion that the SAMDA evaluations performed by TVA are reasonable and sufficient for the license submittal. Although the treatment of SAMDAs for external events was somewhat limited, the likelihood of there being cost-beneficial enhancements in this area was minimized by improvements that have been realized as a result of the IPEEE process, and inclusion of a multiplier to account for external events.

The NRC staff finds acceptable the TVA identified areas in which risk can be reduced in a cost-beneficial manner through the implementation of the identified, potentially cost-beneficial SAMDAs. Given the potential for cost-beneficial risk reduction, the NRC staff finds, subject to the above described dispositions, that implementation of these SAMDAs as committed to by TVA is warranted. Therefore, the NRC staff finds that the TVA analysis meets the requirements of the National Environmental Policy Act.

H.8 References

[BLS] U.S. Bureau of Labor Statistics. 2010. "Bureau of Labor Statistics (BLS) 2010, Inflation Calculator." Available online at http://www.bls.gov/inflation_calculator.htm.

Chanin, D. and M.L. Young. 1998. *Code Manual for MACCS2*. NUREG/CR-6613, Washington, D.C. Accession No. ML110030976.

(a) Because the Third Circuit's opinion in *Limerick Ecology Action, Inc. v. NRC*, 869 F.2d 719, 723 (3d Cir. 1989) is not limited by the scope of license renewal, the relationship of these SAMDAs to aging does not affect the decision to implement.

[EPRI] Electric Power Research Institute. 1991. "A Methodology for Assessment of Nuclear Power Plant Seismic Margin." EPRI NP-6041-SL, Revision 1. Palo Alto, California. August 1991.

[EPRI] Electric Power Research Institute. 1992. "Fire-Induced Vulnerability Evaluation (FIVE)." EPRI TR-100370. Palo Alto, California. April 1992.

[NEI] Nuclear Energy Institute. 2005. "Severe Accident Mitigation Alternatives (SAMA) Analysis Guidance Document." NEI 05-01, Revision A. November 2005. Accession No. ML053500424.

[NRC] U.S. Nuclear Regulatory Commission. 1975. *Standard Review Plan for the Review of Safety Analysis Reports for Nuclear Power Plants.* NUREG-0800, Revision 0. Washington, D.C. November 1975.

[NRC] U.S. Nuclear Regulatory Commission. 1988. "Individual Plant Examination for Severe Accident Vulnerabilities – 10 CFR 50.54(f)." Generic Letter 88-20. Washington, D.C. November 23, 1988.

[NRC] U.S. Nuclear Regulatory Commission. 1990a. Severe *Accident Risks: An Assessment for Five U.S. Nuclear Power Plants.* NUREG-1150. Washington, D.C. Accession No. ML040140729.

[NRC] U.S. Nuclear Regulatory Commission. 1990b. *Evaluation of Severe Accident Risks: Sequoyah Unit 1.* NUREG/CR-4551, SAND86-1390, Vol. 5, Rev. 1, Part 2. Washington, D.C. December 1990.

[NRC] U.S. Nuclear Regulatory Commission. 1990c. *Analysis of Core Damage Frequency: Sequoyah Unit 1.* NUREG/CR-4550, SAND86-2084, Vol. 5, Rev. 1. Washington, D.C. April 1990.

[NRC] U.S. Nuclear Regulatory Commission. 1991. "Individual Plant Examination of External Events (IPEEE) for Severe Accident Vulnerabilities – 10CFR 50.54(f)." Generic Letter No. 88-20, Supplement 4. Washington, D.C. June 28, 1991.

[NRC] U.S. Nuclear Regulatory Commission. 1994. Letter from Peter S. Tam, U.S. NRC, to Oliver D. Kingsley, TVA. Subject: "Watts Bar Nuclear Plant – Completion of Individual Plant Examination Review (TAC No. M74488)." Washington, D.C. October 5, 1994. Accession Nos. ML073230696, ML082320705, ML082320706, ML082320707.

[NRC] U.S. Nuclear Regulatory Commission. 1995. *Final Environmental Statement Related to the Operation of Watts Bar Nuclear Plant, Units 1 and 2.* NUREG-0498, Supplement No. 1. Washington, D.C. April 1995. Accession No. ML081430592.

[NRC] U.S. Nuclear Regulatory Commission. 1997a. *Regulatory Analysis Technical Evaluation Handbook.* NUREG/BR-0184. Washington, D.C.

[NRC] U.S. Nuclear Regulatory Commission. 1997b. *Individual Plant Examination Program: Perspectives on Reactor Safety and Plant Performance.* NUREG-1560. Washington, D.C.

[NRC] U.S. Nuclear Regulatory Commission. 2000. Letter from Robert T. Martin, U.S. NRC, to J.A. Scalice, TVA. Subject: "Watts Bar Nuclear Plant, Unit 1 – Review of Individual Plant Examination of External Events (IPEEE) Submittal (TAC Nos. M83693)." Washington, D.C. May 19, 2000.

[NRC] U.S. Nuclear Regulatory Commission. 2003a. Memorandum from Nilesh C. Chokshi, U.S. NRC, to Ashok C. Thadani, U.S. NRC. Subject: "Results of Initial Screening of Generic Issue 194, 'Implications of Updated Probabilistic Seismic Hazard Estimates.'" September 12, 2003.

[NRC] U.S. Nuclear Regulatory Commission. 2003b. SECPOP *2000: Sector Population, Land Fraction, and Economic Estimation Program.* NUREG/CR-6525, Revision 1. Washington, D.C. August 2003.

[NRC] U.S. Nuclear Regulatory Commission. 2004a. *An Approach for Estimating the Frequencies of Various Containment Failure Modes and Bypass Events.* NUREG/CR-6595 Revision 1. Washington, D.C. October 2004.

[NRC] U.S. Nuclear Regulatory Commission. 2004b. *Regulatory Analysis Guidelines of the U.S. Nuclear Regulatory Commission.* NUREG/BR-0058, Rev. 4. Washington, D.C.

[NRC] U.S. Nuclear Regulatory Commission. 2009. Letter from Joel S. Wiebe, U.S. NRC, to Ashok S. Bhatnagar, Tennessee Valley Authority. Subject: "Watts Bar Nuclear Plant, Unit 2 - Request for Additional Information Regarding Severe Accident Management Alternatives (TAC No. MD8203)." Washington, D.C. November 30, 2009. Accession No. ML092230024.

[NRC] U.S. Nuclear Regulatory Commission. 2010. NRC Information Notice 2010-18: Generic Issue 199 (GI-199), "Implications of Updated Probabilistic Seismic Hazard Estimates in Central and Eastern United States on Existing Plants." Washington, D.C. September 2, 2010. Accession No. ML101970221.

[NRC] U.S. Nuclear Regulatory Commission. 2011a. Letter from Justin C. Poole, U.S. NRC, to Ashok S. Bhatnagar, TVA. Subject: "Watts Bar Nuclear Plant, Unit 2 - Request for Additional Information Regarding Severe Accident Management Alternative Review (TAC No. MD8203)." January 11, 2011. Accession No. ML103470681.

[NRC] U.S. Nuclear Regulatory Commission. 2011b. Letter from Patrick D. Milano, U.S. NRC, to Ashok S. Bhatnagar, TVA. Subject: "Watts Bar Nuclear Plant, Unit 2 – Supplemental Request for Additional Information Regarding Severe Accident Management Design Alternatives Review (TAC No. MD8203)." March 30, 2011. Accession No. ML110820858.

[NRC] U.S. Nuclear Regulatory Commission. 2011c. Letter from Justin C. Poole, U.S. NRC, to Ashok S. Bhatnagar, TVA. Subject: "Watts Bar Nuclear Plant, Unit 2 – Request for Additional Information Regarding Severe Accident Management Design Alternatives Review – June 2011 (TAC No. MD8203)." June 13, 2011. Accession No. ML111530523.

[NRC] U.S. Nuclear Regulatory Commission. 2011d. Letter from Stephen J. Campbell, U.S. NRC, to Ashok S. Bhatnagar, TVA. Subject: "Watts Bar Nuclear Plant, Unit 2 – Review of Individual Plant Examination Submittal – Internal Events and Internal Flood (TAC NO. ME3334)." August 12, 2011. Accession No. ML111960228.

[NRC] U.S. Nuclear Regulatory Commission. 2011e. Letter from Stephen J. Campbell, U.S. NRC, to Ashok S. Bhatnagar, TVA. Subject: "Watts Bar Nuclear Plant, Unit 2 – Review of Individual Plant Examination of External Events Design Report (TAC NO. ME4482)." September 20, 2011. Accession No. ML111960300.

[PLG] PLG, Inc. 1992. *RISKMAN-PRA Workstation Software.* User Manuals I - IV, Version 3.0, Newport Beach, California.

[SNC] Southern Nuclear Operating Company. 2007. *Applicant's Environmental Report-Operating License Renewal Stage for Vogtle Electric Generating Plant Units 1 and 2, Appendix E - Environmental Report.* Southern Nuclear Operating Company, Birmingham, Alabama. June 2007.

[TVA] Tennessee Valley Authority. 1992. Letter from Mark O. Medford, TVA, to U.S. Nuclear Regulatory Commission. Subject: "Watts Bar Nuclear Plant (WBN) Units 1 and 2 – Generic Letter (GL) 88-20 - Individual Plant Examination (IPE) for Severe Accident Vulnerabilities – Response (TAC M74488)." September 1, 1992. Accession Nos. ML080070015, ML080070016, ML080070017, ML080070018, ML080070019.

[TVA] Tennessee Valley Authority. 1994a. Letter from William J. Museler, TVA, to U.S. Nuclear Regulatory Commission. Subject: "Watts Bar Nuclear Plant (WBN) Units 1 and Common – Generic Letter 88-20 - Individual Plant Examination (IPE) – Update of Level 1 and 2 Analysis." May 2, 1994. Accession Nos. ML080070020, ML080070025, ML080070026.

[TVA] Tennessee Valley Authority. 1994b. Letter from Dwight E. Nunn, TVA, to U.S. Nuclear Regulatory Commission Document Control Desk. Subject: "Watts Bar Nuclear Plant (WBN) Units 1 and 2 – Severe Accident Mitigation Design Alternatives (SAMDAs) Evaluation from Updated Individual Plant Examination (IPE) (TAC Nos. M77222 and M77223)." June 30, 1994. Accession No. ML073230670.

[TVA] Tennessee Valley Authority. 1998. Letter from P.L. Pace, TVA, to U.S. Nuclear Regulatory Commission Document Control Desk. Subject: "Watts Bar Nuclear Plant (WBN) Unit 1 – Generic Letter (GL) 88-20, Supplements 4 and 5 - Individual Plant Examination of External Events (IPEEE) for Severe Accident Vulnerabilities (TAC M83693)." February 17, 1998. Accession No. ML073240218.

[TVA] Tennessee Valley Authority. 2006. "Tennessee Multi-Jurisdictional Radiological Emergency Response Plan for the Watts Bar Nuclear Plant," Annex H.

[TVA] Tennessee Valley Authority. 2009a. Letter from Masoud Bajestani, TVA, to U.S. Nuclear Regulatory Commission Document Control Desk. Subject: "Watts Bar Nuclear Plant (WBN) Unit 2 – Final Supplemental Environmental Impact Statement - Severe Accident Management Alternatives (TAC MD8203)." January 27, 2009. Accession Nos. ML090360588 and ML090360589.

Tennessee Valley Authority (TVA). 2009b. Letter from Masoud Bajestani (TVA) to U.S. Nuclear Regulatory Commission dated October 22, 2009 in response to TVA letters dated February 15, 2008, July 2, 2008, and January 27, 2009, "Additional Information in Support of TVA Final Supplemental Environmental Impact Statement (FSEIS) (TAC MD8303)." Accession No. ML093510802.

Tennessee Valley Authority (TVA). 2009c. Letter from Masoud Bajestani (TVA) to U.S. Nuclear Regulatory Commission dated December 23, 2009 in response to NRC letter dated December 3, 2009 and TVA letters dated February 15, 2008, July 2, 2008, and January 27, 2009, "Watts Bar Nuclear Plant (WBN) Unit 2 – Additional Information Regarding Environmental Review (TAC No. MD8203)." Accession No. ML100210350.

[TVA] Tennessee Valley Authority. 2010a. Letter from Masoud Bajestani, TVA, to U.S. Nuclear Regulatory Commission Document Control Desk. Subject: "Watts Bar Nuclear Plant (WBN) Unit 2 –Severe Accident Management Alternatives Using Latest Computer Aided Fault Tree Analysis Model – Additional Information (TAC MD8203)." November 1, 2010. Accession No. ML103080030.

[TVA] Tennessee Valley Authority. 2010b. Letter from Masoud Bajestani, TVA, to U.S. Nuclear Regulatory Commission Document Control Desk. Subject: "Watts Bar Nuclear Plant (WBN) Unit 2 – Probabilistic Risk Assessment Individual Plant Examination." February 9, 2010. Accession No. ML100491535.

[TVA] Tennessee Valley Authority. 2010c. Letter from Masoud Bajestani, TVA, to U.S. Nuclear Regulatory Commission Document Control Desk. Subject: "Watts Bar Nuclear Plant (WBN) Unit 2 – Request for Additional Information Regarding Severe Accident Management Alternatives (TAC No. MD8203)." July 23, 2010. Accession No. ML102100588.

[TVA] Tennessee Valley Authority. 2010d. Letter from Masoud Bajestani, TVA, to U.S. Nuclear Regulatory Commission Document Control Desk. Subject: "Watts Bar Nuclear Plant (WBN) Unit 2 – Response to Request for Additional Information Regarding Severe Accident Management Alternatives SECPOP2000 Errors (TAC MD8203)." September 17, 2010. Accession No. ML102600582.

[TVA] Tennessee Valley Authority. 2010e. Letter from Masoud Bajestani, TVA, to U.S. Nuclear Regulatory Commission Document Control Desk. Subject: "Watts Bar Nuclear Plant (WBN) Unit 2 – Individual Plant Examination for External Events Design Report." April 30, 2010. Accession No. ML101240992.

[TVA] Tennessee Valley Authority. 2011a. Letter from Marie Gillman, TVA, to U.S. Nuclear Regulatory Commission Document Control Desk. Subject: "Watts Bar Nuclear Plant (WBN) Unit 2 – Response to Request for Additional Information Regarding Severe Accident Management Alternative Review (TAC MD8203)." January 31, 2011. Accession No. ML110340040.

[TVA] Tennessee Valley Authority. 2011b. Letter from David Stinson, TVA, to U.S. Nuclear Regulatory Commission Document Control Desk. Subject: "Watts Bar Nuclear Plant (WBN) Unit 2 – Response to Request for Additional Information Regarding Severe Accident Management Alternative Review (TAC MD8203)." May 13, 2011. Accession No. ML11145A088.

[TVA] Tennessee Valley Authority. 2011c. Letter from David Stinson, TVA, to U.S. Nuclear Regulatory Commission Document Control Desk. Subject: "Watts Bar Nuclear Plant (WBN) - Unit 2 – Response to Request for Additional Information Item Numbers 2, 3, 5 and 15 Regarding Severe Accident Management Alternative Review (TAC MD8203)." May 25, 2011. Accession No. ML11147A099.

[TVA] Tennessee Valley Authority. 2011d. Letter from David Stinson, TVA, to U.S. Nuclear Regulatory Commission Document Control Desk. Subject: "Watts Bar Nuclear Plant (WBN) - Unit 2 – Response to Request for Additional Information Regarding Severe Accident Management Alternative Review (SAMDA) (TAC MD8203)." June 17, 2011. Accession No. ML11171A510.

[TVA] Tennessee Valley Authority. 2011e. Letter from David Stinson, TVA, to U.S. Nuclear Regulatory Commission Document Control Desk. Subject: "Watts Bar Nuclear Plant (WBN) - Unit 2 – Response to Request for Additional Information Regarding Severe Accident Management Alternative Review (SAMDA) (TAC MD8203)." June 27, 2011. Accession No. ML11180A008.

[TVA] Tennessee Valley Authority. 2011f. Letter from David Stinson, TVA, to U.S. Nuclear Regulatory Commission Document Control Desk. Subject: "Watts Bar Nuclear Plant (WBN) - Unit 2 – Revised Severe Accident Management Design Alternative Review (SAMDA) Response (TAC MD8203)." September 16, 2011. Accession No. ML11264A052.

[TVA] Tennessee Valley Authority. 2011g. Letter from David Stinson, TVA, to U.S. Nuclear Regulatory Commission Document Control Desk. Subject: "Watts Bar Nuclear Plant (WBN) Unit 2 – Chapter 15.5 Fuel Handling Accident (FHA) Dose Analysis." October 17, 2011. Accession No. ML11294A461.

[TVA] Tennessee Valley Authority. 2012. Letter from Raymond A. Hruby, TVA, to U.S. Nuclear Regulatory Commission Document Control Desk. Subject: "Watts Bar Nuclear Plant (WBN) – Unit 2 – Severe Accident Management Alternative (SAMA) Review." October 17, 2012. Accession No. ML12296A227.

[USGS] U.S. Geologic Survey. 2008. "2008 NSHM Gridded Data, Peak Ground Acceleration." Available online at http://earthquake.usgs.gov/hazards/products/conterminous/2008/data/.

[WCNOC] Wolf Creek Nuclear Operating Corporation. 2006. Letter from Terry J. Garrett, WCNOC to U.S. Nuclear Regulatory Commission Document Control Desk. Subject: "Docket No. 50-482: Application for Renewed Operating License." September 27, 2006.

Appendix I

Supporting Documentation for Radiological Dose Assessment

Appendix I

Supporting Documentation for Radiological Dose Assessment

This appendix contains supporting documentation for the U.S. Nuclear Regulatory Commission (NRC) staff's determinations described in this supplemental final environmental statement (SFES) for the radiological dose assessment.

The staff reviewed and performed an independent dose assessment of the radiological impacts from normal operations of the new nuclear Unit 2 at the Watts Bar Nuclear (WBN) plant in Rhea County, Tennessee. This appendix contains four sections: (1) dose estimates to the public from liquid effluents; (2) dose estimates to the public from gaseous effluents; (3) cumulative dose estimates, and (4) dose estimates to biota from gaseous and liquid effluents.

I.1 Dose Estimate from Liquid Effluents

The NRC staff used the dose assessment approach specified in Regulatory Guide 1.109 (NRC 1977) and the NRC-developed LADTAP II computer code (Strenge et al. 1986) to estimate doses to the maximally exposed individual (MEI) and the population from the liquid effluent pathway of WBN Unit 2. As described in Regulatory Guide 1.109 (NRC 1977), the MEI is characterized as an individual with the "maximum" food consumption, occupancy, and other usages in the vicinity of the plant site and is therefore representative of a member of the public that would receive the maximum dose from all radiological pathways from the site. The NRC staff used the projected radioactive effluents release values from the Tennessee Valley Authority (TVA) final supplemental environmental impact statement (submitted to NRC as the TVA Environmental Report for an Operating License) (TVA 2008a) and responses to Requests for Additional Information (RAIs) submitted by TVA (TVA 2011a, b).

I.1.1 Scope

Doses from proposed WBN Unit 2 to the MEI were calculated and compared to the regulatory criteria for the following:

- Total Body – Dose was the total for the ingestion of aquatic organisms as food and cow meat and external exposure to contaminated sediments deposited along the shoreline (shoreline exposure). Water downstream from the WBN site is not used for irrigation. Refer to Figure 4-2 in Section 4.6.1 for visual representation of the exposure pathway to humans.

- Organ – Dose was the total for each organ for ingestion of aquatic food and cow meat and shoreline exposure with the highest value for adult, teen, child, or infant.

The NRC staff performed calculations for exposure pathways using input parameters and values found in TVA documentation. When site- or design-specific input parameters were not available, staff used default values from Regulatory Guide 1.109 (NRC 1997).

I.1.2   Resources Used

To calculate doses to the public from liquid effluents the NRC staff used a personal computer version of the LADTAP II code titled NRCDOSE Version 2.3.12 (Chesapeake Nuclear Services, Inc. 2006) obtained through the Oak Ridge Radiation Safety Information Computational Center. LADTAP II calculates the radiation exposure to man from potable water, aquatic foods, shoreline deposits, swimming, boating, and irrigated foods, and also the dose to biota. Doses are calculated for both the maximum individual and for the population and are summarized for each pathway by age group and organ. LADTAP II implements the radiological exposure models described in NRC Regulatory Guide 1.109, Rev. 1 (Appendix A) for radioactivity releases in liquid effluent. The usage factors contained in Regulatory Guide 1.109 have been included as standard assumptions but may easily be replaced with site-specific data.

I.1.3   Input Parameters

The population distribution assumed for all NRC staff calculations was obtained from the TVA RAI response letter dated May 26, 2011 and is shown in Table I-1 (TVA 2011b). Table I-2 lists the major parameters used in calculating dose to the public from liquid effluent releases during normal operation. It should be noted that the 80-km (50-mi) population was assumed to be for the year 2040. Section 5.4.1 of the Environmental Standard Review Plan (ESRP) guidance suggests that populations be projected only 5 years out from the date of the licensing action under consideration (NRC 2000). The staff considers that using the population for the year 2040, rather than the recommended 5 years from licensing, is acceptable because it assesses the population dose for a time period that approximates the operating life of WBN Unit 2.

I.1.4   Results of Calculations

Table I-3 shows the results of the calculations of dose to the public from liquid effluent releases. The data in this table indicate fairly good agreement between NRC staff calculations and TVA calculations (TVA 2008b) and therefore the staff can use the TVA calculations for conclusions in Section 4.6 of this SFES-OL.

Table I-4 lists the NRC staff's calculated doses to the MEI from liquid effluent releases from WBN Unit 2, which would include such things as eating the fish, drinking the water, and swimming and other recreational uses of the water.

Table I-1. Projected Population by Sector and Radial Distance Around the WBN Site for the Year 2040.

Sectors	Year	Radii/Distances (mi)											
		0-1	1-2	2-3	3-4	4-5	5-10	0-10	10-20	20-30	30-40	40-50	0-50
North	2040	0	18	0	0	135	2,465	2,619	1,885	2,778	4,798	6,172	18,222
North-Northeast	2040	0	0	18	411	185	1,536	2,150	11,762	18,766	14,502	2,547	49,727
Northeast	2040	0	0	18	308	287	827	1,441	3,783	16,734	29,838	78,334	130,130
East-Northeast	2040	0	0	18	308	287	497	1,110	3,553	29,539	63,798	253,831	351,832
East	2040	0	8	431	308	616	552	1,915	11,352	18,647	30,063	44,013	105,990
East-Southeast	2040	0	0	0	27	41	68	135	6,230	20,120	5,068	3,280	34,833
Southeast	2040	8	0	0	29	39	135	203	19,852	15,185	3,950	7,822	44,012
South-Southeast	2040	21	0	0	246	413	103	783	8,951	12,907	2,918	48,593	74,151
South	2040	16	0	0	0	1,983	3,824	5,823	4,586	42,883	56,430	17,985	127,707
South-Southwest	2040	0	0	21	0	0	546	567	5,725	42,517	46,281	106,392	201,482
Southwest	2040	0	0	0	0	0	1,051	1,051	12,978	14,499	62,307	111,795	202,630
West-Southwest	2040	0	6	36	59	126	711	938	12,791	2,837	2,840	3,372	22,778
West	2040	0	14	22	101	90	710	937	3,406	5,555	2,944	5,474	18,316
West-Northwest	2040	0	0	22	126	79	490	717	2,091	4,372	5,654	20,511	33,345
Northwest	2040	0	108	332	376	526	2,655	3,998	2,889	18,634	10,462	15,956	51,940
North-Northwest	2040	0	0	0	173	123	3,116	3,413	1,536	33,843	11,609	5,890	56,290
Total		45	155	919	2,471	4,930	19,287	27,799	113,368	299,818	353,432	728,968	1,523,385

Source: TVA 2011b

Table I-2. Parameters Used in Calculating Dose to the Public from Liquid Effluent Releases (WBN Unit 2 only)

Parameter	Staff Value		Comments
New unit liquid effluent source term (Ci/yr)[a,b]	Br-84	6.88×10^{-4}	Table 3-16, p. 80 of the TVA ER (TVA 2008a, 2011b, Enclosure 1; p. E1-23).
	I-131	1.16	
	I-132	1.21×10^{-1}	
	I-133	9.10×10^{-1}	
	I-134	3.28×10^{-2}	
	I-135	4.70×10^{-1}	
	Rb-88	7.68×10^{-3}	
	Cs-134	1.98×10^{-1}	
	Cs-136	1.98×10^{-2}	
	Cs-137	2.61×10^{-1}	
	Na-24	1.86×10^{-2}	
	Cr-51	9.98×10^{-2}	
	Mn-54	5.59×10^{-2}	
	Fe-55	8.09×10^{-3}	
	Fe-59	1.15×10^{-2}	
	Co-58	1.66×10^{-1}	
	Co-60	3.16×10^{-2}	
	Zn-65	3.82×10^{-4}	
	Sr-89	4.52×10^{-3}	
	Sr-90	4.10×10^{-4}	
	Sr-91	2.47×10^{-3}	
	Y-91m	1.68×10^{-4}	
	Y-91	3.90×10^{-4}	
	Y-93	1.27×10^{-3}	
	Zr-95	1.34×10^{-2}	
	Nb-95	1.11×10^{-2}	
	Mo-99	1.04×10^{-1}	
	Tc-99M	3.35×10^{-3}	
	Ru-103	5.88×10^{-3}	
	Ru-106	7.63×10^{-2}	
	Te-129M	1.41×10^{-4}	
	Te-129	7.30×10^{-4}	
	Te-131M	8.05×10^{-4}	
	Te-131	2.03×10^{-4}	
	Te-132	3.05×10^{-2}	
	Ba-140	3.58×10^{-1}	
	La-140	5.14×10^{-1}	
	Ce-141	3.41×10^{-4}	
	Ce-143	1.53×10^{-3}	
	Ce-144	1.33×10^{-1}	
	Np-239	1.37×10^{-3}	
	H-3	1.25×10^{3}	

Table I-2. (contd)

Parameter	Staff Value	Comments
Freshwater site	Selected	Discharge is to the freshwater Tennessee River.
Discharge flow rate (cfs)	44.56	Site-specific value. Cooling tower blowdown rate used for dilution from Figure 3-7 of TVA ER (TVA 2008a).
Source-term multiplier	1	For one unit.
Reconcentration model	No impoundment	Site-specific value.
Effluent discharge rate from impoundment system to receiving water body (cfs)	44.56	Matches discharge flow rate for "no impoundment" model (Strenge et al. 1986).
Impoundment total volume (ft^3)	0	Set to zero for "no impoundment" model (Strenge et al. 1986).
Shore-width factor	0.2	Suggested value for river shoreline (NRC 1977; Strenge et al. 1986).
Dilution factors for aquatic food and boating, shoreline and swimming, and drinking water	78	Site-specific value. The quotient of the minimum Tennessee River flow rate to allow release of liquid effluent divided by the cooling tower blowdown used for dilution prior to release into the river.
Transit time (hr)	0	Site-specific value from RAI TVA letter dated May 26, 2011, p. E1-12 (TVA 2011b).
Consumption and usage factors for adults, teens, children, and infants	Shoreline usage (hr/yr) 500 (adult) 500 (teen) 500 (child) 500 (infant) Water usage (L/yr) 730 (adult) 510 (teen) 510 (child) 330 (infant) Fish consumption (kg/yr) 21 (adult) 16 (teen) 6.9 (child) 0 (infant)	Shoreline Usage: Site-specific value from Offsite Dose Calculation Manual (ODCM; TVA 2008b) Water Usage: LADTAP II code default values (NRC 1977; Strenge et al. 1986). Note: for fish consumption, NRC staff used default values rather than site values because site values were for average consumption and these values are for calculating the dose to the MEI.
Total 50-mi population	1,523,385	Site-specific value from RAI TVA letter dated May 26, 2011, p. E1-11. The population was estimated for the year 2040 (TVA 2011b).
50-mi drinking water population	453,296	Site-specific value from April 9, 2010 RAI response (TVA 2010). Note: the population datum provided for this RAI was 1,066,580. In the May 26, 2011 letter, TVA updated the population by 500,000 but did not update the 50-mi drinking water population (TVA 2011b).

Table I-2. (contd)

Parameter	Staff Value	Comments
Total 50-mi sport fishing (kg/yr)	942,296	Site-specific value from WBN FSAR (TVA 2009) and Table 3-15, p. 79, of the TVA ER (TVA 2008a). Note: the population datum provided for this RAI was 1,066,580. In the May 26, 2011 letter, TVA updated the population by 500,000 but TVA did not update the 50-mi sport fishing population (TVA 2011b).
Total 50-mi shoreline usage (person-hr/yr)	6.52×10^7	Site-specific value from Table 3-15 of the TVA ER (TVA 2009) and ODCM Eq. 6-18, p. E1-144 (5 hours per visit) (TVA 2008b). Note: the population datum provided for this RAI was 1,066,580. In the May 26, 2011 letter, TVA updated the population by 500,000 but TVA did not update the 50-mi shoreline usage population (TVA 2011b).
Total 50-mi swimming usage (person-hr/yr)	6.52×10^7	NRC staff assumes that swimming could equal shoreline use. Site-specific value from Table 3-15 of the TVA ER (TVA 2009) and ODCM Eq. 6-18, p. E1-144 (5 hours per visit) (TVA 2008b). Note: the population datum provided for this RAI was 1,066,580. In the May 26, 2011 letter, TVA updated the population by 500,000 but TVA did not update the 50-mi swimming usage population (TVA 2011b).
Total 50-mi boating usage (person-hr/yr)	6.52×10^7	NRC staff assumes that boating could equal shoreline use. Note: the population datum provided for this RAI was 1,066,580. In the May 26, 2011 letter, TVA updated the population by 500,000 but TVA did not update the 50-mi boating usage population (TVA 2011b).
Fraction of crops irrigated	0	Site-specific value from ODCM, p. 71 (TVA 2008b).
Fraction of population using contaminated water for drinking and food production	0	Site-specific value from ODCM, p. 71 (TVA 2008b).
Fraction of agricultural products within 50-mi radius	0	Site-specific value from ODCM, p. 71 (TVA 2008b).
Irrigation rate for food products ($L/m^2/mo$)	0	Site-specific value from ODCM, p. 71 (TVA 2008b).
Fraction of contaminated water not used for feed or drinking water	0	Site-specific value from ODCM, p. 71 (TVA 2008b).
Total production of vegetables within 50-mi radius (kg/yr)	8.07×10^8	Site-specific value from WBN FSAR p. 11.3-9 (vegetable production in each sector annulus = vegetable consumption in that sector annulus) (TVA 2009).

Table I-2. (contd)

Parameter	Staff Value	Comments
Production rate for irrigated vegetables (kg/yr)	0	Site-specific value from ODCM, p. 71 (TVA 2008b).
Total production of leafy vegetables within 50-mi radius (kg/yr)	1.37×10^8	Site-specific value from WBN FSAR p. 11.3-9 (leafy vegetable production in each sector annulus = leafy vegetable consumption in that sector annulus) (TVA 2009).
Production rate for irrigated leafy vegetables (kg/yr)	0	Site-specific value from ODCM p. 71 (TVA 2008b).
Total production of milk within 50-mi radius (L/yr)	4.99×10^8	Site-specific value from WBN FSAR p. 11.3-9 (milk production in each sector annulus = milk consumption in that sector annulus) (TVA 2009).
Production rate for irrigated milk (L/yr)	0	Site-specific value from ODCM p. 71 (TVA 2008b).
Total production of meat within 50-mi radius (kg/yr)	1.37×10^8	Site-specific value from WBN FSAR p. 11.3-9 (meat production in each sector annulus = meat consumption in that sector annulus) (TVA 2009).
Production rate for irrigated meat (kg/yr)	0	Site-specific value from ODCM p. 71 (TVA 2008b).

(a) To convert Ci/yr to Bq/yr, multiply the value by 3.7×10^{10}.
(b) 10 CFR 50; Appendix I. Radionuclides included in Regulatory Guide 1.109 are considered (NRC 1977).

Table I-3. Comparison of Doses to the Public from Liquid Effluent Releases for WBN Unit 2

Type of Dose	TVA ER (2009b)[a]	Staff Calculation[b]
Total body (mrem/yr)	0.72 (adult)	0.64 (adult)
Organ dose (mrem/yr)	0.13 (adult GI tract)	0.86 (teen liver)
Thyroid (mrem/yr)	0.92 (child)	1.91 (infant)
Population dose from liquid pathway (person-rem/yr)	1.6	11.8

(a) TVA 2008a.
(b) 100 × (Staff value – TVA value)/(Staff value).
GI = gastrointestinal

Table I-4. Staff Calculation of Annual Doses to the Maximally Exposed Individual for Liquid Effluent Releases from Unit 2

Pathway	Age Group	Total Body (mrem/yr)	Maximum Organ (mrem/yr)[a]	Thyroid (mrem/yr)
Fish and Other Organisms	Adult	5.82×10^{-1}	7.89×10^{-1} (liver)	2.26×10^{-1}
	Teen	3.31×10^{-1}	8.11×10^{-1} (liver)	2.12×10^{-1}
	Child	1.29×10^{-1}	7.08×10^{-1} (GI-LLI)	2.22×10^{-1}
	Infant	0	0	0
Drinking Water	Adult	2.90×10^{-2}	4.17×10^{-2} (GI-LLI)	5.81×10^{-1}
	Teen	1.89×10^{-2}	3.00×10^{-2} (liver)	4.98×10^{-1}
	Child	3.14×10^{-2}	5.43×10^{-2} (liver)	1.21
	Infant	2.96×10^{-2}	6.23×10^{-2} (liver)	1.89
Direct Radiation (Shoreline)	Adult	2.41×10^{-2}	2.82×10^{-2} (skin)	2.41×10^{-2}
	Teen	2.41×10^{-2}	2.82×10^{-2} (skin)	2.41×10^{-2}
	Child	2.41×10^{-2}	2.82×10^{-2} (skin)	2.41×10^{-2}
	Infant	2.41×10^{-2}	2.82×10^{-2} (skin)	2.41×10^{-2}

(a) Other than thyroid.
To convert mrem/yr to mSv/yr divide by 100.
GI-LLI = gastrointestinal tract – lower large intestine

Table I-5 lists the NRC staff's calculated doses to the population in various locations away from the plan from drinking water and shoreline recreational use such as boating and swimming.

Table I-6 compares the doses to the MEI from liquid effluents as calculated by NRC staff to the same doses calculated by the applicant (TVA 2008b). This table indicates fairly good agreement between the two sets of calculations, despite the fact that the applicant used a site-specific model, approved by NRC, which used some parameter values that were different from the mixture of site-specific and default parameter values used by the NRC staff.

Table I-5. Staff Calculation of Population Doses Due to Liquid Effluent Releases from WBN Unit 2

	Dose to Population (person-rem/yr)							
	Whole Body	Skin	Bone	Liver	Thyroid	Kidney	Lung	GI-LLI
Drinking Water								
Dayton, TN	0.268	(a)	0.146	0.346	4.980	0.254	0.192	0.325
Soddy-Daisy/Falling Water Utility District	0.158	(a)	0.083	0.203	2.230	0.147	0.115	0.176
East Side Utility, TN	0.681	(a)	0.348	0.875	7.490	0.624	0.497	0.735
Chattanooga, TN	3.240	(a)	1.630	4.150	31.400	2.950	2.370	3.460
Shoreline Use								
Chickamauga Reservoir (from WBN to 100 percent mixing point)	0.029	0.034	(b)	(b)	0.029	(b)	(b)	(b)
Chickamauga Reservoir (from 100 percent mixing point to SQN)	0.313	0.366	(b)	(b)	0.313	(b)	(b)	(b)
Chickamauga Reservoir (from SQN to Chickamauga Dam)	1.790	2.090	(b)	(b)	1.790	(b)	(b)	(b)
Nickajack Reservoir (from Chickamauga Dam to WBN 50-mi radius)	0.069	0.080	(b)	(b)	0.069	(b)	(b)	(b)

(a) Skin Dose is not appropriate for drinking water pathway.
(b) Not available for Shoreline Use Pathway.
GI-LLI = gastrointestinal tract – lower large intestine
SQN   = Sequoyah Nuclear
WBN   = Watts Bar Nuclear

Table I-6. Comparison of TVA and NRC Staff Calculations for the Dose to the Maximally Exposed Individual and the Projected 2040 Population from Liquid Effluents Released from WBN Unit 2

| | Dose to the Maximally Exposed Individual (MEI) from Liquid Effluents (mrem/yr) | | | | | | | | Population (2040) Person-rem/yr | |
| | Adult | | Teen | | Child | | Infant | | | |
	TVA	NRC	TVA	NRC	TVA	NRC	TVA	NRC	TVA	NRC
Skin	0.031	0.028	0.031	0.028	0.031	0.028	0.031	0.028	0.315	2.57
Bone	0.56	0.487	0.6	0.511	0.76	0.647	0.036	0.061	1.761	2.22
Liver	0.96	0.847	1	0.862	0.88	0.786	0.036	0.087	2.13	5.60
Thyroid	0.88	0.832	0.8	0.735	0.92	1.460	0.264	1.910	15.336	48.31
Kidney	0.352	0.316	0.356	0.315	0.312	0.292	0.034	0.064	1.392	3.99
Lung	0.136	0.130	0.152	0.141	0.128	0.132	0.032	0.051	1.037	3.19
GI-LLI	0.132	0.144	0.104	0.111	0.06	0.086	0.033	0.059	1.420	4.71
Total Body	0.72	0.635	0.44	0.374	0.188	0.185	0.032	0.055	1.619	6.57

Source: TVA 2007
GI-LLI = gastrointestinal tract – lower large intestine

I.2   Dose Estimates to the Public from Gaseous Effluents

The NRC staff used the dose assessment approach specified in Regulatory Guide 1.109 (NRC 1977) and the GASPAR II computer code (Strenge et al. 1987) to estimate doses to the MEI and to the public within 80 km (50 mi) of the WBN Unit 2 site from the gaseous effluent pathway for the proposed units. GASPAR II calculates radiation exposure to humans from routine air releases from nuclear reactor effluents.

I.2.1   Scope

The NRC staff calculated the MEI dose at 3.8 km (2.38 mi) northeast of WBN Unit 2. Pathways included were plume, ground, inhalation, and ingestion of locally produced milk, meat, and vegetables. Refer to Figures 4-2 in Section 4.6.1 for visual representation of the exposure pathway to humans.

The site parameters listed in Table I-7 were the basis for the doses calculated by the NRC staff.

Joint frequency distribution data of wind speed and wind direction by atmospheric stability class for the WBN site provided in the ODCM (TVA 2010) were used as input to the XOQDOQ code (Sagendorf et al. 1982) to calculate the average χ/Q and D/Q values for routine releases. A summary of XOQDOQ provided by Sagendorf (2010) states, "XOQDOQ was designed for meteorological evaluation of continuous and anticipated intermittent releases from commercial nuclear power reactors. It calculates annual relative effluent concentrations and average relative deposition values at locations specified by the user and at various standard radial distances and segments for downwind sectors. It also calculates these values at the specified locations for anticipated intermittent (e.g., containment or purge) releases, which occur during routine operation. The program computes an effective plume height that accounts for physical release height, aerodynamic downwash, plume rise, and terrain features. The user may optionally select additional plume dispersion due to building wakes, plume depletion via dry deposition, and plume radioactive decay, or specify adjustments to represent non-straight line trajectories (recirculation or stagnation)."

The NRC staff performed a comparative review of χ/Q and D/Q values calculated by TVA against the values calculated by the NRC staff. The χ/Q and D/Q values calculated by the NRC staff, using joint frequency data from the applicant's ODCM based on meteorological data from January 2004 through December 2006, are slightly lower (e.g., provides more atmospheric dispersion) than χ/Q and D/Q values calculated by TVA (TVA 2011b), using joint frequency tables based on meteorological data from January 1986 through December 2005. However, the differences in χ/Q and D/Q values are not significant. Furthermore, because the NRC χ/Q and D/Q values are lower than the TVA values, the NRC staff's projected dose to members of the public from the of operation of WBN Unit 2 are slightly lower than the doses calculated by TVA. The differences do not affect the NRC staff's conclusions regarding the radiological evaluation for the operation of WBN Unit 2 contained in Chapter 4 of this draft SFES.

Table I-7. Parameters Used in Calculating Dose to Public from Gaseous Effluent Releases

Parameter	Staff Value		Comments
Single new unit gaseous effluent source term (Ci/yr)	H-3	1.39×10^{-2}	TVA ER (TVA 2008a) Table 3-20 p. 87; TVA letter dated May 20, 2011; Enclosure 2, Attachment 4; Proposed Markups for final SEIS, Chapter 3, p. 87 (TVA 2011a)
	Br-84	5.07×10^{-2}	
	I-131	1.53×10^{-1}	
	I-132	6.73×10^{-1}	
	I-133	4.57×10^{-1}	
	I-134	1.07	
	I-135	8.42×10^{-1}	
	Cr-51	5.92×10^{-4}	
	Mn-54	4.31×10^{-4}	
	Co-57	8.20×10^{-6}	
	Co-58	2.32×10^{-2}	
	Co-60	8.74×10^{-3}	
	Fe-59	7.70×10^{-5}	
	Sr-89	2.98×10^{-3}	
	Sr-90	1.14×10^{-3}	
	Zr-95	1.00×10^{-3}	
	Nb-95	2.45×10^{-3}	
	Ru-103	7.70×10^{-5}	
	Ru-106	7.50×10^{-5}	
	Sb-125	6.09×10^{-5}	
	Cs-134	2.27×10^{-3}	
	Cs-136	8.01×10^{-5}	
	Cs-137	3.48×10^{-3}	
	Ba-140	4.00×10^{-4}	
	Ce-141	3.95×10^{-5}	
	C-14	7.30	
	Kr-85m	9.48	
	Kr-85	6.78×10^{2}	
	Kr-87	5.81	
	Kr-88	1.32	
	Xe-131m	1.09×10^{3}	
	Xe-133m	4.31×10^{1}	
	Xe-133	2.90×10^{3}	
	Xe-135m	4.68	
	Xe-135	8.88×10^{3}	
	Xe-137	1.23	
	Xe-138	4.34	
	Ar-41	3.40×10^{1}	
Population distribution	Updated population data was provided by TVA in letter dated May 26, 2011 p. E1-11 (TVA 2011b)		Population distribution used by the staff was for year 2040
Wind speed and direction	Site-specific data		Site-specific data for Jan 04 through Dec. 06 (hourly data obtained from file wb0408)
Joint frequency distribution of wind speed and direction by stability class	Site-specific data		Site-specific data for Jan 04 through Dec. 06 (hourly data obtained from file wb0408)
Atmospheric dispersion factors (sec/m^3)	Calculated using XOQDOQ		Site-specific data for Jan 04 through Dec. 06
Ground deposition factors	Calculated using XOQDOQ		Site-specific data for Jan 04 through Dec. 06

Table I-7. (contd)

Parameter	Staff Value	Comments
Vegetable production rate within 50 mi of WBN site	8.07×10^8 kg/yr	Site-specific value from WBN FSAR, p. 11.3-9 (TVA 2009) (leafy vegetable production in each sector annulus equals leafy vegetable consumption in that sector annulus)
Meat production rate within 50 mi of WBN site	1.37×10^6 kg/yr	Site-specific value from WBN FSAR, p. 11.3-9 (TVA 2009) (meat production in each sector annulus equals the consumption in that sector annulus)
Milk production rate within 50 mi of WBN site	4.99×10^8 L/yr	Site-specific value from WBN FSAR, p. 11.3-9 (TVA 2009) (milk production in each sector annulus equals milk consumption in that sector annulus)
Pathway receptor locations (direction and distance), nearest site boundary, MEI location	Table 3-19 of the ER (TVA 2008a)	
Consumption factors for milk, meat, leafy vegetables, and vegetables	Milk (L/yr) 310 (adult) 400 (teen) 330 (child) 330 (infant) Meat (kg/yr) 110 (adult) 65 (teen) 41 (child) 0 (infant) Leafy Vegetable (kg/yr) 64 (adult) 42 (teen) 26 (child) 0 (infant) Vegetable (kg/yr) 520 (adult) 630 (teen) 520 (child) 0 (infant)	Default value of GASPAR II code (Strenge et al. 1987)
Fraction of leafy vegetables grown	1	Site-specific value from WBN FSAR, p. 11.3-9 (TVA 2009) (leafy vegetable production in each sector annulus equals leafy vegetable consumption in that sector annulus)
Fraction of year that milk cows are on pasture	0.65	TVA RAI response letter dated May 20, 2011, p. E1-1 (TVA 2011a)
Fraction of MEI vegetable intake from own garden	1	Site-specific value from WBN FSAR, p. 11.3-9 (TVA 2009) (vegetable production in each sector annulus equals vegetable consumption in that sector annulus)
Fraction of year beef cattle are on pasture	1	Default value of GASPAR II code (Strenge et al. 1987).
Fraction of year beef cattle intake is from pasture while on pasture	1	Default value of GASPAR II code (Strenge et al. 1987).

Population doses were calculated for all types of releases (i.e., noble gases, particulates, iodines H-3 and C-14) using the GASPAR II code for the following: plume immersion, direct radiation from radionuclides deposited on the ground, inhalation, ingestion of vegetables, milk, and meat.

I.2.2   Resources Used

To calculate doses to the public from gaseous effluents, the NRC staff used a personal computer version of the XOQDOQ and GASPAR II computer codes entitled NRCDOSE Version 2.3.12 (Chesapeake Nuclear Services, Inc. 2006) obtained through the Oak Ridge Radiation Safety Information Computational Center.

I.2.3   Input Parameters

Table I-7 lists the major parameters used by NRC staff to calculate the doses to the public from gaseous effluents during normal operation. It should be noted that the 80 km (50-mi) population was assumed to be for the year 2040. Section 5.4.1 of the ESRP guidance suggests that populations be projected only 5 years out from the date of the licensing action under consideration (NRC 2000). The staff considers that using the population for the year 2040, rather than the recommended 5 years from licensing is acceptable because it assesses the population dose for a time period that approximates the operating life of WBN Unit 2.

I.2.4   Results

Table I-8 lists the doses to the public at the exclusion area boundary from gaseous effluent releases from WBN Unit 2. Table I-9 lists the doses to the MEI, a child, 3.8 km (2.38 mi) northeast of Unit 2.

Table I-8. Comparison of Doses to the Public from Noble Gas Releases from WBN Unit 2

Type of Dose[a]	WBN Calculations (TVA ER 2008)[b]	Staff Calculation
Gamma air dose at exclusion area boundary[c] – noble gases only (mrad/yr)	0.801	0.829
Beta air dose at exclusion area boundary[c] – noble gases only (mrad/yr)	2.71	2.53
Total body dose at exclusion area boundary[c] – noble gases only (mrem/yr)	0.571	0.499
Skin dose at exclusion area boundary[c] – noble gases only (mrem/yr)	1.54	1.78

(a) To convert from mrad/yr or mrem/yr to mGy/yr or mSv/yr divide by 100.
(b) Taken from Table 3-21 of the TVA ER; data is for MEI or maximum residence (TVA 2011a).
(c) At the exclusion area boundary, 1.3 km (0.8 mi) east.

Table I-9. Staff Calculation for Annual Doses to the Maximally Exposed Individual from Gaseous Effluent Releases from WBN Unit 2[a]

Pathway (Location)	Age Group	Total Body (mrem/yr)	Max. Organ (mrem/yr)	Skin Dose (mrem/yr)	Thyroid Dose (mrem/yr)
Plume (0.85 mi SE)	All	0.269	0.269 (lung)	0.958	0.269
Ground (0.85 mi SE)	All	0.079	0.079	0.093	0.079
Inhalation (0.85 mi SE)	Adult	0.025	0.041 (lung)	[a]	0.720
	Teen	0.026	0.05 (lung)	[a]	0.91
	Child	0.034	0.06 (lung)	[a]	1.61
	Infant	0.014	0.27 (lung)	[a]	0.983
Vegetable (2.08 mi NE)	Adult	0.159	0.783 (bone)	[a]	0.969
	Teen	0.257	1.3 (bone)	[a]	1.4
	Child	0.601	3.12 (bone)	[a]	2.83
	Infant	[b]	[b]	[a]	[b]
Cow Milk (1.42 mi SSW)	Adult	0.033	0.138 (bone)	[a]	0.751
	Teen	0.056	0.25 (bone)	[a]	1.2
	Child	0.128	0.616 (bone)	[a]	2.39
	Infant	0.26	1.19 (bone)	[a]	5.77
Meat (1.42 mi SSW)	Adult	0.026	0.120 (bone)	[a]	0.081
	Teen	0.021	0.101 (bone)	[a]	0.061
	Child	0.039	0.189 (bone)	[a]	0.1
	Infant	[b]	[b]	[a]	[b]

(a) Skin dose is not applicable for these exposure pathways.
(b) Infant dose is not applicable for this pathway.
To convert person-rem to person-Sv, divide by 100.
To convert miles (mi) to kilometers (km), multiply by 1.6.

I.3 Cumulative Dose Estimates

Based on parameters shown for the liquid pathway and the gaseous pathway, Table I-2 and Table I-7, respectively, NRC staff calculated doses from the WBN Unit 2 using LADTAP II and GASPAR II to the MEI and the population within 80 km (50 mi) of the WBN Unit 2 site. It should be noted that the 80-km (50-mi) population was assumed to be for the year 2040. Section 5.4.1 of the ESRP guidance suggests that populations be projected only 5 years out from the date of the licensing action under consideration (NRC 2000). The staff considers that using the population for the year 2040, rather than the recommended 5 years from licensing is acceptable because it assesses the population dose for a time period that approximates the operating life of WBN Unit 2.

As stated in Section 4.6, there are no regulatory requirements for population doses, but the comparison to population dose and dose from natural background demonstrates that the annual estimated population doses from WBN Unit 2 are not significant when compared to the population dose from natural background (0.236 person-Sv/yr [23.6 person-rem/yr] and 4,738 person-Sv/yr [473,800 person-rem/yr], respectively) Table I-10 lists the staff's calculation of cumulative dose rates to the population for the year 2040 from WBN Unit 2.

Table I-11 compares the NRC staff's results for cumulative dose estimates to the MEI with Title 40 of the Federal Code of Regulations (CFR) Part 190 criteria. All dose estimates are within the 40 CFR Part 190 criteria.

Table I-10. Population Total Body Doses (person-rem) for the Year 2040

Pathway	Gaseous Effluent	Liquid Effluent	Total
Noble Gases	1.06	-	1.06
Iodines and particulates	0.30	7.91	8.21
Tritium and C-14	4.73	4.12	8.85
Total	6.09	12	18.1

To convert person-rem to person-Sv, divide by 100.

Table I-11. Comparison of Maximally Exposed Individual Annual Dose Estimates with 40 CFR Part 190(a) Criteria (Staff Calculations)

	Annual Dose Estimate (mrem/yr)							
	Total Body	GI-LLI	Bone	Liver	Kidney	Thyroid	Lung	Skin
	Gaseous Effluent							
Adult	0.59	0.6	1.4	0.6	0.59	2.87	0.61	1.05
Teen	0.71	0.71	2.01	0.72	0.72	3.92	0.73	1.05
Child	1.15	1.14	4.29	1.18	1.17	7.28	1.17	1.05
Infant	0.62	0.61	1.55	0.66	0.64	7.1	0.64	1.05
	Liquid Effluent							
Adult	0.63	0.13	0.49	0.85	0.31	0.83	0.13	0.03
Teen	0.37	0.10	0.51	0.86	0.31	0.74	0.14	0.03
Child	0.18	0.08	0.64	0.78	0.29	1.46	0.13	0.03
Infant	0.05	0.06	0.06	0.08	0.06	1.91	0.05	0.03
	Total (may not sum due to rounding)							
Adult	1.23	0.73	1.88	1.44	0.91	3.7	0.74	1.08
Teen	1.208	0.81	2.52	1.59	1.03	4.66	0.87	1.08
Child	1.33	1.22	4.94	1.97	1.46	8.74	1.3	1.08
Infant	0.67	0.67	1.6	0.75	0.7	9.01	0.69	1.08
	40 CFR Part 190(a) Criteria							
	25	25	25	25	25	75	25	25

To convert person-rem to person-Sv, divide by 100.
GI-LLI = gastrointestinal tract – lower large intestine

I.4 Biota Doses

To calculate doses to the biota from liquid effluents, the NRC staff used personal computer versions of the NRC-developed LADTAP II and GASPAR II that are integrated into NRCDOSE Version 2.3.12 (Chesapeake Nuclear Services, Inc. 2006). NRC staff obtained NRCDOSE through the Oak Ridge Radiation Safety Information Computational Center.

The LADTAP II input parameters are specified in Section I.2.2, above, to include the source term, the discharge flow rate to the receiving freshwater system, the shore-width factor, and fractions of radionuclides in the liquid effluent reaching offsite bodies of water. The transit time from the effluent release location to the exposure location was zero hours.

NRC staff assessed dose to terrestrial biota from the gaseous effluent pathway using GASPAR II by assuming doses for raccoons and ducks were equivalent to adult human doses for inhalation, vegetation ingestion, plume and twice the ground pathways at a location 1.29 km (0.8 mi) east. The doubling of doses from ground deposition reflects the closer proximity of these organisms to the ground. Muskrats and herons do not consume terrestrial vegetation, so that pathway was not included for those organisms.

As stated in Section 4.6, the NRC does not have a regulatory framework for the protection of biota from radioactive discharges from nuclear power reactors. The focus of NRC regulatory framework is for the protection of human beings (NRC 2009). The International Commission on Radiological Protection (ICRP 1977, 1991, 2007) states that if humans are adequately protected, other living things are also likely to be sufficiently protected. Table I-12 lists the results of the NRC staff's biota dose calculations. The results are within the International Atomic Energy Agency/National Council on Radiation Protection and Measurements guidelines for protection of biota (IAEA 1992; NCRP 1991).

Table I-12. Doses to Biota (mrem/yr) Due to Liquid and Gaseous Releases from WBN Unit 2

Biota	Liquid Releases	Gaseous Releases	Total
Fish	4.3	-	4.30
Invertebrate	11.4	-	11.4
Algae	19.2	-	19.2
Muskrat	10.8	0.77	11.57
Raccoon	4.83	1.53	6.36
Heron	55.5	0.77	56.27
Duck	10.3	1.53	11.83
To convert person-rem to person-Sv, divide by 100.			

I.5   References

10 CFR Part 50. Code of Federal Regulations, Title 10, *Energy*, Part 50, "Domestic Licensing of Production and Utilization Facilities."

40 CFR 190. Code of Federal Regulations. Title 40, *Protection of Environment*, Part 190, "Environmental Radiation Protection Standards for Nuclear Power Operations."

Chesapeake Nuclear Services, Inc. 2006. *NRCDOSE for Windows*. Radiation Safety Information Computational Center, Oak Ridge, Tennessee.

[IAEA] International Atomic Energy Agency. 1992. *Effects of Ionizing Radiation on Plants and Animals at Levels Implied by Current Radiation Protection Standards*. Technical Report Series No. 332, Vienna, Austria.

[ICRP] International Commission on Radiological Protection. 2007. *The 2007 Recommendations of the International Commission on Radiological Protection*. ICRP Publication No. 103. Annals of the ICRP 37(2-4). Accession No. ML080790301

[NCRP] National Council on Radiation Protection and Measurements. 1991. *Effects of Ionizing Radiation on Aquatic Organisms*. NCRP Report No. 109, Bethesda, Maryland.

Sagendorf, J.F., J.T. Goll, and W.F. Sandusky. 1982. *XOQDOQ: Computer Program for the Meteorological Evaluation of Routine Effluent Releases at Nuclear Power Stations*. NUREG/CR-2919, Pacific Northwest National Laboratory, Richland, Washington.

Sagendorf, J.F., J.T. Goll, W.F. Sandusky, and L.R. Eyberger. 2010. *XOQDOQ - Meteorological Evaluation of Atmospheric Nuclear Power Plant Effluents*. Version 2.0. Accessed at http://www.oecd-nea.org/tools/abstract/detail/nesc0964.

Strenge, D.L., R.A. Peloquin, and G. Whelan. 1986. *LADTAP II – Technical Reference and User Guide*. NUREG/CR-4013, Pacific Northwest Laboratory, Richland, Washington.

Strenge, D.L., T.J. Bander, and J.K. Soldat. 1987. *GASPAR II – Technical Reference and User Guide*. NUREG/CR-4653, Pacific Northwest Laboratory, Richland, Washington.

[TVA] Tennessee Valley Authority. 2008a. *Final Supplemental Environmental Impact Statement; Completion and Operation of Watts Bar Nuclear Plant Unit 2, Rhea County, Tennessee*. Submitted to NRC as the TVA Environmental Report for an Operating License, Knoxville, Tennessee.

[TVA] Tennessee Valley Authority. 2008b. "Offsite Dose Calculation Manual." Attachment 3 in *Watts Bar Nuclear Plant Effluent and Waste Disposal Annual Report*. Knoxville, Tennessee.

[TVA] Tennessee Valley Authority. 2009. *Watts Bar Nuclear Plant (WBN) – Unit 2 – Final Safety Analysis Report (FSAR), Amendment 94.* Spring City, Tennessee.

[TVA] Tennessee Valley Authority. 2010. Letter from Masoud Bajestani (TVA, Watts Bar Unit 2 Vice President) to U.S. Nuclear Regulatory Commission dated April 9, 2010, in response to NRC letter dated December 3, 2009 and TVA letters dated February 15, 2008, July 2, 2008, January 27, 2009, December 23, 2009, and February 25, 2010, "Watts Bar Nuclear Plant (WBN) Unit 2 – Response to U.S. Nuclear Regulatory Commission (NRC) Request for Additional Information Regarding Environmental Review; Enclosure 1 (TAC No. MD8203)." Accession No. ML101130393.

[TVA] Tennessee Valley Authority. 2011a. Letter from Dave Stinson (TVA, Watts Bar Unit 2 Vice President) to U.S. Nuclear Regulatory Commission dated May 20, 2011, in response to NRC letter dated April 13, 2011, TVA letters dated February 15, 2008, December 17, 2010, February 25, 2011, and an e-mail from Justin C. Poole to William D. Crouch dated March 4, 2011, "Watts Bar Nuclear Plant (WBN) Unit 2 – Response to Final Safety Analysis Report (FSAR), Chapter 11 and Final Supplemental Environmental Impact Statement (FEIS) Request for Additional Information." Accession No. ML11146A044.

[TVA] Tennessee Valley Authority. 2011b. Letter from Dave Stinson (TVA, Watts Bar Unit 2 Vice President) to U.S. Nuclear Regulatory Commission dated May 26, 2011, in response to an e-mail from Justin C. Poole to William D. Crouch dated March 4, 2011 and TVA letter dated May 20, 2011, "Watts Bar Nuclear Plant (WBN) Unit 2 – Response to Request for Additional Information from Public Meeting on May 11, 2011." Accession No. ML11152A160.

[NRC] U.S. Nuclear Regulatory Commission. 1977. *Calculation of Annual Doses to Man from Routine Releases of Reactor Effluents for the Purpose of Evaluating Compliance with 10 CFR Part 50, Appendix I.* Regulatory Guide 1.109, Washington, D.C. Accession No. ML003740384.

[NRC] U.S. Nuclear Regulatory Commission. 2000. *Environmental Standard Review Plan — Standard Review Plans for Environmental Reviews for Nuclear Power Plants.* NUREG-1555, Vol. 1, Washington, D.C. Includes 2007 updates.

[NRC] U.S Nuclear Regulatory Commission. 2009. Memorandum To: R.W. Borchardt. From: Annette L. Vietti-Cook. Subject: "Staff Requirements – SECY-08-0197 – Options to Revise Radiation Protection Regulations and Guidance with Respect to the 2007 Recommendations of the International Commission on Radiological Protection. Accession No. ML090920103.

NRC FORM 335 (12-2010) NRCMD 3.7	U.S. NUCLEAR REGULATORY COMMISSION	1. REPORT NUMBER (Assigned by NRC, Add Vol., Supp., Rev., and Addendum Numbers, if any.)
BIBLIOGRAPHIC DATA SHEET *(See instructions on the reverse)*		NUREG-0498 Supplement No. 2 Vol. 2

2. TITLE AND SUBTITLE	3. DATE REPORT PUBLISHED	
Final Environmental Statement Supplement 2, Vol. 2 Related to the Operation of Watts Bar Nuclear Plant, Unit 2	MONTH May	YEAR 2013
	4. FIN OR GRANT NUMBER	

5. AUTHOR(S)	6. TYPE OF REPORT
J. Poole, et al.	Technical
	7. PERIOD COVERED (Inclusive Dates)

8. PERFORMING ORGANIZATION - NAME AND ADDRESS (If NRC, provide Division, Office or Region, U. S. Nuclear Regulatory Commission, and mailing address; if contractor, provide name and mailing address.)

U.S. Nuclear Regulatory Commission
Office of Nuclear Reactor Regulation
Division of Operating Reactor Licensing
Washington, DC 20555-001

9. SPONSORING ORGANIZATION - NAME AND ADDRESS (If NRC, type "Same as above", if contractor, provide NRC Division, Office or Region, U. S. Nuclear Regulatory Commission, and mailing address.)

Same as above

10. SUPPLEMENTARY NOTES
Docket No. 50-391

11. ABSTRACT (200 words or less)
The U.S. Nuclear Regulatory Commission (NRC) prepared this supplemental final environmental statement related to the operating license in response to its review of the Tennessee Valley Authority's (TVA's) application for a facility operating license submitted on March 4, 2009. The proposed action requested is for the NRC to is sue an operating license for a second light-water nuclear reactor at the Watts Bar Nuclear (WBN) Plant in Rhea County, TN.

The NRC staff evaluated a full scope of environmental topics, including land and water use, air quality and meteorology, terrestrial and aquatic ecology, radiological and nonradiological impacts on humans and the environment, historic and cultural resources, socioeconomics, alternatives, and environmental justice. The alternatives considered were natural gas-fired power generation, and a combination of energy alternatives.

The staffs evaluations are based on (1) the application submitted by TVA, including the environmental report and previous environmental impact statements and historical documents, (2) consultation with other Federal, State, Tribal, and local agencies, (3) the staff's independent review, and (4) the staff's consideration of comments related to the environmental review received during the public scopinq process and received during the public comment period for the draft document.

12. KEY WORDS/DESCRIPTORS (List words or phrases that will assist researchers in locating the report.)	13. AVAILABILITY STATEMENT
Final Environmental Statement Watts Bar Nuclear Plant Tennessee Valley Authority TVA Docket No. 50-391 National Environmental Policy Act NEPA	unlimited
	14. SECURITY CLASSIFICATION
	(This Page) unclassified
	(This Report) unclassified
	15. NUMBER OF PAGES
	16. PRICE

UNITED STATES
NUCLEAR REGULATORY COMMISSION
WASHINGTON, DC 20555-0001

OFFICIAL BUSINESS

NUREG-0498
Supplement 2, Vol. 2
Final

Final Environmental Statement Related to the Operation of
Watts Bar Nuclear Plant, Unit 2

May 2013